AF454031

Physical Activity, Wellness and Health: Challenges, Benefits and Strategies

Physical Activity, Wellness and Health: Challenges, Benefits and Strategies

Editors

Luciana Zaccagni
Emanuela Gualdi-Russo

MDPI • Basel • Beijing • Wuhan • Barcelona • Belgrade • Manchester • Tokyo • Cluj • Tianjin

Editors
Luciana Zaccagni
University of Ferrara
Italy

Emanuela Gualdi-Russo
University of Ferrara
Italy

Editorial Office
MDPI
St. Alban-Anlage 66
4052 Basel, Switzerland

This is a reprint of articles from the Special Issue published online in the open access journal *International Journal of Environmental Research and Public Health* (ISSN 1660-4601) (available at: https://www.mdpi.com/journal/ijerph/special_issues/Physical_Activity_Wellness_Health_Challenges_Benefits_Strategies).

For citation purposes, cite each article independently as indicated on the article page online and as indicated below:

LastName, A.A.; LastName, B.B.; LastName, C.C. Article Title. *Journal Name* **Year**, *Volume Number*, Page Range.

ISBN 978-3-0365-2402-3 (Hbk)
ISBN 978-3-0365-2403-0 (PDF)

Contents

About the Editors

Luciana Zaccagni (Prof.) is an associate professor of Methods and Didactics of sporting activities in the Department of Neuroscience and Rehabilitation at the University of Ferrara (Italy). Her research interests are in the field of sport sciences, with an emphasis on sports performance, biomechanics and kinanthropometry. She has authored over 40 original articles. Her current h-index is 13 (Scopus) and her ResearchGate-score is 28.51.

Emanuela Gualdi-Russo (Prof.) is an Eminent Scholar at University of Ferrara (Italy), Department of Neuroscience and Rehabilitation. Her areas of expertise are in the fields of Anthropology and Anthropometry, with an emphasis on body composition, nutritional status, and body image perception. She has over 150 original journal publications and many book chapters. Her current h-index is 21 (Scopus) and ResearchGate-score is 37.74.

International Journal of
Environmental Research and Public Health

Editorial

Physical Activity for Health and Wellness

Emanuela Gualdi-Russo [1,*] and **Luciana Zaccagni [1,2]**

[1] Department of Neuroscience and Rehabilitation, Faculty of Medicine, Pharmacy and Prevention, University of Ferrara, 44121 Ferrara, Italy; luciana.zaccagni@unife.it
[2] Center for Exercise Science and Sport, University of Ferrara, 44121 Ferrara, Italy
* Correspondence: emanuela.gualdi@unife.it

check for
updates

Citation: Gualdi-Russo, E.; Zaccagni, L. Physical Activity for Health and Wellness. *Int. J. Environ. Res. Public Health* **2021**, *18*, 7823. https://doi.org/10.3390/ijerph18157823

Received: 19 July 2021
Accepted: 21 July 2021
Published: 23 July 2021

Publisher's Note: MDPI stays neutral with regard to jurisdictional claims in published maps and institutional affiliations.

1. Introduction

Regular physical activity (PA) is both a preventive measure and a cure for non-communicable diseases (NCDs). Moreover, PA improves mental health, quality of life, and well-being [1]. Conversely, physical inactivity and sedentary lifestyles have negative impacts on individuals, families, and society, as evidenced in particular by the spread of the obesity epidemic [2–6].

PA has proven to be a low-cost alternative for the treatment and prevention of disease. Therefore, interventions to prevent avoidable diseases by increasing the proportion of physically active people are fundamental.

The Special Issue "Physical Activity, Wellness and Health: Challenges, Benefits and Strategies" was intended to collect research articles on anthropometric determinants of health and performance, PA and healthy habits, exercise and diet, exercise and body composition, interventions to promote PA for people of all ages, strategies for the implementation of an active life, and the beneficial effects of exercise on metabolic syndrome. Finally, 20 articles covering a wide range of information were published, indicating the interest generated by this call. Below we will provide a summary of the main contents of this Special Issue, highlighting proposals for future research that potentially contribute to the health benefits of being physically active.

Topics included in this Special Issue fall mainly into the following three areas: anthropometry, health, and sport; health benefits of exercise; population studies and strategies for an active life.

2. Anthropometry, Health, and Sport

Anthropometric characteristics are important factors of a person's physical performance and health status. Four studies included in this Special Issue evaluated the contribution of these variables. Matias et al. [7] found that phase angle derived from bioelectrical impedance spectroscopy is predictive of maximal isometric forearm strength in cancer patients. Its relevance as a clinical indicator of disease-related function in breast cancer survivors was suggested. Handgrip strength was particularly influenced by body composition parameters and handedness according to Zaccagni et al. [8], so much so that the authors recommended it as a proxy for unhealthy conditions with impaired muscle mass, taking into account laterality. Further research should also provide evidence for the effectiveness and clinical relevance of hand strength testing in the assessment and prediction of critical health conditions. Barbieri et al. [9] investigated the efficacy and accuracy of a data mining methodology in predicting cardiovascular risk based on anthropometric, demographic, and biomedical data from a very large sample of the population involved in competitive sports practice. The procedure was conducted using a decision tree and logistic regression to classify individuals as at-risk or not. In addition, the authors used the receiver operating characteristic curve to assess classification performance, achieving satisfactory results. The fourth study by Rinaldo et al. [10] departs from the previous themes to deal with injuries that can occur in sporting activities, focusing on the relationship between anthropometric

1

traits and injury occurrence. Their findings pointed out that an increased body mass index, decreased calf muscle area, and being closer to the age of peak height velocity are significant risk factors for injuries in elite soccer players aged 9–13 years. Consistently with these findings, the authors claim that body composition and anthropometric characteristics should be monitored to reduce the risk of injury in young soccer players. Furthermore, training programs must be adapted to both the chronological age and the maturity status of the players.

3. Health Benefits of Exercise

PA contributes to preventing and treating a wide range of NCDs and can improve mental health, while also enhancing the quality of life and well-being. A total of seven studies in the Special Issue were conducted in this area.

Two studies concern, in particular, the implications of regular exercise for disease prevention and treatment. Kanai et al. [11] reported that the health utility score was 0.77 in stroke survivors and was associated with the number of steps; the more stroke survivors walked, the higher their health utility score. Turning to multiple sclerosis, it is well known that physical inactivity reduces cardiorespiratory capacity, promotes physical deconditioning, and leads to comorbidities such as obesity, metabolic syndrome, and osteoporosis. In this field, Pau et al. [12] examined possible sex-related differences in the amount and intensity of PA performed by people with multiple sclerosis and showed that the pattern for women was characterized by greater sedentariness and less activity of light intensity than for men. Both studies [11,12] quantitatively assessed PA (moderate-to-vigorous physical activity, MVPA) using accelerometers.

Five studies focused, in particular, on mental health and PA. PA promotes different kinds of positive psychological responses. Regular exercise has a beneficial impact on depression and anxiety. It reduces stress and improves overall well-being. The first study starts from the evidence that poor sleep quality, common in young people, increases the risk of morbidity and mortality. In this area, Zhai et al. [13] highlighted that regular PA can improve poor sleep quality among college students. PA could enhance sleep by helping individuals cope with stress, indicating that stress management could be a non-pharmaceutical treatment for sleep improvement. Considering the mental health of young people, Usán Supervía et al. [14] examined the relationships between the constructs of goal orientations, emotional intelligence, and burnout in high school students. The authors outlined that the psychological profile arising from these features could be important for academic performance and school participation. Bíró et al. [15] examined gender, as a socio-economic determinant of health, by testing the validity of the biopsychosocial model of health with a limited life course perspective on a very large sample of students from Hungarian universities and colleges. Their findings suggested that determinants of male health included fewer variables focused on physical activity, and were less influenced by social relationships, in contrast to female health, which was influenced by age and social support. Kim and Ahn [16] showed that exercise participation for six weeks led to positive changes in the self-esteem and mental health of college students. In a narrative review, Belvederi Murri et al. [17] investigated the beneficial effects of PA on depressed populations. A specific public health problem is the premature mortality of depressed individuals. This is mainly caused by increased cardiovascular risk, as depression leads to the development or exacerbation of unhealthy lifestyles. According to their findings, PA can reduce depression severity and directly address cardiovascular risk factors. In the field of public health, the development and dissemination of initiatives promoting exercise-based interventions in depressed populations are recommended, focusing on their cost-effectiveness.

4. Population Studies and Strategies for an Active Life Implementation

Nine articles in the Special Issue deal with this topic.

Two studies took into account the multiple negative effects of physical inactivity on health and the factors involved. In a South African adult population, Chifaku et al. [18] assessed the levels and correlates of PA. They found that gender, marital status, and health awareness were significant predictors, pointing out a high prevalence of insufficient PA in some vulnerable groups, particularly the elderly and obese, and a general lack of participation in sports and recreational activities. As PA plays a fundamental role in the process of growth and development, Baqal et al. [19] analyzed data from a national study, "Jeeluna", on a large sample of adolescents living in the Kingdom of Saudi Arabia. The authors found that 67% of adolescents who did not exercise led a sedentary lifestyle. Males and adolescents aged 10–14 years were significantly more likely to engage in PA than females and adolescents aged 15–19 years. Among the factors contributing to high rates of inactivity among adolescents, the authors include the lack of PA programs in schools, hot weather conditions, poor family and peer support, and socio-cultural barriers, which have a particular impact on girls.

Despite the known benefits of regular PA, there is a high percentage of physically inactive adults worldwide. Increased national attention on PA as a tool for health promotion and disease prevention is therefore required [20]. Five studies in this Special Issue examine different approaches and strategies that aim to increase PA. The first article, by Potter et al. [21], is a pilot study on activities that naturally involve PA, considering a stealth health approach to increase PA among inactive dog owners. The approach tested in this study showed that dog obedience training could have, as a side effect, a positive impact on both PA and sedentary behavior among dog owners; dog owners are induced to walk more and sit less. Given the large number of dog owners, this new approach to promoting PA may have a significant impact on public health and merits further investigation. In Latin America, the prevalence of obesity and overweight is increasing in all countries, despite the efforts of governments to promote healthy lifestyles. In this context, Farías [22] analyzed which emotions out of fear and hope are most effective in stimulating individuals to make health-related decisions, showing that these appeals in health advertisements do not have any main effect on PA intention, although this effect is positively moderated by perceived body weight and past healthy eating behavior, and is negatively moderated by subjective norms in diet and exercise. Another study conducted by Shi et al. [23] on university students indicated that the combination of insufficient physical activity levels with mobile phone addiction is significantly linked to high levels of irrational procrastination. To improve efficiency and reduce irrational procrastination, it would be necessary to increase physical activity and reducing mobile phone addiction. A systematic review by Zaccagni et al. [24] reported the consequences on physical activity and health of the general lockdown implemented in Italy from March to May 2020 due to the COVID-19 pandemic. Their analysis of 23 studies showed that there has been a general reduction in PA and unhealthy dietary habits as a result of this lockdown in Italy, with a deterioration of the health status in both the general population and people with chronic diseases. According to the authors, individual outdoor exercise should be promoted, especially during daylight hours, while maintaining physical distance in the case of another lockdown to contain current and future pandemics. Particularly in older people, sedentary behavior is a serious public health problem. Monteagudo et al. [25] examined the impact of overground walking interval training in sedentary older adults by comparing two different dose distributions during a longitudinal study. Both training protocols led to a significant overall improvement of physical function in older adults. As regards the strategy to be used in the elderly, Monteagudo showed that the bout length is not a determinant of the functional health effects associated with exercise; splitting a single exercise into two sets during the day can be beneficial for autonomy, agility, and health-related quality of life. In particular, the accumulative strategy is to be recommended when health-related quality of life is the main goal, whereas the continuous strategy is to be recommended when weakness may be a short- or medium-term threat.

The last two studies of this section concern the fitness sector and the spread of sports venues. The research of Moustakas et al. [26] aimed to define the drivers of change in the fitness sector and to identify the skills needed by the fitness workforce to navigate these changes. The main finding was that technology, health needs, and customer loyalty are critical drivers of change in the fitness industry. Fitness professionals must therefore respond by improving both their professional skills, especially in providing services for special populations, and their soft skills, stressing the particular importance of engaging with technology and having an understanding of specific health issues. Mainland China, one of the most populous upper-middle-income countries, also has to deal with a prevalence of NCDs and physical inactivity. Analyzing the relevant characteristics of sports venues associated with leisure-time PA in China, Wang et al. [27] identified the number and area of sports venues as the most important indicators. The number of sports venues, which increased between 2000 and 2013, is still comparatively small compared to the United States and Japan. The urban–rural gap in sports venues exemplifies just a few aspects of the 'urban–rural dual structure' in Chinese society.

5. Conclusions

The 20 manuscripts included differ in subject matter and methodologies applied, and we consider this variability to be an enrichment for the Special Issue. According to the previous subdivision, the studies included in this Special Issue dealt mainly with interventions to promote PA for persons of all age groups and implementation strategies for active living in different populations.

In general, the studies made important suggestions for planning targeted interventions for specific diseases, ages, or population groups, but also for providing guidelines for a healthy lifestyle, tailored to the requirements of individuals to achieve maximum effectiveness. PA interventions are needed to reduce the treatment costs of chronic morbidity that may result from a lower prevalence and better control of CVD and its risk factors. PA-based interventions have also been shown to be effective as additional interventions in mental health. In this respect, it should be emphasized that exercise is still underprescribed for depressed individuals. It is therefore important to eliminate the barriers that are currently restricting this prescription by clinicians.

The findings of several studies support the relevance of specific anthropometric variables as potential health indicators, suggesting that anthropometric characteristics and growth rates should be monitored in younger athletes. To improve clinical decision making by reducing the number of unnecessary examinations, the application of data mining to biomedical data, including anthropometric data, may be effective.

The importance to ensure the application of appropriate methodologies of measuring quantitative traits (PA, strength, body composition measurements, etc.) was often emphasized in the articles.

All of the studies support strategies to promote PA and reduce sedentary behavior among adolescents, adults, and the elderly. There is no doubt that regular exercise is beneficial to health, but the general population should be encouraged to engage in more of it.

With the support of all the contributing authors, we are confident that we have provided a significant contribution to the knowledge of the topic addressed in this Special Issue.

Author Contributions: Conceptualization, E.G.-R., L.Z.; writing—original draft preparation, E.G.-R.; writing—review and editing, E.G.-R., L.Z. Both authors have read and agreed to the published version of the manuscript.

Funding: This research received no external funding.

Institutional Review Board Statement: Not applicable.

Informed Consent Statement: Not applicable.

Data Availability Statement: Not applicable.

Acknowledgments: We sincerely thank all authors and reviewers for their valuable contribution to this Special Issue.

Conflicts of Interest: The authors declare no conflict of interest.

References

1. World Health Organization (WHO). *Global Action Plan on Physical Activity 2018–2030: More Active People for a Healthier World*; World Health Organization: Geneva, Switzerland, 2018; Licence: CC BY-NC-SA 3.0 IGO.
2. NCD Risk Factor Collaboration (NCD-RisC). Worldwide trends in body-mass index, underweight, overweight, and obesity from 1975 to 2016: A pooled analysis of 2416 population-based measurement studies in 128·9 million children, adolescents, and adults. *Lancet* **2017**, *390*, 2627–2642. [CrossRef]
3. NCD Risk Factor Collaboration (NCD-RisC). Rising rural body-mass index is the main driver of the global obesity epidemic in adults. *Nature* **2019**, *569*, 260–264. [CrossRef] [PubMed]
4. NCD Risk Factor Collaboration (NCD-RisC). Height and body-mass index trajectories of school-aged children and adolescents from 1985 to 2019 in 200 countries and territories: A pooled analysis of 2181 population-based studies with 65 million participants. *Lancet* **2020**, *396*, 1511–1524. [CrossRef]
5. NCD Risk Factor Collaboration (NCD-RisC). Heterogeneous contributions of change in population distribution of body mass index to change in obesity and underweight. *Elife* **2021**, *10*, e60060. [CrossRef] [PubMed]
6. Gualdi-Russo, E.; Rinaldo, N.; Toselli, S.; Zaccagni, L. Associations of Physical Activity and Sedentary Behaviour Assessed by Accelerometer with Body Composition among Children and Adolescents: A Scoping Review. *Sustainability* **2021**, *13*, 335. [CrossRef]
7. Matias, C.N.; Cavaco-Silva, J.; Reis, M.; Campa, F.; Toselli, S.; Sardinha, L.; Silva, A.M. Phase Angle as a Marker of Muscular Strength in Breast Cancer Survivors. *Int. J. Environ. Res. Public Health* **2020**, *17*, 4452. [CrossRef] [PubMed]
8. Zaccagni, L.; Toselli, S.; Bramanti, B.; Gualdi-Russo, E.; Mongillo, J.; Rinaldo, N. Handgrip Strength in Young Adults: Association with Anthropometric Variables and Laterality. *Int. J. Environ. Res. Public Health* **2020**, *17*, 4273. [CrossRef]
9. Barbieri, D.; Chawla, N.; Zaccagni, L.; Grgurinović, T.; Šarac, J.; Čoklo, M.; Missoni, S. Predicting Cardiovascular Risk in Athletes: Resampling Improves Classification Performance. *Int. J. Environ. Res. Public Health* **2020**, *17*, 7923. [CrossRef]
10. Rinaldo, N.; Gualdi-Russo, E.; Zaccagni, L. Influence of Size and Maturity on Injury in Young Elite Soccer Players. *Int. J. Environ. Res. Public Health* **2021**, *18*, 3120. [CrossRef]
11. Kanai, M.; Izawa, K.P.; Kubo, H.; Nozoe, M.; Mase, K.; Shimada, S. Association of Health Utility Score with Physical Activity Outcomes in Stroke Survivors. *Int. J. Environ. Res. Public Health* **2020**, *18*, 251. [CrossRef]
12. Pau, M.; Porta, M.; Coghe, G.; Frau, J.; Lorefice, L.; Cocco, E. Does Multiple Sclerosis Differently Impact Physical Activity in Women and Man? A Quantitative Study Based on Wearable Accelerometers. *Int. J. Environ. Res. Public Health* **2020**, *17*, 8848. [CrossRef] [PubMed]
13. Zhai, X.; Wu, N.; Koriyama, S.; Wang, C.; Shi, M.; Huang, T.; Wang, K.; Sawada, S.S.; Fan, X. Mediating Effect of Perceived Stress on the Association between Physical Activity and Sleep Quality among Chinese College Students. *Int. J. Environ. Res. Public Health* **2021**, *18*, 289. [CrossRef]
14. Usán Supervía, P.; Salavera Bordás, C.; Murillo Lorente, V. Psychological Analysis among Goal Orientation, Emotional Intelligence and Academic Burnout in Middle School Students. *Int. J. Environ. Res. Public Health* **2020**, *17*, 8160. [CrossRef] [PubMed]
15. Bíró, É.; Kovács, S.; Veres-Balajti, I.; Ádány, R.; Kósa, K. Modelling Health in University Students: Are Young Women More Complicated Than Men? *Int. J. Environ. Res. Public Health* **2021**, *18*, 7310. [CrossRef]
16. Kim, I.; Ahn, J. The Effect of Changes in Physical Self-Concept through Participation in Exercise on Changes in Self-Esteem and Mental Well-Being. *Int. J. Environ. Res. Public Health* **2021**, *18*, 5224. [CrossRef] [PubMed]
17. Belvederi Murri, M.; Folesani, F.; Zerbinati, L.; Nanni, M.G.; Ounalli, H.; Caruso, R.; Grassi, L. Physical Activity Promotes Health and Reduces Cardiovascular Mortality in Depressed Populations: A Literature Overview. *Int. J. Environ. Res. Public Health* **2020**, *17*, 5545. [CrossRef]
18. Chikafu, H.; Chimbari, M.J. Levels and Correlates of Physical Activity in Rural Ingwavuma Community, uMkhanyakude District, KwaZulu-Natal, South Africa. *Int. J. Environ. Res. Public Health* **2020**, *17*, 6739. [CrossRef]
19. Baqal, O.J.; Saleheen, H.; AlBuhairan, F.S. Urgent Need for Adolescent Physical Activity Policies and Promotion: Lessons from "Jeeluna". *Int. J. Environ. Res. Public Health* **2020**, *17*, 4464. [CrossRef]
20. World Health Organization (WHO). *Steps to a Health: A European Framework to Promote Physical Activity for Health*; WHO Regional Office for Europe: Copenhagen, Denmark, 2007.
21. Potter, K.; Masteller, B.; Balzer, L.B. Examining Obedience Training as a Physical Activity Intervention for Dog Owners: Findings from the Stealth Pet Obedience Training (SPOT) Pilot Study. *Int. J. Environ. Res. Public Health* **2021**, *18*, 902. [CrossRef]
22. Farías, P. The Use of Fear versus Hope in Health Advertisements: The Moderating Role of Individual Characteristics on Subsequent Health Decisions in Chile. *Int. J. Environ. Res. Public Health* **2020**, *17*, 9148. [CrossRef] [PubMed]
23. Shi, M.; Zhai, X.; Li, S.; Shi, Y.; Fan, X. The Relationship between Physical Activity, Mobile Phone Addiction, and Irrational Procrastination in Chinese College Students. *Int. J. Environ. Res. Public Health* **2021**, *18*, 5325. [CrossRef] [PubMed]
24. Zaccagni, L.; Toselli, S.; Barbieri, D. Physical Activity during COVID-19 Lockdown in Italy: A Systematic Review. *Int. J. Environ. Res. Public Health* **2021**, *18*, 6416. [CrossRef] [PubMed]

25. Monteagudo, P.; Roldán, A.; Cordellat, A.; Gómez-Cabrera, M.C.; Blasco-Lafarga, C. Continuous Compared to Accumulated Walking-Training on Physical Function and Health-Related Quality of Life in Sedentary Older Persons. *Int. J. Environ. Res. Public Health* **2020**, *17*, 6060. [CrossRef]
26. Moustakas, L.; Szumilewicz, A.; Mayo, X.; Thienemann, E.; Grant, A. Foresight for the Fitness Sector: Results from a European Delphi Study and Its Relevance in the Time of COVID-19. *Int. J. Environ. Res. Public Health* **2020**, *17*, 8941. [CrossRef]
27. Wang, K.; Wang, X. Providing Sports Venues on Mainland China: Implications for Promoting Leisure-Time Physical Activity and National Fitness Policies. *Int. J. Environ. Res. Public Health* **2020**, *17*, 5136. [CrossRef] [PubMed]

International Journal of
*Environmental Research
and Public Health*

Article

Phase Angle as a Marker of Muscular Strength in Breast Cancer Survivors

Catarina N. Matias [1,2], Joana Cavaco-Silva [3], Mafalda Reis [1], Francesco Campa [4,*],
Stefania Toselli [5], Luís Sardinha [1] and Analiza M. Silva [1]

[1]   Exercise and Health Laboratory, CIPER, Faculdade de Motricidade Humana, Universidade de Lisboa,
      1499-002 Lisboa, Portugal; cmatias@fmh.ulisboa.pt (C.N.M.); mafaldamtreis@gmail.com (M.R.);
      lsardinha@fmh.ulisboa.pt (L.S.); analiza@fmh.ulisboa.pt (A.M.S.)
[2]   Physiology and Biochemistry of Exercise Laboratory, CIPER, Faculdade de Motricidade Humana,
      Universidade Lisboa, 1499-002 Lisboa, Portugal
[3]   ScienceCircle-Scientific and Biomedical Consulting, 1600 544 Lisboa, Portugal; jo.cvsilva@gmail.com
[4]   Department for Life Quality Studies, University of Bologna, 47921 Rimini, Italy
[5]   Department of Biomedical and Neuromotor Science, University of Bologna, 40126 Bologna, Italy;
      stefania.toselli@unibo.it
*   Correspondence: francesco.campa3@unibo.it; Tel.: +39-3450-031-080

Received: 22 May 2020; Accepted: 18 June 2020; Published: 21 June 2020

Abstract: Background: accurate prognostic tools are relevant for decision-making in cancer care. Objective measures, such as bioelectrical impedance (BI), have the potential to improve prognostic accuracy for these patients. This cross-sectional study aimed to investigate whether phase angle (PhA) derived from the electrical properties of the body tissues is a predictor of muscular strength in breast cancer survivors (BCS). Methods: a total of 41 BCS (age 54.6 ± 9.2 years) were evaluated. PhA, obtained at frequency 50 kHz, was assessed with BI spectroscopy, and muscular strength with a handgrip dynamometer. Moderate-to-vigorous physical activity (MVPA) was assessed using the International Physical Activity Questionnaire (IPAQ). Measurements were performed in the morning after an overnight fast. Results: linear regression analysis showed that PhA accounted for 22% ($r^2 = 0.22$) of muscular strength variance. PhA remained a borderline predictor of muscular strength variance independently of age and MVPA. Conclusions: the findings of this study suggest that PhA is a significant predictor of maximal forearm isometric strength and a potential indicator of disease-related functionality in BCS.

Keywords: body composition; breast cancer; bioimpedance; handgrip strength

---

1. Introduction

Although muscular strength has received far less attention than cardiorespiratory fitness, recent studies support the hypothesis that low muscular strength in adulthood also predicts all-cause mortality, as well as mortality due to cardiovascular disease and cancer [1–10].

As most cancer patients do not engage in physical activity programs [11], muscle dysfunction—characterized by an impairment in muscle strength—may occur. Growing evidence suggests that exercise has the ability to ameliorate and/or reverse muscle dysfunction in cancer patients [11]. As shown by Christensen et al. [11], early-stage breast cancer patients increased muscle strength by 25–35% after a 17-week resistance training program. In this regard, the maximal isometric forearm strength test is a reliable and valid method to assess muscle function [12,13]. Recent studies of breast cancer survivors (BCS) show that surgical procedures and cancer cachexia, defined by severe muscle wasting, systemic inflammation, and malnutrition [11,14], are associated with muscle strength reduction and represent risk factors for all-cause mortality [11,13,14].

Lymphedema related to surgery or radiotherapy of the axillary area is a common complication of breast cancer treatment [15], characterized by the accumulation of protein-rich extracellular fluid resulting from damaged or blocked vessels [13], and leading to a significant increase in the volume of the affected limb that results from an impairment in the ability of the lymphatic system to drain the proteins and macromolecules of the interstitium [15]. This edema formation, resulting from the fluid redistribution between extracellular water (ECW) and intracellular water (ICW) spaces, can compromise all cell functions [16,17]. Hence, an ECW/ICW ratio seems to be of great help in detecting the early onset of lymphedema. This ratio, easily obtained with bioelectrical impedance analysis, is a validated method and appears to have equal or better sensitivity than other techniques for detecting lymphedema [18].

Recently, another simple measure obtained from bioelectrical impedance analysis has received far more attention: the phase angle (PhA), which is a noninvasive simple measure directly retrieved from resistance (R) and reactance (Xc) raw data [16]. From a biophysical point of view, PhA is calculated from the arctangent of the ratio between the R and Xc, where R arises from ECW and ICW distribution, and, conversely, Xc arises from the cell membrane's ability to take an electric load and liberate it at a later moment, after a brief delay. Therefore, it could be compared to a vessel-capacitance-like property.

Hence, PhA is considered a valuable indicator of cellular health and, as it is derived purely from electrical properties of the tissue, it avoids the typical concerns associated with prediction equations [19,20]. PhA is one of the indicators for cell membrane structure, and a lower PhA suggests decreased cellular integrity [21]. Thus PhA is considered a prognostic marker in several clinical conditions, including cancer [17], as it represents either cell death or malnutrition, both of which are characterized by changes in cellular membrane integrity. Due to this characteristic, PhA seems to be a useful predictor of impaired muscle function [16,17,22].

Gupta et al. [17] reported that PhA is a strong predictor of survival in breast cancer patients after controlling for confounders, such as stage at diagnosis and prior treatment history. Additionally, an association seems to exist between PhA and muscle strength in cancer patients, as both represent prognostic measures [23,24].

The present study emerged from the unmet need to develop a simple, easily applicable, and noninvasive tool to be used in the clinical setting to assess muscular strength. Although clinical studies using these parameters have been previously conducted, none have explored the relationship between both PhA and muscular strength in BCS after adjusting for potential confounders. Therefore, the aim of this study was to determine if PhA is a predictor of muscular strength in BCS after considering the mediation effect of lymphedema.

2. Methods

2.1. Participants

Forty-one BCS were recruited from Viva Mulher Viva Association and assessed in this observational, cross-sectional study developed at Hospital de São José, in Lisbon. Before providing written informed consent to participate, each participant was informed about the study's goals and potential benefits.

All procedures were approved by the Ethics Committee (approval code: 27012016) of the Faculty of Human Kinetics of the University of Lisbon and conducted in accordance with the World Medical Association's Declaration of Helsinki for human studies [25].

2.2. Inclusion/Exclusion Criteria

Inclusion criteria comprised patients who survived breast cancer and were currently in follow-up for their disease. Since breast cancer treatments cause body composition and muscular strength changes until six months after surgery, according to Gomes et al. [13], exclusion criteria included BCS who performed a recent (<6 months) mammary tissue removal surgery.

Recruited subjects could not be participating in other studies or have any kind of dependent relationship with study investigators.

2.3. Measurements

Before the morning visit to Hospital de São José, each participant was instructed to perform measurements after an overnight fast and to wear minimal clothing. Participants were further asked to remove all objects that could interfere with the bioelectrical impedance assessment.

2.4. Anamnesis

All patients answered a general health questionnaire, in which questions about the type of surgery were included.

2.5. Anthropometric Data

Participants' weight and height were measured using a stadiometer with an incorporated scale (SECA, Hamburg, Germany) according to standardized procedures [26]. Body mass index (BMI) was calculated as body mass (kg) divided by the stature (m) squared. Waist circumference was measured at the top of the iliac crest according to the United States National Institute of Health protocol [27]. Hip circumference measurement was performed at the widest portion of the buttocks [28]. Waist to hip (WHR) ratio was calculated accordingly.

2.6. Phase Angle and the Ratio of Extra to Intracellular Water Compartments

Whole-body R and Xc data were obtained with bioelectrical impedance spectroscopy (BIS) (model 4200B, Xitron Technologies, San Diego, CA, USA), where participants adopted a supine position with their arms and legs abducted at a 45° angle, and right hand and foot dorsal surfaces were cleaned with alcohol. After a 10-min rest, four electrodes were placed on the cleaned surfaces and measurements were performed.

Data collection was performed with a 5- to 1-MHz spectrum, from which the software was programmed to perform biophysical modelling of the impedance data, fitting spectral data to a Cole–Cole cell suspension model [29]. This procedure derives a theoretical impedance at zero and infinity frequencies based on a non-linear curve fitting, and produces general Cole model terms, namely Re (resistance associated with ECW), Ri (resistance associated with ICW), Cm (cell membrane capacitance), and exponent α. The aforementioned Cole terms are automatically applied to equations derived from the Hannai mixture theory [30], and ECW and ICW are individually calculated based on the assumption that R0 represents the R of ECW and R∞ represent the R of the intracellular and extracellular fluid sum. Accordingly, the ECW/ICW ratio was calculated and the presence of lymphedema classified [31].

PhA was estimated by recording the voltage drop between the current applied and the two output sites, and measuring phase shift at frequency 50 kHz.

2.7. Maximal Isometric Forearm Strength Test

Maximal isometric forearm strength was assessed with a dynamometer (Jamar, Sammons Preston, Inc., Bolingbrook, IL, USA). Handgrip was measured on the right and left sides, with the dominant side selected for further analysis. The assessment protocol for maximal isometric forearm strength was conducted with the subject sitting in a straight-backed chair with their feet flat on the floor, their shoulder adducted and neutrally rotated, and their elbow flexed at 90° with the forearm in neutral position [32]. A 3-second contraction time was used to obtain the maximal isometric forearm strength reading.

2.8. Physical Activity

Physical activity was evaluated with the short form of the International Physical Activity Questionnaire (IPAQ).

From the data treatment obtained in the questionnaire, several variables were accounted for in scoring the domain of activity: minutes of physical activity (total), MET/minute/week, time spent in vigorous/moderate/light physical activity, and time spent sitting over the week or weekend. The IPAQ short form is a seven-item measure of four domains of activity: vigorous-intensity physical activity, moderate-intensity physical activity, walking, and sitting [33].

In the present study, physical activity was calculated as the sum of the days, hours, and minutes of vigorous-intensity and moderate-intensity physical activity (MVPA), presented in minutes [34].

2.9. Statistical Analysis

Sample size was calculated while considering a large (>0.15) Cohen's $f2$ effect size (appropriate for calculating effect size within a multiple regression model with continuous independent and dependent variables), with a 5% type I error, 80% power, and 4 predictors (independent variables: PhA, age, MVPA, and TPS as confounding variable). The estimated sample size was 40 participants.

Statistical analysis was performed using IBM SPSS Statistics for Mac OS version 22.0, 2010 (SPSS Inc., IBM Company, Chicago, IL, USA). Descriptive statistics (mean ± standard deviation) were performed for all measurements. All variables were checked for normality using the Shapiro-Wilk test. All the variables that resulted were normally distributed, with the exception of MVPA. Bivariate correlations were conducted in a preliminary analysis. A potential confounding factor in BCS is edema of the distal extremities, which may result from lymphedema and may potentially affect impedance measurement [35]. A univariate general linear model test was performed to investigate whether lymphedema had an impact on the relationship between PhA and strength. If this association was nonsignificant ($p < 0.05$), the sample was further analyzed as a whole. Multiple regression analysis was used to determine whether PhA was a significant predictor of muscular strength after adjusting for confounding variables (age, MVPA).

$p < 0.05$ was established as significant, except in the multiple regression analysis, in which a $p < 0.1$ indicated that the independent variable was a significant model predictor [36].

3. Results

A total of 41 BCS were included in this study. Anamnesis showed that 22 women underwent mastectomy (15 of which included axillary dissection) and 19 underwent conservative surgery (10 of which included axillary dissection). Additionally, five women had lymphedema.

Table 1 summarizes the characteristics of the study sample ($n = 41$).

A univariate general linear test was performed to investigate whether PhA was associated with muscular strength regardless of the presence or absence of lymphedema. Since the analysis of the interaction term lymphedema by PhA was nonsignificant ($\beta = 0.006$; $p = 0.556$) in explaining muscular strength, the whole sample was used to test the association of PhA as a predictor of muscular strength.

A multiple regression analysis was performed while adjusting the relationship between PhA and muscular strength for confounding variables (Table 2). PhA alone explained ~22% of muscular strength variance ($\beta = 0.47$; $p < 0.001$). After adjusting for age, PhA remains a significant predictor in the model ($\beta = 0.32$; $p = 0.03$). After adjusting for age and MVPA, PhA remains a significant predictor in the model ($\beta = 0.30$; $p = 0.05$).

Figure 1 represents the association between forearm muscular strength and PhA.

Table 1. Participant characteristics (*N* = 41).

Characteristic	Mean ± Std. Deviation
Age (y)	54.6 ± 9.2
Body mass (kg)	68.0 ± 11.7
Stature (cm)	159.9 ± 6.7
BMI (kg/m^2)	26.6 ± 4.6
Waist circumference (cm)	87.1 ± 10.9
Hip circumference (cm)	100.3 ± 8.9
WHR	0.9 ± 0.1
PhA (°)	5.5 ± 0.7
ECW (L)	13.9 ± 1.7
ICW (L)	16.1 ± 3.0
E/I ratio	0.87 ± 0.09
MIF strength (kg)	25.0 ± 5.5
MVPA (min/week)	44.4 ± 49.7

Abbreviations: BMI: body mass index; WHR: waist hip ratio; PhA: phase angle; ECW: extracellular water; ICW: intracellular water; E/I ratio: ratio of extracellular and intracellular water; MIF strength: maximal isometric forearm strength; MVPA: moderate-to-vigorous physical activity.

Table 2. Unadjusted and adjusted models using phase angle (PhA) as the independent variable for determining maximal isometric forearm strength.

Model	Unstandardized Coefficients		Standardized Coefficients	*p*-Value
	β	Std. Error	β	
Model 1 (R^2 = 0.22; SEE = 4.9 kg)				
(Constant)	3.51	6.57		0.60
PhA (°)	3.91	1.19	0.47	0.00
Model 2 (R^2 = 0.35; SEE = 4.53 kg)				
(Constant)	22.60	9.23		0.02
PhA (°)	2.71	1.19	0.32	0.03
Age	−0.23	0.08	−0.39	0.01
Model 3 (R^2 = 0.36; SEE = 4.57 kg)				
(Constant)	24.39	9.77		0.02
PhA (°)	2.48	1.26	0.30	0.05
Age	−0.25	0.09	−0.42	0.01
MVPA (min/week)	0.00	0.01	0.09	0.06

Abbreviations: PhA: phase angle; MVPA: moderate-to-vigorous physical activity.

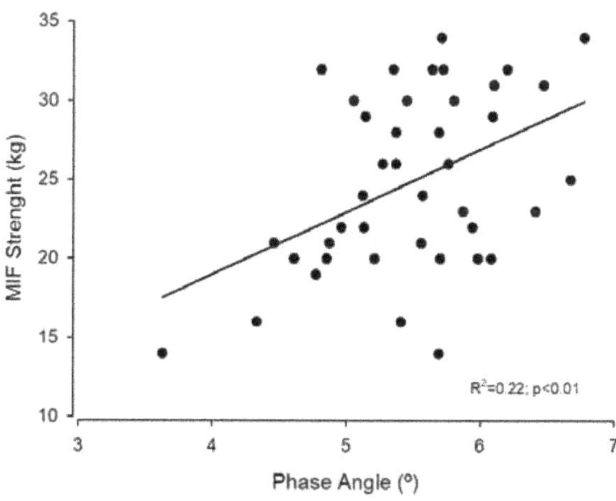

Figure 1. Regression plot of phase angle and maximal isometric forearm (MIF) strength.

4. Discussion

This study revealed that PhA accounted for 22% of muscular strength variability, remaining a significant predictor regardless of age and MVPA. PhA represents a novel and not yet fully understood marker of cellular function [20]. A high PhA represents good cell integrity, and low PhA represents cell death or decreased cell integrity [17].

However, despite its prognostic relevance, it is still necessary to define valid cut-offs to use this marker as a clinical indicator of disease-related malnutrition in several conditions [37].

Indeed, there is an absence of consensus on the best cut-off to use, with different authors using different cut-offs according to their study population, precluding their applicability to other populations [38]. Gupta et al. [17], Hui et al. [23], and Lee et al. [39] reported longer survival in cancer patients with a PhA higher than 5.6°. In the current study, the mean PhA value (5.5 ± 0.7) may be regarded as low when considering the cut-off values presented in the above-mentioned studies. Nevertheless, all those studies were performed using single-frequency analysis equipment, while the present study used a bioelectrical impedance spectroscopy approach. With that in mind, data from single- and multi-frequency devices should not be used interchangeably, as previously noted, since a lack of agreement exists between devices in determining individual R, Xc, Z, and PhA values, with methodological and biological factors pointed out as potential justifications for the observed differences [40].

Gupta et al. [17] took a step further and established that breast cancer patients with PhA <5.6° had a median survival of 23.1 months, while those with PhA >5.6° had a median survival of 49.9 months. Importantly, the aforementioned study was conducted in cancer patients, while the present investigation only included breast cancer survivors. Although the PhA values of our sample were <5.6°, the higher mean age of our sample and methodological differences between impedance instruments may help explain the lower than expected PhA values.

PhA is also one of the best indicators of cell membrane function related to the ECW/ICW ratio [19,20,41]. Significant alterations in body fluid hydration, fluid distribution, and the ECW/body cell mass (BCM) ratio caused by certain medical conditions can affect impedance measurement and are probably associated with PhA changes [42]. According to Schwenk et al. [43], a low PhA corresponds to a high ECW/ICW ratio in systemic illnesses due to ECW expansion and ICW loss. One of the most significant confounding factors is edema of the distal extremities, which may result in lymphedema [35]. In the present study, a univariate general linear model was employed to explore whether lymphedema moderated the relationship between PhA and muscular strength. Our results failed to find any significant effect, extending the findings of Gomes et al. [13] that reported similar maximal forearm strength values, regardless of the presence or absence of lymphedema.

Despite the existence or even the influence of lymphedema, loss of function and muscle strength occurs with both disease and malnutrition and is of major clinical significance [38]. Further studies using a longitudinal approach are required to investigate whether a decrease in muscle strength in BCS is mediated by lymphedema.

Data from the present investigation suggest that PhA is a predictor of maximal isometric forearm strength measured on the dominant side using the handgrip test. The gold standard for muscle strength assessment is the force exerted in a maximum voluntary contraction, with force output measured by a dynamometer [1,11,12]. This test reflects the maximum strength derived from the combined contraction of extrinsic and intrinsic hand muscles, leading to hand joint flexion [44]. The Jamar dynamometer has been reported as the most reliable, valid, fast, and easy-to-apply instrument, with the highest calibration accuracy for maximal isometric forearm strength measurement. This tool has been recently validated in advanced cancer patients [45–49].

In a study conducted by Norman et al. [24], a significant association was found between PhA and forearm muscular strength. Another study [23] conducted in advanced cancer patients found that PhA was a significant predictor of survival and that lower forearm muscular strength demonstrated a trend towards shorter survival. A systematic review conducted by Neil-Sztramko et al. [50] pooled grip-strength data from 26 studies of breast cancer and reported a 22.8 kg mean value (95% CI 20.6–25.1). However, stratifying by age groups, the mean muscular strength for the 50–59-year age group was 27.7 kg. Considering that the mean age of the present BCS cohort is within this age group, the mean forearm muscular strength retrieved (24.9 kg) was below the expected. This emphasizes the need to monitor physical activity after treatment (as well as during) to help health professionals identify functional declines and promote functional outcome improvements in these patients.

Physical activity has been used as a therapeutic auxiliary in the treatment of various pathophysiological conditions [51,52]. In healthy populations, a muscular endurance decrease was found after a short period of physical inactivity [53]. In cancer patients, physical inactivity has more exacerbated consequences [54–56], as moderate-intensity exercise can provide a sufficient physiological stimulus to improve muscular strength in cancer survivors. One finding from the present study is that most BCS do not sufficiently engage in physical activity programs (moderate-to-vigorous physical activity (MVPA): 99.39 ± 118.67 min), not meeting minimal recommendations for weekly physical activity (150 min). Besides the cancer treatment itself, this fact may be one of the reasons for muscular strength loss.

Despite the encouraging findings of this study, some limitations must be addressed. Firstly, the study sample is too small ($n = 41$) to detect a small-to-moderate effect size, and hence the power to detect an association may have been compromised. Secondly, the sample was not age balanced, with the youngest participant being 36 years old and the oldest 76 years old. This necessarily influences body composition parameters assessed by bioelectrical impedance equipment, such as PhA, as age is an important PhA determinant (lower PhA values are observed in older people due to a reactance reduction that parallels muscle mass loss) [57,58]. Considering the above, the authors adjusted the analysis for age as a potential confounder of the association between PhA and muscular strength. Finally, given the study's cross-sectional nature, a causal-effect relationship could not be established, and therefore a reverse causality should not be discarded. In addition, future studies should consider evaluating changes in the ECW/ICW ratio using the segmental bioimpedance analysis approach by assessing the PhA measured in the arms.

5. Conclusions

The findings of this study highlight phase angle as a predictor of maximal isometric forearm strength, regardless of age and level of physical activity, suggesting its usefulness as a clinical indicator of disease-related functionality in breast cancer survivors. PhA represents a novel and not-yet-fully-understood marker of cellular function, but despite its prognostic relevance, it is still necessary to define valid cut-offs to use this marker as a clinical indicator.

Author Contributions: Data curation, C.N.M. and J.C.-S.; funding acquisition, C.N.M.; investigation, C.N.M. and M.R.; methodology, J.C.-S., F.C., and A.M.S.; supervision, L.S.; writing—original draft, M.R. and A.M.S.; writing—review and editing, M.R., S.T., L.S., and A.M.S. All authors have read and agreed to the published version of the manuscript.

Funding: This research received no external funding.

Acknowledgments: The authors especially acknowledge all breast cancer survivors who contributed to this research.

Conflicts of Interest: The authors declare no conflict of interest.

References

1. Al Snih, S.; Markides, K.S.; Ray, L.; Ostir, G.V.; Goodwin, J.S. Handgrip strength and mortality in older Mexican Americans. *J. Am. Geriatr. Soc.* **2002**, *50*, 1250–1256. [CrossRef] [PubMed]
2. Fujita, Y.; Nakamura, Y.; Hiraoka, J.; Kobayashi, K.; Sakata, K.; Nagai, M.; Yanagawa, H. Physical-strength tests and mortality among visitors to health-promotion centers in Japan. *J. Clin. Epidemiol.* **1995**, *48*, 1349–1359. [CrossRef]

3. Gale, C.R.; Martyn, C.N.; Cooper, C.; Sayer, A.A. Grip strength, body composition, and mortality. *Int. J. Epidemiol.* **2007**, *36*, 228–235. [CrossRef] [PubMed]

4. Katzmarzyk, P.T.; Craig, C.L. Musculoskeletal fitness and risk of mortality. *Med. Sci. Sports Exerc.* **2002**, *34*, 740–744. [CrossRef] [PubMed]

5. Metter, E.J.; Talbot, L.A.; Schrager, M.; Conwit, R. Skeletal muscle strength as a predictor of all-cause mortality in healthy men. *J. Gerontol. A Biol. Sci. Med. Sci.* **2002**, *57*, 359–365. [CrossRef] [PubMed]

6. Sasaki, H.; Kasagi, F.; Yamada, M.; Fujita, S. Grip strength predicts cause-specific mortality in middle-aged and elderly persons. *Am. J. Med.* **2007**, *120*, 337–342. [CrossRef]

7. Ruiz, J.R.; Sui, X.; Lobelo, F.; Morrow, J.R.; Jackson, A.W.; Sjöström, M.; Blair, S.N. Association between muscular strength and mortality in men: Prospective cohort study. *BMJ* **2008**, *337*, a439. [CrossRef]

8. Rantanen, T.; Harris, T.; Leveille, S.G.; Visser, M.; Foley, D.; Masaki, K.; Guralnik, J.M. Muscle strength and body mass index as long-term predictors of mortality in initially healthy men. *J. Gerontol. A Biol. Sci. Med. Sci.* **2000**, *55*, M168–M173. [CrossRef]

9. Rantanen, T.M.K.; He, Q.; Ross, G.W.; Willcox, B.J.; White, L. Midlife muscle strength and human longevity up to age 100 years: A 44 year prospective study among a decent cohort. *Age (Dordr.)* **2012**, *34*, 564–570. [CrossRef]

10. Artero, E.G.; Lee, D.C.; Ruiz, J.R.; Sui, X.; Ortega, F.B.; Church, T.S.; Lavie, C.J.; Castillo, M.J.; Blair, S.N. A prospective study of muscular strength and all-cause mortality in men with hypertension. *J. Am. Coll. Cardiol.* **2011**, *57*, 1831–1837. [CrossRef]

11. Christensen, J.F.; Jones, L.W.; Andersen, J.L.; Daugaard, G.; Rorth, M.; Hojman, P. Muscle dysfunction in cancer patients. *Ann. Oncol.* **2014**, *25*, 947–958. [CrossRef] [PubMed]

12. Ortega, F.B.; Silventoinen, K.; Tynelius, P.; Rasmussen, F. Muscular strength in male adolescents and premature death: Cohort study of one million participants. *BMJ* **2012**, *345*, e7279. [CrossRef] [PubMed]

13. Gomes, P.R.; Freitas, I.F.; da Silva, C.B.; Gomes, I.C.; Rocha, A.P.; Salgado, A.S.; do Carmo, E.M. Short-term changes in handgrip strength, body composition, and lymphedema induced by breast cancer surgery. *Rev. Bras. Ginecol. Obstet.* **2014**, *36*, 244–250. [CrossRef] [PubMed]

14. Rantanen, T.; Volpato, S.; Ferrucci, L.; Heikkinen, E.; Fried, L.P.; Guralnik, J.M. Handgrip strength and cause-specific and total mortality in older disabled women: Exploring the mechanism. *J. Am. Geriatr. Soc.* **2003**, *51*, 636–641. [CrossRef]

15. Hayes, S.C.; Janda, M.; Cornish, B.; Battistutta, D.; Newman, B. Lymphedema after breast cancer: Incidence, risk factors, and effect on upper body function. *J. Clin. Oncol.* **2008**, *26*, 3536–3542. [CrossRef]

16. Matias, C.N.; Santos, D.A.; Martins, F.; Silva, A.M.; Laires, M.J.; Sardinha, L.B. Magnesium and phase angle: A prognostic tool for monitoring cellular integrity in judo athletes. *Magnes. Res.* **2015**, *28*, 92–98. [CrossRef]

17. Gupta, D.; Lammersfeld, C.A.; Vashi, P.G.; King, J.; Dahlk, S.L.; Grutsch, J.F.; Lis, C.G. Bioelectrical impedance phase angle as a prognostic indicator in breast cancer. *BMC Cancer* **2008**, *8*, 249. [CrossRef]

18. Rockson, S. Bioimpedance analysis in the assessment of lymphoedema diagnosis and management. *J. Lymphoedema* **2007**, *2*, 44–48.

19. Campa, F.; Matias, C.N.; Marini, E.; Heymsfield, S.B.; Toselli, S.; Sardinha, L.B.; Silva, A.M. Identifying Athlete Body-Fluid Changes During a Competitive SeasonWith Bioelectrical Impedance Vector Analysis. *Int. J. Sports Physiol. Perform.* **2020**, *15*, 361–367. [CrossRef]

20. Marini, E.; Campa, F.; Buffa, R.; Stagi, S.; Matias, C.N.; Toselli, S.; Sardinha, L.B.; Silva, A.M. Phase angle and bioelectrical impedance vector analysis in the evaluation of body composition in athletes. *Clin. Nutr.* **2020**, *39*, 447–454. [CrossRef]

21. Castizo-Olier, J.; Irurtia, A.; Jemni, M.; Carrasco-Marginet, M.; Fernandez-Garcia, R.; Rodriguez, F.A. Bioelectrical impedance vector analysis (BIVA) in sport and exercise: Systematic review and future perspectives. *PLoS ONE* **2018**, *13*, e0197957. [CrossRef] [PubMed]

22. Yamada, M.; Kimura, Y.; Ishiyama, D.; Nishio, N.; Otobe, Y.; Tanaka, T.; Ohji, S.; Koyama, S.; Sato, A.; Suzuki, M.; et al. Phase Angle Is a Useful indicator for Muscle Function in Older Adults. *J. Nutr. Health Aging* **2019**, *23*, 251–255. [CrossRef] [PubMed]

23. Hui, D.; Bansal, S.; Morgado, M.; Dev, R.; Chisholm, G.; Bruera, E. Phase angle for prognostication of survival in patients with advanced cancer: Preliminary findings. *Cancer* **2014**, *120*, 2207–2214. [CrossRef] [PubMed]

24. Norman, K.; Pirlich, M.; Sorensen, J.; Christensen, P.; Kemps, M.; Schütz, T.; Lochs, H.; Kondrup, J. Bioimpedance vector analysis as a measure of muscle function. *Clin. Nutr.* **2009**, *28*, 78–82. [CrossRef] [PubMed]

25. Association, W.M. World Medical Association Declaration of Helsinki: Ethical principles for medical research involving human subjects. *JAMA* **2013**, *310*, 2191–2194.

26. Lohman, T.G.; Roche, A.F.; Martorell, R. *Antrophometric Standardization Reference Manual*, 3–8th ed.; Human Kinetics Publishers: Champaign, IL, USA, 1988.

27. World Health Association. *Waist Circumference and Waist Hip Ratio: Report of a WHO Expert Consultation*; World Health Association: Geneva, Switzerland, 2008.

28. Pacholczak, R.; Klimek-Piotrowska, W.; Kuszmiersz, P. Associations of anthropometric measures on breast cancer risk in pre- and postmenopausal women-a case-control study. *J. Physiol. Anthropol.* **2016**, *35*, 7. [CrossRef]

29. Cole, K.S.; Cole, R.H. Dispersion and absorption in dielectrics. *J. Chem. Physiol.* **1941**, *9*. [CrossRef]

30. Hanai, T. Electrical properties of emulsions. In *Emulsion Science*; Sherman, P., Ed.; Academic Press: New York, NY, USA, 1968; pp. 354–477.

31. Wang, Z.; Deurenberg, P.; Wang, W.; Pietrobelli, A.; Baumgartner, R.N.; Heymsfield, S.B. Hydration of fat-free body mass: New physiological modeling approach. *Am. J. Physiol.* **1999**, *276*, E995–E1003. [CrossRef]

32. Innes, E. Handgrip strength testing: A review of the literature. *Aust. Occup. Ther. J.* **1999**, *46*, 120–140. [CrossRef]

33. IPAQ (2005) Guidelines for Data Processing and Analysis of the International Physical Activity Questionnaire (IPAQ)—Short and Long Forms, revised on November 2005. Available online: http://www.ipaq.ki.se/scoring.pdf (accessed on 15 March 2010).

34. Craig, C.L.; Marshall, A.L.; Sjostrom, M.; Bauman, A.E.; Booth, M.L.; Ainsworth, B.E.; Pratt, M.; Ekelund, U.; Yngve, A.; Sallis, J.F.; et al. International physical activity questionnaire: 12-country reliability and validity. *Med. Sci. Sports Exerc.* **2003**, *35*, 1381–1395. [CrossRef]

35. Dehghan, M.; Merchant, A.T. Is bioelectrical impedance accurate for use in large epidemiological studies? *Nutr. J.* **2008**, *7*, 26. [CrossRef] [PubMed]

36. Lang, T. Documenting research in scientific articles: Guidelines for authors: 3. Reporting multivariate analyses. *Chest* **2007**, *131*, 628–632. [CrossRef] [PubMed]

37. Norman, K.; Stobaus, N.; Pirlich, M.; Bosy-Westphal, A. Bioelectrical phase angle and impedance vector analysis—Clinical relevance and applicability of impedance parameters. *Clin. Nutr.* **2012**, *31*, 854–861. [CrossRef]

38. Norman, K.; Stobaus, N.; Zocher, D.; Bosy-Westphal, A.; Szramek, A.; Scheufele, R.; Smoliner, C.; Pirlich, M. Cutoff percentiles of bioelectrical phase angle predict functionality, quality of life, and mortality in patients with cancer. *Am. J. Clin. Nutr.* **2010**, *92*, 612–619. [CrossRef]

39. Lee, S.Y.; Lee, Y.J.; Yang, J.H.; Kim, C.M.; Choi, W.S. The Association between Phase Angle of Bioelectrical Impedance Analysis and Survival Time in Advanced Cancer Patients: Preliminary Study. *Korean J. Fam. Med.* **2014**, *35*, 251–256. [CrossRef]

40. Silva, A.M.; Matias, C.N.; Nunes, C.L.; Santos, D.A.; Marini, E.; Lukaski, H.C.; Sardinha, L.B. Lack of agreement of in vivo raw bioimpedance measurements obtained from two single and multi-frequency bioelectrical impedance devices. *Eur. J. Clin. Nutr.* **2019**, *73*, 1077–1083. [CrossRef]

41. Francisco, R.; Matias, C.N.; Santos, D.A.; Campa, F.; Minderico, C.S.; Rocha, P.; Heymsfield, S.B.; Lukaski, H.; Sardinha, L.B.; Silva, A.M. The Predictive Role of Raw Bioelectrical Impedance Parameters in Water Compartments and Fluid Distribution Assessed by Dilution Techniques in Athletes. *Int. J. Environ. Res. Public Health* **2020**, *17*, 759. [CrossRef] [PubMed]

42. Selberg, O.; Selberg, D. Norms and correlates of bioimpedance phase angle in healthy human subjects, hospitalized patients, and patients with liver cirrhosis. *Eur. J. Appl. Physiol.* **2002**, *86*, 509–516. [CrossRef] [PubMed]

43. Schwenk, A.; Beisenherz, A.; Römer, K.; Kremer, G.; Salzberger, B.; Elia, M. Phase angle from bioelectrical impedance analysis remains an independent predictive marker in HIV-infected patients in the era of highly active antiretroviral treatment. *Am. J. Clin. Nutr.* **2000**, *72*, 496–501. [CrossRef]

44. Norman, K.; Stobäus, N.; Gonzalez, M.C.; Schulzke, J.D.; Pirlich, M. Hand grip strength: Outcome predictor and marker of nutritional status. *Clin. Nutr.* **2011**, *30*, 135–142. [CrossRef]

45. Peolsson, A.; Hedlund, R.; Oberg, B. Intra- and inter-tester reliability and reference values for hand strength. *J. Rehabil. Med.* **2001**, *33*, 36–41. [CrossRef] [PubMed]

46. Mathiowetz, V.; Weber, K.; Volland, G.; Kashman, N. Reliability and validity of grip and pinch strength evaluations. *J. Hand Surg Am.* **1984**, *9*, 222–226. [CrossRef]

47. Spruit, M.A.; Sillen, M.J.; Groenen, M.T.; Wouters, E.F.; Franssen, F.M. New normative values for handgrip strength: Results from the UK Biobank. *J. Am. Med. Dir. Assoc.* **2013**, *14*, e5–e11. [CrossRef] [PubMed]

48. Kilgour, R.D.; Vigano, A.; Trutschnigg, B.; Lucar, E.; Borod, M.; Morais, J.A. Handgrip strength predicts survival and is associated with markers of clinical and functional outcomes in advanced cancer patients. *Support. Care Cancer* **2013**, *21*, 3261–3270. [CrossRef]

49. Trutschnigg, B.; Kilgour, R.D.; Reinglas, J.; Rosenthall, L.; Hornby, L.; Morais, J.A.; Vigano, A. Precision and reliability of strength (Jamar vs. Biodex handgrip) and body composition (dual-energy X-ray absorptiometry vs. bioimpedance analysis) measurements in advanced cancer patients. *Appl. Physiol. Nutr. Metab.* **2008**, *33*, 1232–1239. [CrossRef]

50. Neil-Sztramko, S.E.; Kirkham, A.A.; Hung, S.H.; Niksirat, N.; Nishikawa, K.; Campbell, K.L. Aerobic capacity and upper limb strength are reduced in women diagnosed with breast cancer: A systematic review. *J. Physiother.* **2014**, *60*, 189–200. [CrossRef]

51. Schneider, C.M.; Hsieh, C.C.; Sprod, L.K.; Carter, S.D.; Hayward, R. Cancer treatment-induced alterations in muscular fitness and quality of life: The role of exercise training. *Ann. Oncol.* **2007**, *18*, 1957–1962. [CrossRef]

52. Toselli, S.; Badicu, G.; Bragonzoni, L.; Spiga, F.; Mazzuca, P.; Campa, F. Comparison of the Effect of Different Resistance Training Frequencies on Phase Angle and Handgrip Strength in ObeseWomen: A Randomized Controlled Trial. *Int. J. Environ. Res. Public Health* **2020**, *17*, 1163. [CrossRef]

53. Bogdanis, G.C. Effects of physical activity and inactivity on muscle fatigue. *Front. Physiol.* **2012**, *3*, 142. [CrossRef]

54. Phillips, S.M.; Dodd, K.W.; Steeves, J.; McClain, J.; Alfano, C.M.; McAuley, E. Physical activity and sedentary behavior in breast cancer survivors: New insight into activity patterns and potential intervention targets. *Gynecol Oncol.* **2015**, *138*, 398–404. [CrossRef]

55. Ligibel, J.A.; Engquist, K.B.; Bea, J.W. Weight Management and Physical Activity for Breast Cancer Prevention and Control. *Am. Soc. Clin. Oncol. Educ. Book* **2019**, *39*, e22–e33. [CrossRef] [PubMed]

56. Reis, A.D.; Pereira, P.; Diniz, R.R.; de Castro Filha, J.; Dos Santos, A.M.; Ramallo, B.T.; Filho, F.; Navarro, F.; Garcia, J. Effect of exercise on pain and functional capacity in breast cancer patients. *Health Qual. life Outcomes* **2018**, *16*, 58. [CrossRef] [PubMed]

57. Campa, F.; Silva, A.M.; Toselli, S. Changes in Phase Angle and Handgrip Strength Induced by Suspension Training in Older Women. *Int. J. Sports Med.* **2018**, *39*, 442–449. [CrossRef] [PubMed]

58. Kyle, U.G.; Bosaeus, I.; De Lorenzo, A.D.; Deurenberg, P.; Elia, M.; Gómez, J.M.; Heitmann, B.L.; Kent-Smith, L.; Melchior, J.C.; Pirlich, M.; et al. Bioelectrical impedance analysis—Part I: Review of principles and methods. *Clin. Nutr.* **2004**, *23*, 1226–1243. [CrossRef] [PubMed]

International Journal of
*Environmental Research
and Public Health*

Article

Handgrip Strength in Young Adults: Association with Anthropometric Variables and Laterality

Luciana Zaccagni [1,2], **Stefania Toselli** [3,*], **Barbara Bramanti** [1,4,*], **Emanuela Gualdi-Russo** [1,*], **Jessica Mongillo** [1] **and Natascia Rinaldo** [1]

[1]   Department of Biomedical and Specialty Surgical Sciences, Faculty of Medicine, Pharmacy and Prevention, University of Ferrara, 44121 Ferrara, Italy; luciana.zaccagni@unife.it (L.Z.); mngjsc@unife.it (J.M.); natascia.rinaldo@unife.it (N.R.)
[2]   Biomedical Sport Studies Center, University of Ferrara, 44123 Ferrara, Italy
[3]   Department of Biomedical and Neuromotor Science, University of Bologna, 40126 Bologna, Italy
[4]   University Center for Studies on Gender Medicine, University of Ferrara, 44121 Ferrara, Italy
[*]   Correspondence: stefania.toselli@unibo.it (S.T.); brmbbr@unife.it (B.B.); emanuela.gualdi@unife.it (E.G.-R.)

Received: 5 May 2020; Accepted: 12 June 2020; Published: 15 June 2020

Abstract: The measurement of handgrip strength (HGS) is an indicator of an individual's overall strength and can serve as a predictor of morbidity and mortality. This study aims to investigate whether HGS is associated with handedness in young adults and if it is influenced by anthropometric characteristics, body composition, and sport-related parameters. We conducted a cross-sectional study on a sample of 544 young Italian adults aged 18–30 years. We measured HGS using a dynamometer and collected data on handedness and physical activity, along with anthropometric measurements. In both sexes, the HGS of the dominant side was significantly greater than that of the non-dominant side. Furthermore, in ambidextrous individuals, the right hand was stronger than the left. A comparison between the lowest and the highest tercile of HGS highlighted its significant association with anthropometric and body composition parameters in both sexes. Moreover, sex, dominant upper arm muscle area, arm fat index, fat mass, and fat-free mass were found to be significant predictors of HGS by multiple regression analysis. Our findings suggest that HGS is especially influenced by body composition parameters and handedness category. Therefore, HGS can be used as a proxy for unhealthy conditions with impairment of muscle mass, provided that the dominance in the laterality of the subject under examination is taken into account.

Keywords: handgrip strength; anthropometry; handedness; body composition; physical activity; sports practice

---

1. Introduction

Handgrip strength (HGS) is a fundamental parameter in biomechanical modeling, which has found many important applications in the development of ergonomic tools, in the design of equipment and consumer products, and in sports practices [1–8]. Grip strength is crucial for the human body when performing prehensile and precision hand functions [9], and it is used as one of the main indicators for testing muscle power. Moreover, it is a low-cost tool for predicting an individual's overall strength, which can reflect general health conditions and level of physical activity [10]. In fact, muscle weakness and low grip strength have been related to disability and are considered predictors of sarcopenia, higher recovery time, and higher mortality, especially in hospitalized patients [11–14]. Nevertheless, several populations lack the HGS reference values that are necessary to ensure the health, safety, comfort, and productivity of workers and consumers, as well as for clinical purposes and to monitor recovery after post-injury treatment [15].

Reference values are essential for the correct interpretation of acquired data, for the definition of appropriate treatments, and for the evaluation of the effectiveness of interventions [16–18]. Furthermore, reference values can also be used as motivation for patients during rehabilitation interventions. One of the goals of this study is to contribute new mean HGS values obtained from healthy young Italian individuals disaggregated by sex, which can be used as a reference for equivalent populations.

Many studies have reported the relationships between grip strength and various variables [9,19,20]; however, the predominant factor which influences hand grip strength remains unclear. From a summary of the principal findings in the literature, some relations can be identified. Age has an important influence on grip strength, for which a curvilinear relationship has been observed, resulting in an increase in grip strength with age that reaches a peak between 30 and 40 years of age, and decreases thereafter. The non-linear relation differs between the sexes and is more pronounced in females. Males maintain their grip strength for at least another decade longer than females [15,21–23]. The difference between the sexes is evident at all ages and males generally have greater grip strength than females [15,17,23–25]. This is the consequence of the larger size and related muscle mass in males, as muscle strength is a function of these characteristics. Males generally feature bigger arm muscles and are more involved in more activities which require strength than females. Strong correlations have also been reported between grip strength and body dimensions (e.g., weight, height, and hand length) [26–32]. With respect to body mass index, the results are controversial: some authors have reported that static grip strength is positively related to body mass index (BMI), considering it to be a predictor for grip strength, while others found no significant association, concluding that BMI does not influence handgrip strength [15,20]. Forearm circumference, wrist joint circumference, palm circumference, hand length, and middle finger length play major roles in influencing the dominant hand's grip strength [15,20,33–35]. Working activity can also influence handgrip strength, where manual workers have a higher average grip strength than non-manual workers [15]. In light of the above, it is necessary to find the best predictors of HGS values among different anthropometric characteristics, as well as establish the impact of physical and sporting activities on hand strength.

Hand grip differs according to the laterality, where the dominant hand has a greater capability for grip strength than the non-dominant hand [15]. This is probably related to the fact that the dominant hand is used forcefully more often than the non-dominant one, such that the muscles of the dominant side are bigger and, thus, stronger. Nevertheless, this is true for people with a dominant right hand; while for people with a dominant left hand, this difference is smaller and sometimes not even significant [15,20,36]. In this respect, it is important to remember that humans demonstrate functional differences in the right and left of each bilaterally symmetrical body part [37–41]. Laterality is particularly evident in the functions of the fingers, such as for spoon use or letter writing, and has been considered to be due to the more preferential and frequent use of one hand in daily activities. Until now, lateral dominance of muscle function has frequently been reported. Laterality represents a multidimensional trait [42] and it is well-known that, in the adult population, 90% of people prefer to use their right hand for common manual tasks, whereas about 10% of the population is left-handed [43–46]. Another important observation is that, throughout human life, the development of laterality is a very active process, influenced by both genetic and environmental factors [46–51].

Handedness can be defined as "the individual's preference to use one hand predominantly for unimanual tasks and the ability to perform these tasks more efficiently with one hand" [52]. Significant differences in the consistency of handedness between males and females have been observed, and a significantly higher number of ambidextrous individuals among males has been detected [51,53–55].

Another significant aim of this study is to assess the influence of laterality on hand muscle strength in young adults subdivided by sex, in order to evaluate whether the level of laterality assessed as the difference in handgrip skillfulness between the preferred and the non-preferred limb differs in males and females. Finally, considering the limited information on dominant handgrip strength in the literature [56], especially for youth, this study is designed to identify anthropometric and activity-related determinants of dominant hand strength in healthy young subjects.

2. Materials and Methods

2.1. Participants and Procedure

We carried out a cross-sectional survey which included young subjects, aged 18–30, in a convenience sample. Given the ethnic variability of the anthropometric traits [57,58], only individuals of Italian origin were included in the sample. They were recruited among the students of the Universities of Ferrara and Bologna (Northern Italy), according to the inclusion and exclusion criteria specified below.

The inclusion criteria for participation were: (1) being of Italian origin; (2) being aged ≥18 years; (3) having signed the written informed consent; and (4) having declared to be healthy. Those aged >30 years were excluded.

For this study, we excluded 16 foreign students, 20 students over 30 years of age, and 50 subjects with incomplete data on handedness. The result was a final sample of 544 young Italian adults (356 males and 188 females) with an average age of 21.6 ± 2.9 years for males and 21.3 ± 2.0 years for females.

The study was approved by the Ethics Committee of the University of Bologna (registration number: 253035, 17 October 2019) and followed the rules and principles of the Helsinki Declaration.

2.2. Data Collection

The collection of data was carried out through self-administered questionnaires on hand laterality and physical activity, as well as through the direct anthropometric measurements described in detail below.

Questionnaires: First, we collected personal demographic information (sex, place and date of birth, citizenship), health status (absence/presence of illness or infirmity), and the physical/sporting activity of the participants. With regards to sporting activity, participants were asked whether they had practiced any sport, which sport, for how many years, and how many hours they practiced per week. Sports were categorized according to their metabolic equivalent of task (MET) intensity values [59]. Physical activities were assessed by the International Physical Activity Questionnaire (IPAQ) and classified as light, moderate, or intense. We used the Edinburgh Handedness Inventory (EHI) as a method to assess handedness on a quantitative scale [60]. Participants examined the 10-item inventory by choosing the answer column ("left hand" or "right hand") for each item and indicating their weak (1) or strong (2) preference for the left or right hand in performing the given action specified in the item. If there was no specific preference for one hand over the other in that item, the participant indicated 1 in both columns. At the end of the procedure, a handedness score (R) between −100 (strongly left) and +100 (strongly right) was obtained. An ambidextrous condition was indicated by scores between −40 and +40 [61,62].

Measurements: Anthropometric measurements were collected by trained anthropometrists using specific instruments (i.e., Raven anthropometer, Seca weighing scale, Lange plicometer, GPM measuring tape, and Takei handgrip dynamometer). This study took into consideration seven measurements—stature, weight, mid-upper arm circumference (MUAC), waist circumference (WC), triceps skinfold thickness, and right and left handgrip strength (HGS)—collected from each subject according to standard procedures [63–65], as well as five anthropometric indices (reported below) obtained from previous measurements. Moreover, the body density (BD), estimated according to the formula developed by Durnin and Womersley [66], was converted to percentage fat mass (%F) using the Siri [67] equation: $((4.95/BD) - 4.50) \times 100$. The fat mass (FM, kg) was calculated as (%F × weight)/100, and the fat-free mass (FFM, kg) as weight − FM.

With regards to anthropometric measurements, stature was measured from each subject while standing barefoot. The subject's head was oriented according to the Frankfurt plane. The weight of each participant was measured, while dressed in underwear or light clothing, using a mechanical scale. The maximal HGS of both hands was measured in kg to five decimal points by a hand-grip dynamometer. The participants carried out three maximum voluntary contractions per side with a 60 s rest break between each test. In the statistical analyses, the highest strength value obtained in the three tests of each hand was used. MUAC and triceps skinfold thicknesses were measured on both sides of the body at the midpoint between the acromion process of the scapula and the olecranon process of the ulna. WC was measured at the narrowest point between the lowest ribs and the iliac crest.

For the anthropometric indices, the body mass index (BMI) was computed as weight (kg)/stature (m^2). Subjects were classified as underweight, normal weight, overweight, or obese, according to the World Health Organization (WHO) cut-points [68]. Based on the %F value, subjects were classified as under-fat, normal fat, over-fat, or very over-fat, according to the cut-points of Gallagher et al. [69]. Using MUAC and triceps skinfold, we also calculated the following indices of nutritional status: total upper arm area (TUA, cm^2), upper arm muscle area (UMA, cm^2), upper arm fat area (UFA, cm^2), and arm fat index (AFI, %) [70]. For all bilateral characteristics and related indices, we reported separately the mean values of the right, left, and dominant side (the last value was replaced by the mean of the two sides for ambidextrous subjects).

2.3. Statistical Analysis

Quantitative variables were described using means and standard deviations, and qualitative variables described by frequencies. Due to the fact that all the variables (except for skinfold thickness) had a normal univariate distribution (skewness and kurtosis were within acceptable ranges), parametric statistical tests were applied. Comparisons between two independent samples (males vs. females; lowest vs. highest tercile) were performed by Student's *t*-test. These comparisons were carried out using the logarithmic transformation of skinfold thickness to normalize this trait, which is usual practice [66,71,72].

The paired samples (dominant vs. non-dominant handgrip strength) were compared with the *t*-test for dependent samples, or with the Wilcoxon signed-rank test when the subsample size for a certain trait was too small (i.e., comparison of HGS within some category of handedness). Then, one-way analysis of variance (ANOVA) was applied to detect any difference in HGS means of the dominant side among the three handedness groups (i.e., right-handed, left-handed, and ambidextrous individuals).

Comparisons among categorical data (frequencies of fat status categories; weight status categories) were performed with the Chi-squared test.

The Pearson correlation coefficient was used to measure associations of anthropometric traits with grip strength. Backward multiple regression analysis was carried out to assess possible predictors of dominant HGS. Anthropometric variables (of the dominant side for characters detectable on both sides), metabolic equivalent of task (MET), and fat status were included in the regression models in the continuous scale, while sex was included in the model as a binary variable with females as the reference group. Predictors input into the model were those found to have significant associations with HGS (i.e., $p < 0.05$), while those with $p > 0.05$ were removed from the model. The multicollinearity of the data was assessed by variance inflation factors (VIFs), assuming VIF values between 0.10 and 10 as acceptable [23].

A value of $p < 0.05$ was accepted as the level of significance for all statistical tests.

All analyses were performed using the software STATISTICA, version 11 (StatSoft, Tulsa, OK, USA).

3. Results

Table 1 shows descriptive statistics of the sample by sex. As expected, males were significantly taller, heavier, stronger in both hands in absolute and relative terms (handgrip strength/weight), and with larger circumferences (waist and upper arms), on average, but with less skinfold thickness than females.

Table 1. Anthropometric characteristics, sports habits, and fat and weight status by sex (R = right; L = left; D = dominant side).

Anthropometric Traits	Males		Females		
	Mean	SD	Mean	SD	*p*
Stature (cm)	178.0	7.0	163.6	6.1	<0.0001
Weight (kg)	74.9	11.1	58.2	8.0	<0.0001
BMI (kg/m^2)	23.6	2.9	21.7	2.6	<0.0001
WC (cm)	79.7	7.8	69.3	5.4	<0.0001
L MUAC (cm)	30.4	3.5	26.5	2.8	<0.0001
R MUAC (cm)	30.7	3.3	26.6	2.7	<0.0001
D MUAC (cm)	30.6	3.3	26.6	2.8	<0.0001
L Triceps skinfold (mm)	9.8	4.8	17.0	5.1	<0.0001 [a]
R Triceps skinfold (mm)	9.9	5.0	16.9	5.2	<0.0001 [a]
D Triceps skinfold (mm)	9.9	4.9	16.9	5.1	<0.0001 [a]
D TUA (cm^2)	75.4	16.6	56.5	12.4	<0.0001
D UMA (cm^2)	61.2	14.8	36.1	7.6	<0.0001
D UFA (cm^2)	14.4	7.2	20.4	7.5	<0.0001
D AFI (%)	18.9	7.7	35.5	8.0	<0.0001
%F	14.4	4.4	25.8	4.5	<0.0001
FM (kg)	11.0	4.5	15.2	4.3	<0.0001
FFM (kg)	63.9	8.3	43.0	4.9	<0.0001
L HGS (kg)	43.9	8.1	27.5	5.0	<0.0001
R HGS (kg)	45.9	8.3	28.8	4.8	<0.0001
D HGS (kg)	45.7	8.2	28.9	4.7	<0.0001
D HGS/weight	0.6	0.1	0.5	0.1	<0.0001
Sports and PA					
Sport amount (h/week)	7.0	4.0	6.1	4.1	0.0169
Sport practice (years)	9.2	5.2	8.9	5.2	0.5458
PA (METs)	4827.6	3268.7	3621.4	3300.5	0.0005
	N	%	N	%	*p*
Weight status					<0.0001
Under weight	5	1.4	16	8.5	
Normal weight	269	75.6	155	82.4	
Overweight	70	19.7	15	8.0	
Obese	12	3.4	2	1.1	
Fat status					0.0065
Under fat	23	6.5	26	13.8	
Normal fat	295	82.8	152	80.9	
Overfat	34	9.6	10	5.3	
Very overfat	4	1.1	0	0.0	
Distribution by categories of sports by METs					0.9595
METs < 2	11	3.1	6	3.3	
2 ≤ METs < 4	8	2.3	3	1.6	
4 ≤ METs < 6.5	74	20.8	40	21.2	
METs ≥ 6.5	262	73.8	139	73.9	
Distribution by categories of PA by METs					0.0463
Light	13	3.6	11	5.6	
Moderate	70	19.8	54	28.9	
Intense	273	76.7	123	65.5	

[a] Comparisons were performed using Log skinfolds. Abbreviations: BMI, body mass index; WC, waist circumference; MUAC, mid-upper arm circumference; TUA, total upper arm area; UMA, upper arm muscle area; UFA, upper arm fat area; AFI, arm fat index; %F, percent fat mass; FM, fat mass; FFM, fat-free mass; HGS, handgrip strength; PA, physical activity; MET, metabolic equivalent of task.

Mean BMI values fell into the normal weight status, according to WHO weight status categories, but the sexes differed in their distribution of weight status: underweight and normal weight categories were more represented in females, while overweight and obese were more represented in males.

Concerning body composition, males were on average leaner than females: the muscle area of the upper arm and FFM in males were higher than in females, but the fat area of the upper arm, arm fat index, %F, and FM in males were lower than in females.

The fat status distribution also differed significantly by sex: the under-fat category was more represented in females (almost twice as often in females than in males), and over-fat and very over-fat categories were more represented in males (over-fat: almost twice as often in males as in females; very over-fat: no females).

In both sexes, the number of years of sports practice (about nine) were similar, but males practiced a significantly higher weekly amount of sport than females. Both sexes chose a similar sports discipline, according to the classification of METs: nearly three-quarters chose intense sports and almost one-fifth of them chose moderate sports.

With regards to physical activity (PA), as assessed with the IPAQ, males were significantly more active than females and practiced more intense activities than females.

Table 2 shows the results obtained from the administration of the EHI. Females had, on average, a significantly higher R score than males. The distribution of handedness differed significantly by sex, with more ambidextrousness among males and more right-handers among females.

Table 2. R scores (mean and SD) and frequencies of hand preference (according to Edinburgh Handedness Inventory (EHI)) by sex.

Handedness	Males		Females		
	Mean	SD	Mean	SD	p
R score	45.0	42.7	54.2	37.7	0.0347
Frequencies	**N**	**%**	**N**	**%**	0.0075
Right-handed	212	59.6	137	72.9	
Left-handed	18	5.1	8	4.3	
Ambidextrous	126	35.4	43	22.9	

Table 3 shows the results of the HGS by handedness category. In both sexes, the HGS of the dominant side was significantly greater than that of the non-dominant side, apart from left-handed females; moreover, in ambidextrous subjects, the right hand was stronger than the left hand.

Table 3. Comparison between sides in handgrip strength for each handedness category by sex.

Handedness Category	Right Handgrip Strength		Left Handgrip Strength		
	Mean	SD	Mean	SD	p
Males					
Right-handed	46.0	8.8	43.6	8.4	<0.0001 [a]
Left-handed	40.9	6.3	43.3	6.3	0.0468 [b]
Ambidextrous	46.6	7.9	44.6	8.3	<0.0001 [a]
Females					
Right-handed	29.0	4.9	27.4	5.2	<0.0001 [a]
Left-handed	25.3	5.8	26.6	4.9	0.2489 [b]
Ambidextrous	28.8	4.1	28.0	4.5	0.0693 [a]

[a] Comparison was performed using *t*-test for dependent sample; [b] comparison was performed using Wilcoxon non-parametric test.

No difference was detected in the comparison of the mean HGS of the dominant side among the three independent handedness groups (disaggregated by sex) by ANOVA (males: F = 1.010, df = 2; 358, p = 0.3653; females: F = 1.330, df = 2; 191, p = 0.2669).

Tables 4 and 5 report the results of the comparisons between the lowest and the highest terciles of the dominant HGS, separately for each sex.

Table 4. Comparison between first and last tercile of dominant handgrip strength in males (D = dominant side).

Anthropometric Traits (Males)	1st Tercile (Strength ≤ 42.0 kg)		3rd Tercile (Strength ≥ 49.5 kg)		
	Mean	SD	Mean	SD	*p*
Stature (cm)	176.0	6.6	180.5	6.6	0.0000
Weight (kg)	69.5	8.8	80.9	10.9	0.0000
BMI (kg/m^2)	22.4	2.5	24.8	2.9	0.0000
WC (cm)	76.8	6.8	82.1	8.6	0.0000
D Triceps skinfold (mm)	9.7	4.6	9.9	5.8	0.1969 [a]
D MUAC (cm)	28.9	2.8	32.5	3.2	0.0000
D TUA (cm^2)	67.5	13.1	85.0	17.3	0.0000
D UMA (cm^2)	54.5	11.1	70.2	15.9	0.0000
D UFA (cm^2)	13.3	6.1	15.1	8.6	0.0647
D AFI (%)	19.3	7.2	17.6	8.3	0.0963
%F	14.0	4.3	14.4	4.7	0.5185
FM (kg)	9.9	3.8	11.9	5.2	0.0008
FFM (kg)	59.6	6.5	69.2	7.8	0.0000
Sports and PA					
Sport amount (h/week)	6.6	3.7	7.4	4.3	0.1149
Sport practice (years)	10.1	5.1	8.2	5.2	0.0062
PA (METs)	4431.9	2912.9	4853.0	3050.2	0.3625
		%		%	*p*
Weight status					0.0001
Underweight	4	3.2	0	0.0	
Normal weight	103	83.2	75	63.6	
Overweight	16	12.8	36	30.5	
Obese	1	0.8	7	5.9	
Fat status					0.8168
Under fat	10	8.0	7	6.0	
Normal fat	102	82.4	95	81.2	
Overfat	11	8.8	14	12.0	
Obese	1	0.8	1	0.9	
Distribution by categories of sports by METs					0.0038
METs < 2	4	3.2	4	3.4	
2 ≤ METs < 4	0	0.0	4	3.4	
4 ≤ METs < 6.5	18	14.5	35	29.7	
METs ≥ 6.5	102	82.3	75	63.6	
Distribution by categories of PA by METs					0.7627
Light	3	2.7	2	2.1	
Moderate	28	22.7	23	19.1	
Intense	93	74.7	93	78.7	
Handedness categories					0.2947
Right-handed	75	60.8	76	64.4	
Left-handed	10	8.0	4	3.4	
Ambidextrous	39	31.2	38	32.2	

[a] Comparisons were performed using Log skinfolds. Abbreviations: BMI, body mass index; WC, waist circumference; MUAC, mid-upper arm circumference; TUA, total upper arm area; UMA, upper arm muscle area; UFA, upper arm fat area; AFI, arm fat index; %F, percent fat mass; FM, fat mass; FFM, fat-free mass; PA, physical activity; MET, metabolic equivalent of task.

Table 5. Comparison between the lowest and the highest tercile of dominant handgrip strength in females (D = dominant side).

Anthropometric Traits (Females)	1st Tercile (Strength ≤ 26.8 kg)		3rd Tercile (Strength ≥ 30.5 kg)		
	Mean	SD	Mean	SD	*p*
Stature (cm)	162.0	6.4	165.1	6.2	0.0062
Weight (kg)	55.6	6.1	61.8	8.7	0.0000
BMI (kg/m^2)	21.2	2.1	22.6	2.8	0.0015
WC (cm)	68.2	4.4	70.7	5.2	0.0037
D Triceps skinfold (mm)	15.5	4.8	18.0	4.9	0.0026 [a]
D MUAC (cm)	25.7	2.1	27.5	2.8	0.0002
D TUA (cm^2)	52.9	8.9	60.5	13.4	0.0007
D UMA (cm^2)	34.2	6.4	38.4	8.3	0.0036
D UFA (cm^2)	18.7	6.0	22.1	7.8	0.0107
D AFI (%)	35.0	8.4	36.2	7.3	0.4105
%F	25.2	4.5	26.6	4.0	0.0597
FM (kg)	14.1	3.4	16.7	4.5	0.0006
FFM (kg)	41.5	4.3	45.1	5.1	0.0000
Sports and PA					
Sport amount (h/week)	5.4	3.6	7.0	3.7	0.0153
Sport practice (yrs)	9.0	5.6	9.1	5.2	0.9536
PA (METs)	2910.4	3080.9	3908.1	3373.6	0.1571
	%		%		*p*
Weight status					0.3834
Under weight	4	6.9	4	6.0	
Normal weight	51	87.9	54	80.6	
Overweight	3	5.2	7	10.4	
Obese	0	0.0	2	3.0	
Fat status					0.1797
Under fat	11	18.3	5	7.5	
Normal fat	45	78.3	59	88.1	
Overfat	2	3.3	3	4.5	
Obese	0	0.0	0	0.0	
Distribution by categories of sports by METs					0.4536
METs < 2	1	1.8	2	3.0	
2 ≤ METs < 4	2	3.5	0	0.0	
4 ≤ METs < 6.5	9	15.8	12	17.9	
METs ≥ 6.5	46	78.9	53	79.1	
Distribution by categories of PA by METs					0.0948
Light	3	5.9	4	6.8	
Moderate	22	38.2	14	20.3	
Intense	32	55.9	49	72.9	
Handedness categories					0.1770
Right-handed	39	67.2	49	73.1	
Left-handed	5	8.6	1	1.5	
Ambidextrous	14	24.1	17	25.4	

[a] Comparisons were performed using Log skinfolds. Abbreviations: BMI, body mass index; WC, waist circumference; MUAC, mid-upper arm circumference; TUA, total upper arm area; UMA, upper arm muscle area; UFA, upper arm fat area; AFI, arm fat index; %F, percent fat mass; FM, fat mass; FFM, fat-free mass; PA, physical activity; MET, metabolic equivalent of task.

As shown in Table 4, males with higher values of HGS (highest tercile) were taller and heavier, with higher mean BMI and waist circumference values. Analyzing the body composition of the participants in detail, we found higher values of strength in males with higher values of upper arm muscularity, MUAC, FM, and FFM. However, the triceps skinfold, %F, and AFI did not differ between the lowest and the highest terciles. Consistently with the mean values of BMI, the distribution of weight status categories was significantly different between the two terciles, as we found high percentages of overweight and obese males in the highest tercile. In contrast, the fat status category distribution was similar between the two groups. Regarding sport and physical activity, only the years of sports practice and the intensity of the sport resulted in a significant difference between the two terciles. Males from the lowest tercile, who showed more years of experience and practiced sports classified as more intense by METs, had lower values of handgrip strength. There were no significant differences in the handedness distribution between the two terciles (Table 4).

Comparing the anthropometric and body composition characteristics between the lowest and the highest tercile in females (Table 5), the results were similar to those of the male group. Females with higher values of HGS were taller and heavier, with higher values of BMI, WC, triceps skinfold, MUAC, FM, FFM, UFA, and UMA. As in the male group, %F and AFI did not differ between the two terciles. Moreover, the sample distribution in the weight and fat status categories was similar between the two terciles. Considering the practice of sports and PA, only the hours of sports practice resulted in a significant difference between the two terciles, with females with higher values of HGS practicing more h/week. The handedness distribution was similar between the two terciles.

Table 6 shows the results of the correlation between the anthropometric variables and the dominant HGS for each sex. Almost all the traits were positively correlated with HGS in both sexes. HGS was positively correlated with dominant UFA in females and negatively correlated with AFI in males. Both FM and FFM proved to be positively correlated with HGS in both sexes.

Table 6. Pearson correlation coefficients between anthropometric variables and dominant handgrip strength by sex (asterisks denote statistical significance at the $p < 0.05$ level; D = dominant side).

Dominant Handgrip Strength	Males	Females
Stature	0.229 *	0.245 *
Weight	0.436 *	0.392 *
BMI	0.372 *	0.293 *
WC	0.272 *	0.270 *
D MUAC	0.502 *	0.319 *
D Triceps skinfold (Log)	−0.047	0.201 *
D TUA	0.501 *	0.317 *
D UMA	0.522 *	0.275 *
D UFA	0.089	0.232 *
D AFI	−0.125 *	0.076
%F	0.037	0.162
FM	0.176 *	0.300 *
FFM	0.490 *	0.375 *

Abbreviations: BMI, body mass index; WC, waist circumference; MUAC, mid-upper arm circumference; TUA, total upper arm area; UMA, upper arm muscle area; UFA, upper arm fat area; AFI, arm fat index; %F, percent fat mass; FM, fat mass; FFM, fat-free mass.

Table 7 shows the multiple linear regression analysis for the whole sample, conducted with all the significant predictors input into the model with a backward method. In the multivariate model, sex, dominant UMA, AFI, FM (negative coefficient), and FFM were significant predictors of HGS. These independent variables accounted for 74.6% of the variance of HGS in young adults.

Table 7. Results of multiple linear regression with backward regression analysis.

Predictor Variables	Dominant Handgrip Strength		
	β	t	p
Sex (male)	0.1980	3.4323	0.0007
D UMA	0.3463	5.9100	0.0000
D AFI	0.1345	2.1342	0.0335
FM	−0.1784	−3.6116	0.0003
FFM	0.4420	6.2158	0.0000
R^2	0.7491		
R^2 adjusted	0.7457		
p	0.0000		

β: standardized regression coefficient. Abbreviations: D UMA, upper arm muscle area (dominant side); D AFI, arm fat index (dominant side); FM, fat mass; FFM, fat-free mass.

4. Discussion

The main purpose of this study was to analyze the relationship of HGS with anthropometric characteristics and laterality, taking physical activity and sports practice of the participants into account, in order to identify the determinants of dominant HGS by multivariate analysis. Moreover, we intended to analyze the differences in the values of dominant and non-dominant handgrip strength among three categories of handedness, comparing the results between the two sexes.

While the curvilinear association between HGS and age is well-known [73,74], the relationships between HGS and several anthropometric characteristics or indices have not been well documented in samples, nor selected for a specific sport, which has led to inconsistent or incomplete results in different studies. The association between BMI and strength, for example, has been widely debated in the literature; while some studies have recognized their positive association in both sexes, others have denied it [75–77]. In addition, previous analyses have not generally taken into account manual dominance, assessed according to defined criteria, as indicated by Oldfield [60]. Therefore, we conducted the present study on a sample of healthy young Italian adults of both sexes to examine these aspects in depth, taking into account their handedness through the EHI questionnaire. In particular, we observed significant differences in HGS between sides in the sample's paired comparison after its division into three handedness categories by sex. The vast majority of the sample was right-handed, but there was a significantly different distribution of handedness categories between sexes: more males were left-handed (+0.8%) or ambidextrous (+12.5%), while more females were right-handed (+13.3%). This finding was consistent with previous studies [78,79].

Within the categories with different handedness, the results of this study indicate a significant difference in HGS between the sexes in favor of males, as well as a trend for the dominant hand to be significantly stronger in both sexes (this difference was not significant in left-handed females, probably due to the small sample size). Ambidextrous subjects of both sexes showed higher HGS of the right hand (significant only in the male sex). This latter trend is likely due to the greater pressure of our society to use the right hand in various occupational tasks from childhood on [47,80,81].

The HGS mean values of the dominant hand obtained in our study for young healthy Italian adults were 45.7 ± 8.2 kg for males and 28.9 ± 4.7 kg for females. A detailed comparison of our HGS results with previous studies is difficult, especially due to the different methods applied (if any) in the evaluation of handedness. In general, we observed that male and female HGS values were, on average, within the range of healthy young adults (not selected for sports) found in the literature (i.e., in subjects aged 20, in Innes [82]; 18–33, in Nicolay and Walker [83]; 20–24, in Werle et al. [84]; 18–25, in Koley [85]; 20, in McGrath et al. [86], 50th percentile values) with mean values ranging from 20.4 to 35.6 kg for the dominant hand among females, and from 39.5 to 58.8 kg for the dominant hand among males. A majority of studies reporting reference values for HGS were conducted by examining participants living in developed countries, mainly Europe, Australia, and the US [82,84,86]. Only a few studies have reported reference HGS values for developing countries (i.e., Werle et al. [84] for the Indian population). According to Dopsaj et al. [87], differences in HGS between sexes may be explained by various factors, such as the cross-sectional area, muscle fiber characteristics, amount of skeletal muscle mass, distribution of muscle mass in the upper limbs, and common anatomical differences.

According to the scientific literature [88], there is a strong positive correlation between HGS and weight. As body mass can be considered a confounding factor in the evaluation and analysis of HGS, we decided to adjust HGS to the body mass, in order to reduce this bias. Despite the fact that the importance of the adjustment of HGS for weight has been proven, especially for the association of this parameter with metabolic syndrome and sarcopenia [89], the results of our study did not show particular differences between the sexes using the two parameters of HGS; males had a stronger handgrip strength than females, regardless of their weight.

The dominant hand is used more often than the non-dominant one, such that the muscles of the dominant arm are bigger and stronger. The general rule assumes that the dominant hand is 5–10% stronger, compared to the non-dominant [90–93]. According to Adam et al. [93], there is a higher percentage of recruitment of motor units at lower absolute force levels in the dominant hand; whereas, in the non-dominant hand, there is a spread-out recruitment pattern. In our study, the difference was observed for the lowest percentages, shifting from 5.1% for left-handed females (5.9% in males) to a maximum of 5.8% for right-handed females (5.5% in males). These results are in line with the study of Ekşioğlu [15], and the possible differences may be due to the different methodologies used. In some studies, in fact, the dominance of the hand was established simply on the basis of the best performance obtained in the measurement of HGS and not on the basis of the function of the hands in different daily actions. In the examined ambidextrous subjects, the right hand was, on average, stronger than the left hand, with a difference of 2.9% in females and 4.5% in males. To explain this feature, it is important to take into account that most everyday tools and appliances are designed for right-handed individuals. A left-hander who does not have access to left-handed instruments often has to adapt by making his/her right hand the dominant hand [94]. As a result, the right hand is exercised more often than the left, on a daily basis [91]. In addition, Ozcan et al. [95] reported that the laterality of manual function is less pronounced in left-handed people than in right-handed people.

The results of the comparison between the lowest and the highest tercile of the dominant handgrip strength for each anthropometric variable showed a significant difference for almost all considered traits. Individuals of both sexes with higher values of HGS (highest tercile) were taller and heavier, with higher mean BMI, WC, D triceps skinfold (only in females), dominant (D) MUAC, D TUA, D UMA, D UFA (only in females), FM, and FFM. The larger sizes of the strongest individuals were confirmed by the significant correlation of anthropometric characteristics with HGS. As the HGS increased, height, weight, BMI, WC, D MUAC, D triceps (in females only), D TUA, D UMA, D UFA (in females only), FM, and FFM increased. In males, D AFI decreased as HGS increased. Previous studies have generally considered and confirmed only the associations of HGS with stature, weight, and BMI [96–98].

The hypothesis of a positive association between HGS and lean body mass, which has been formulated in other studies only on the basis of an increase in BMI (especially in males) [99,100], was confirmed in this study, through the assessment of body composition parameters and indices.

As evidence that BMI is an unreliable measure of body fat and does not differentiate between weight changes due to an increase or decrease in muscularity or adiposity [101], we found higher percentages of overweight and obese males in the highest tercile of HGS than in the lowest one (36.4% vs. 13.6%). As further confirmation, neither D triceps skinfold, D UFA, D AFI, nor %F differed between terciles and were not significantly correlated with HGS. As a consequence, underweight and normal weight males were more frequent in the lowest HGS tercile than in the highest one (86.4% vs. 63.6%). This trend was consistent with the results in the literature, reporting that grip strength is lower in underweight individuals and increases across normal and overweight categories, but plateaus through obese BMI ranges (although obese individuals are still stronger than normal weight individuals), suggesting that strength declines only in extremely obese individuals [41,86].

This was also consistent with the positive association we found between HGS and FM. In females, the trend was similar but more shaded. Therefore, underweight and normal weight females were more frequent in the lowest HGS tercile (94.8% vs. 86.6%) and overweight and obese females in the highest HGS tercile (13.4% vs. 5.2%), without any statistical significance. This finding could be related to the lower representation of overweight and obese categories in females. Again, however, D AFI and %F did not differ between terciles and showed no significant correlation with HGS.

The comparison between the parameters of physical activity and sport practice between the two terciles showed different results in the two sexes. According to the literature, intensity and years of sports practice are both important for increasing HGS [102,103]. Surprisingly, we observed significantly fewer years of sports practice in the stronger males, in addition to a lower weekly amount of sports activity and a lower number of METs, although not significant. A possible explanation could be that the HGS of the participants in this study was associated more with the specificity of each sport, rather than with the intensity of the sport (as categorized by METs). This is an important result, as there is a paucity of data in the literature regarding the general influence of sport intensity on HGS, as the majority of the studies have generally focused on a specific sport. Among females, there was a significantly higher amount of sport in the highest tercile of HGS. Thus, basically, a greater amount of sport does not lead to higher HGS in men, while a greater amount of sport in women similar in entity to males can lead to an increase in strength which is closer to the maximum strength, due to their relatively smaller muscle mass [104].

Considering the complexity of HGS associations with anthropometric variables, weight status, body composition, and sex, we applied multiple regression analysis in such a way that the final model retained only the most significant factors after a stepwise exclusion procedure. According to the model, body composition parameters and indices have proven to be the most effective, jointly with sex, in explaining the variability of D HGS, which was best predicted by sex, D UMA, D AFI, FM, and FFM. These factors account for about 75% of the HGS variability of the dominant side in our sample of healthy Italian young adults.

In general, it can be deduced that HGS differs significantly between sexes and is directly influenced by body composition. Consequently, we stress the impact that severe undernutrition, with subsequent deficits in FFM, can have on HGS. Based on the results obtained, although we did not have severely undernourished subjects in our sample, we can confirm that hand grip strength is an essential parameter for indicating a deteriorating nutritional status [105,106]. Norman et al. [107] suggested that a reduction in nutritional intake translates into a compensatory loss of whole-body protein, which is preferably lost from the muscle mass; that is, the largest protein reserve in the body. As a result of this process, cellular changes lead to a decrease in protein synthesis and an increase in proteolysis, thus causing fiber atrophy. This, in turn, leads to a decrease in muscle strength and muscle function. This underlines the importance of HGS, at all ages, as a marker of nutritional status which can be used to identify patients at risk of undernutrition and malnutrition and to control them after nutritional interventions. Moreover, it has been proven that the measurement of HGS could motivate malnourished subjects to improve their nutritional status and may increase their perceived quality of life [104,108].

The use of grip strength estimates is a robust method to assess muscle weakness and can be of very practical use in evaluating ageing populations, patients with neuromuscular disorders, and patients during rehabilitation or return to work or sport. Further longitudinal studies may provide relevant indications on HGS decline with body composition changes during the ageing process.

This study has some strengths and limitations. Our sample of healthy university students cannot be considered representative of the whole population, even though it has proved adequate to draw relevant conclusions. Moreover, the sample was homogeneous with respect to age, preventing us from assessing changes in HGS with the age of the participants. It must also be considered that our sample consisted only of Italians. Further investigations could be carried out in other populations, in order to gain new data and generate more reference values. The strengths of our study included the direct measurement of several anthropometric characters collected by expert operators and the quantitative assessment of handedness using recognized methods of investigation.

5. Conclusions

In this study, we successfully analyzed HGS according to laterality, as well as anthropometric parameters and indices, in a sample of healthy young adults, in order to obtain an appropriate model of the relationships between these variables. With regards to laterality, the dominant hand was significantly stronger in both sexes, with ambidextrous subjects showing higher HGS in the right hand. While the parameters connected with physical activity did not seem to show any association with HGS, we identified body composition parameters and sex to be the major predictors of strength in the dominant hand. Our findings are of particular relevance, suggesting that HGS can be associated with unhealthy conditions—in particular, those related to low values of FFM—and can be used to monitor changes in body composition, given that the dominant side is always measured. In conclusion, we have highlighted the importance of the application of appropriate measurement methodologies and the objective assessment of handedness, although further research is needed to provide evidence of the effectiveness and clinical relevance of HGS testing in the evaluation and prediction of critical health conditions.

Author Contributions: Conceptualization, B.B., E.G.-R., N.R., and L.Z.; methodology, N.R., L.Z., E.G.-R., and S.T.; formal analysis, L.Z. and N.R.; investigation, N.R., J.M., and S.T.; data curation, L.Z., J.M., and N.R.; writing—original draft preparation, S.T., E.G.-R., L.Z., N.R., and J.M.; writing—review and editing, E.G.-R., S.T., L.Z., N.R., J.M., and B.B.; visualization, S.T.; supervision, L.Z. and E.G.-R.; project administration, N.R. and S.T. All authors have read and agreed to the published version of the manuscript.

Funding: This research received no external funding.

Acknowledgments: The authors would like to thank all the participants in this research.

Conflicts of Interest: The authors declare no conflict of interest.

References

1. Yu, A.; Yick, K.-L.; Ng, S.; Yip, J. Case study on the effects of fit and material of sports gloves on hand performance. *Appl. Ergon.* **2019**, *75*, 17–26. [CrossRef] [PubMed]
2. Leyk, D.; Gorges, W.; Ridder, D.; Wunderlich, M.; Rüther, T.; Sievert, A.; Essfeld, D. Hand-grip strength of young men, women and highly trained female athletes. *Eur. J. Appl. Physiol.* **2007**, *99*, 415–421. [CrossRef] [PubMed]
3. Turnes, T.; Silva, B.A.; Kons, R.L.; Detanico, D. Is bilateral deficit in handgrip strength associated with performance in specific judo tasks? *J. Strength Cond. Res.* **2019**. [CrossRef] [PubMed]
4. Tan, Y.H.; Ng, P.K.; Saptari, A.; Jee, K.S. Ergonomics aspects of knob designs: A literature review. *TIES* **2015**, *16*, 86–98. [CrossRef]
5. Toselli, S.; Badicu, G.; Bragonzoni, L.; Spiga, F.; Mazzuca, P.; Campa, F. Comparison of the effect of different resistance training frequencies on phase angle and handgrip strength in obese women: A randomized controlled trial. *IJERPH* **2020**, *17*, 1163. [CrossRef]
6. Campa, F.; Silva, A.; Toselli, S. Changes in phase angle and handgrip strength induced by suspension training in older women. *Int. J. Sports Med.* **2018**, *39*, 442–449. [CrossRef]
7. Trivic, T.; Eliseev, S.; Tabakov, S.; Raonic, V.; Casals, C.; Jahic, D.; Jaksic, D.; Drid, P. Somatotypes and hand-grip strength analysis of elite cadet sambo athletes. *Medicine* **2020**, *99*. [CrossRef]
8. Lee, K.-S.; Jung, M. Ergonomic evaluation of biomechanical hand function. *Saf. Health Work* **2014**, *6*, 9–17. [CrossRef]
9. AlAhmari, K.; Silvian, S.P.; Reddy, R.S.; Kakaraparthi, V.N.; Ahmad, I.; Alam, M.M. Hand grip strength determination for healthy males in Saudi Arabia: A study of the relationship with age, body mass index, hand length and forearm circumference using a hand-held dynamometer. *J. Int. Med. Res.* **2017**, *45*, 540–548. [CrossRef]

10. Leong, D.P.; Teo, K.K.; Rangarajan, S.; López-Jaramillo, P.; Avezum, A.; Orlandini, A.; Serón, P.; Ahmed, S.H.; Rosengren, A.; Kelishadi, R.; et al. Prospective Urban Rural Epidemiology (PURE) study investigators. Prognostic value of grip strength: Findings from the Prospective Urban Rural Epidemiology (PURE) study. *Lancet* **2015**, *386*, 266–273. [CrossRef]

11. Oliveira, V.H.F.; Wiechmann, S.L.; Narciso, A.M.; Deminice, R. Knee extension and flexion strength asymmetry in Human Immunodeficiency Virus positive subjects: A cross-sectional study. *Braz. J. Phys. Ther.* **2017**, *21*, 434–439. [CrossRef] [PubMed]

12. Oksuzyan, A.; Demakakos, P.; Shkolnikova, M.; Thinggaard, M.; Vaupel, J.W.; Christensen, K.; Shkolnikov, V.M. Handgrip strength and its prognostic value for mortality in Moscow, Denmark, and England. *PLoS ONE* **2017**, *12*, e0182684. [CrossRef] [PubMed]

13. Rabelo, N.D.D.A.; Lucareli, P.R.G. Do hip muscle weakness and dynamic knee valgus matter for the clinical evaluation and decision-making process in patients with patellofemoral pain? *Braz. J. Phys. Ther.* **2017**, *22*, 105–109. [CrossRef] [PubMed]

14. Santos, C.; Matos, S.; Pitanga, F.; Maia, H. Handgrip strength as discriminator of sarcopenia and sarcopenia obesity in adults of the ELSA-Brasil. *Eur. J. Public Health* **2018**, *28* (Suppl. 4). [CrossRef]

15. Ekşioğlu, M. Normative static grip strength of population of Turkey, effects of various factors and a comparison with international norms. *Appl. Ergon.* **2016**, *52*, 8–17. [CrossRef] [PubMed]

16. Meldrum, D.; Cahalane, E.; Conroy, R.; Fitzgerald, D.; Hardiman, O. Maximum voluntary isometric contraction: Reference values and clinical application. *Amyotroph. Lateral Scler.* **2007**, *8*, 47–55. [CrossRef]

17. Mathiowetz, V.; Kashman, N.; Volland, G.; Weber, K.; Dowe, M.; Rogers, S. Grip and pinch strength: Normative data for adults. *Arch. Phys. Med. Rehabil.* **1985**, *66*, 69–74.

18. Portney, L.G.; Watkins, M.P. *Foundations of Clinical Research: Applications to Practice*, 3rd ed.; Prentice-Hall: Upper Saddle River, NJ, USA, 2008.

19. Sağiroğlu, I.; Kurt, C.; Kurt Ömürlü, I.; Çatikkaş, F. Does hand grip strength change with gender? The traditional method vs. the allometric normalisation method. *EJPESS* **2016**, *2*, 84–93.

20. Chelliah, K.; Rizam, R. Association of dominant hand grip strength with anthropometric measurements and Body Mass Index. *Int. J. Pharm. Sci. Res.* **2018**, *10*, 525–531.

21. Andersen-Ranberg, K.; Petersen, I.; Frederiksen, H.; MacKenbach, J.P.; Christensen, K. Cross-national differences in grip strength among 50+ year-old Europeans: Results from the SHARE study. *Eur. J. Ageing* **2009**, *6*, 227–236. [CrossRef]

22. Bohannon, R. Hand-grip dynamometry predicts future outcomes in aging adults. *J. Geriatr. Phys. Ther.* **2008**, *31*, 3–10. [CrossRef] [PubMed]

23. Amo-Setién, F.J.; Leal-Costa, C.; Abajas-Bustillo, R.; González-Lamuño, D.; Redondo-Figuero, C. Factors associated with grip strength among adolescents: An observational study. *J. Hand Ther.* **2020**, *33*, 96–102. [CrossRef] [PubMed]

24. Kamon, E.; Goldfuss, A.J. In-plant evaluation of the muscle strength of workers. *Am. Ind. Hyg. Assoc. J.* **1978**, *39*, 801–807. [CrossRef] [PubMed]

25. Bohannon, R.W.; Peolsson, A.; Massy-Westropp, N.; Desrosiers, J.; Bear-Lehman, J. Reference values for adult grip strength measured with a Jamar dynamometer: A descriptive meta-analysis. *Physiotherapy* **2006**, *92*, 11–15. [CrossRef]

26. Malina, R.M.; Zavaleta, A.N.; Little, B.B. Body size, fatness, and leanness of Mexican American children in Brownsville, Texas: Changes between 1972 and 1983. *Am. J. Public Health* **1987**, *77*, 573–577. [CrossRef]

27. Häger, C.K.; Rösblad, B. Norms for grip strength in children aged 4-16 years. *Acta Paediatr.* **2002**, *91*, 617–625. [CrossRef]

28. Singh, A.P.; Koley, S.; Sandhu, J.S. Association of hand grip strength with some anthropometric traits in collegiate population of Amritsar. *Orient. Anthr. A Bi-Annual Int. J. Sci. Man* **2009**, *9*, 99–110. [CrossRef]

29. Koley, S.; Yadav, M.K. An association of hand grip strength with some anthropometric variables in Indian Cricket players. *Phys. Educ. Sport* **2009**, *7*, 113–123.

30. Koley, S.; Singh, A.P. Pal an association of dominant hand grip strength with some anthropometric variables in indian collegiate population. *Anthr. Anz.* **2009**, *67*, 21–28. [CrossRef]

31. Jürimäe, T.; Hurbo, T.; Jürimäe, J. Relationship of handgrip strength with anthropometric and body composition variables in prepubertal children. *HOMO* **2009**, *60*, 225–238. [CrossRef]

32. Kaur, M. Age-related changes in hand grip strength among rural and urban Haryanvi Jat females. *HOMO* **2009**, *60*, 441–450. [CrossRef] [PubMed]
33. Kallman, D.A.; Plato, C.C.; Tobin, J.D. The Role of muscle loss in the age-related decline of grip strength: Cross-sectional and longitudinal perspectives. *J. Gerontol.* **1990**, *45*, 82–88. [CrossRef] [PubMed]
34. Fraser, A.; Vallow, J.; Preston, A.; Cooper, R.G. Predicting 'normal' grip strength for rheumatoid arthritis patients. *Rheumatology* **1999**, *38*, 521–528. [CrossRef] [PubMed]
35. Anakwe, R.E.; Huntley, J.; McEachan, J.E. Grip strength and forearm circumference in a healthy population. *J. Hand Surg. Eur. Vol.* **2007**, *32*, 203–209. [CrossRef] [PubMed]
36. Incel, N.A.; Ceceli, E.; Durukan, P.B.; Erdem, H.R.; Yorgancioglu, Z.R. Grip strength: Effect of hand dominance. *Singap. Med. J.* **2002**, *43*, 234–237.
37. Dolcos, F.; Rice, H.J.; Cabeza, R. Hemispheric asymmetry and aging: Right hemisphere decline or asymmetry reduction. *Neurosci. Biobehav. Rev.* **2002**, *26*, 819–825. [CrossRef]
38. Bohannon, R.W. Grip strength: A summary of studies comparing dominant and nondominant limb measurements. *Percept. Mot. Ski.* **2003**, *96*, 728–730. [CrossRef]
39. Roy, E.A.; Bryden, P.; Cavill, S. Hand differences in pegboard performance through development. *Brain Cogn.* **2003**, *53*, 315–317. [CrossRef]
40. Noguchi, T.; Demura, S.; Aoki, H. Superiority of the dominant and nondominant hands in static strength and controlled force exertion. *Percept. Mot. Ski.* **2009**, *109*, 339–346. [CrossRef]
41. Aoki, H.; Demura, S. Laterality and accuracy of force exertion in elbow flexion. *Adv. Phys. Educ.* **2017**, *7*, 101–106. [CrossRef]
42. Corballis, M.C. Handedness and cerebral asymmetry: An evolutionary perspective. In *The Two Halves of the Brain: Information Processing in the Cerebral Hemispheres*; Hugdahl, K., Westerhausen, R., Eds.; MIT Press: Cambridge, MA, USA, 2010. [CrossRef]
43. Annett, M. Handedness as a continuous variable with dextral shift: Sex, generation, and family handedness in subgroups of left- and right-handers. *Behav. Genet.* **1994**, *24*, 51–63. [CrossRef] [PubMed]
44. Raymond, M.; Pontier, M.; Dufour, A.-B.; Møller, A.P. Frequency-dependent maintenance of left handedness in humans. *Proc. R. Soc. B Boil. Sci.* **1996**, *263*, 1627–1633. [CrossRef]
45. Bryden, M.; Roy, E.; Manus, I.; Mc Bulman-Fleming, M. On the Genetics and Measurement of Human Handedness. *Laterality* **1997**, *2*, 317–336. [CrossRef] [PubMed]
46. McManus, C. *Right Hand Left Hand: The Origins of Asymmetry in Brains, Bodies, Atoms and Culture*; Harvard University Press: Cambridge, MA, USA, 2002.
47. Porac, C.; Coren, S.; Duncan, P. Life-span Age Trends in Laterality. *J. Gerontol.* **1980**, *35*, 715–721. [CrossRef] [PubMed]
48. Geschwind, N.; Galaburda, A.M. *Cerebral Lateralization: Biological Mechanisms, Associations, and Pathology*; MIT Press: Cambridge, MA, USA, 1987.
49. Halpern, D.F.; Coren, S. Handedness and life span. *N. Engl. J. Med.* **1991**, *324*, 998. [CrossRef] [PubMed]
50. Annett, M. *Handedness and Brain Asymmetry: The Right Shift Theory*; Psychology Press: Hove, UK, 2002.
51. Musalek, M. Skilled performance tests and their use in diagnosing handedness and footedness at children of lower school age 8–10. *Front. Psychol.* **2015**, *5*, 5. [CrossRef]
52. Brown, S.; Roy, E.; Rohr, L.; Bryden, P. Using hand performance measures to predict handedness. *Laterality* **2006**, *11*, 1–14. [CrossRef]
53. Whittington, J.E.; Richards, P.N. The stability of children's laterality prevalences and their relationship to measures of performance. *Br. J. Educ. Psychol.* **1987**, *57*, 45–55. [CrossRef]
54. Sommer, I.E.C.; Aleman, A.; Somers, M.; Boks, M.P.M.; Kahn, R.S. Sex differences in handedness, asymmetry of the Planum Temporale and functional language lateralization. *Brain Res.* **2008**, *1206*, 76–88. [CrossRef]
55. Johnston, D.W.; Nicholls, M.E.R.; Shah, M.; Shields, M.A. Nature's experiment? Handedness and early childhood development. *Demography* **2009**, *46*, 281–301. [CrossRef]
56. Shahida, M.N.; Zawiah, M.S.; Case, K.; Shalahim, N.S.M. The relationship between anthropometry and hand grip strength among elderly Malaysians. *Int. J. Ind. Ergon.* **2015**, *50*, 17–25. [CrossRef]
57. Rønn, P.F.; Andersen, G.S.; Lauritzen, T.; Christensen, D.L.; Aadahl, M.; Carstensen, B.; Jørgensen, M.E. Ethnic differences in anthropometric measures and abdominal fat distribution: A cross-sectional pooled study in Inuit, Africans and Europeans. *J. Epidemiol. Community Health* **2017**, *71*, 536–543. [CrossRef] [PubMed]

58. Mongraw-Chaffin, M.; Golden, S.H.; Allison, M.; Ding, J.; Ouyang, P.; Schreiner, P.J.; Szklo, M.; Woodward, M.; Young, J.H.; Anderson, C.A.M.; et al. The sex and race specific relationship between anthropometry and body fat composition determined from computed tomography: Evidence from the multi-ethnic study of atherosclerosis. *PLoS ONE* **2015**, *10*. [CrossRef] [PubMed]

59. Ainsworth, B.; Haskell, W.L.; Herrmann, S.D.; Meckes, N.; Bassett, D.R.; Tudor-Locke, C.; Greer, J.L.; Vezina, J.; Whitt-Glover, M.C.; Leon, A.S.; et al. 2011 compendium of physical activities: A second update of codes and MET values. *Med. Sci. Sports Exerc.* **2011**, *43*, 1575–1581. [CrossRef]

60. Oldfield, R. The assessment and analysis of handedness: The Edinburgh inventory. *Neuropsychologia* **1971**, *9*, 97–113. [CrossRef]

61. Stöckel, T.; Vater, C. Hand preference patterns in expert basketball players: Interrelations between basketball-specific and everyday life behavior. *Hum. Mov. Sci.* **2014**, *38*, 143–151. [CrossRef]

62. Gualdi-Russo, E.; Rinaldo, N.; Pasini, A.; Zaccagni, L. Hand preference and performance in basketball tasks. *Int. J. Environ. Res. Public Health* **2019**, *16*, 4336. [CrossRef]

63. Lohman, T.J.; Roache, A.F.; Martorell, R. *Anthropometric Standardization Reference Manual*; Human Kinetics: Champaign, IL, USA, 1988.

64. Rinaldo, N.; Gualdi-Russo, E. Anthropometric techniques. *Annali Online Università Ferrara Sez. Didattica Formazione Docente* **2015**, *10*, 275–289.

65. Centers for Disease Control and Prevention. National Youth fitness Survey (NYFS) Muscle Strength (grip) Procedures Manual. 2012. Available online: www.cdc.gov/nchs/data/nnyfs/Handgrip_Muscle_Strength.pdf (accessed on 7 October 2019).

66. Durnin, J.V.G.A.; Womersley, J. Body fat assessed from total body density and its estimation from skinfold thickness: Measurements on 481 men and women aged from 16 to 72 Years. *Br. J. Nutr.* **1974**, *32*, 77–97. [CrossRef]

67. Siri, W.E. *Body Composition from Fluid Spaces and Density: Analysis of Methods*; Lawrence Radiation Laboratory: Berkeley, CA, USA, 1956.

68. James, W.P.T.; Leach, R.; Kalamara, E.; Shayeghi, M. The worldwide obesity epidemic. *Obes. Res.* **2001**, *9*, 228S–233S. [CrossRef]

69. Gallagher, D.; Heymsfield, S.B.; Heo, M.; Jebb, S.A.; Murgatroyd, P.R.; Sakamoto, Y. Healthy percentage body fat ranges: An approach for developing guidelines based on body mass index. *Am. J. Clin. Nutr.* **2000**, *72*, 694–701. [CrossRef] [PubMed]

70. Frisancho, A. *Anthropometric Standards: An. Interactive Nutritional Reference of Body Size and Body Composition for Children and Adults*, 2nd ed.; University of Michigan Press: Ann Arbor, MI, USA, 2008.

71. Zaccagni, L.; Barbieri, D.; Gualdi-Russo, E. Body composition and physical activity in Italian university students. *J. Transl. Med.* **2014**, *12*, 120. [CrossRef] [PubMed]

72. Toselli, S.; Gualdi-Russo, E.; Mazzuca, P.; Campa, F. Ethnic differences in body composition, sociodemographic characteristics and lifestyle in people with type 2 diabetes mellitus living in Italy. *Endocrine* **2019**, *65*, 558–568. [CrossRef] [PubMed]

73. Lunaheredia, E.; Martinpena, G.; Ruiz-Galiana, J. Handgrip dynamometry in healthy adults. *Clin. Nutr.* **2005**, *24*, 250–258. [CrossRef] [PubMed]

74. Frederiksen, H.; Hjelmborg, J.; Mortensen, J.; McGue, M.; Vaupel, J.W.; Christensen, K. Age trajectories of grip strength: Cross-sectional and longitudinal data among 8,342 danes aged 46 to 102. *Ann. Epidemiol.* **2006**, *16*, 554–562. [CrossRef]

75. Chilima, D.M.; Ismail, S.J. Nutrition and handgrip strength of older adults in rural Malawi. *Public Health Nutr.* **2001**, *4*, 11–17. [CrossRef]

76. Koley, S.; Kaur, N.; Sandhu, J. A study on hand grip strength in female labourers of Jalandhar, Punjab, India. *J. LIFE Sci.* **2009**, *1*, 57–62. [CrossRef]

77. Massy-Westropp, N.; Gill, T.K.; Taylor, A.W.; Bohannon, R.; Hill, C.L. Hand grip strength: Age and gender stratified normative data in a population-based study. *BMC Res. Notes* **2011**, *4*, 127. [CrossRef]

78. Vuoksimaa, E.; Koskenvuo, M.; Rose, R.J.; Kaprio, J. Origins of handedness: A nationwide study of 30161 adults. *Neuropsychologia* **2009**, *47*, 1294–1301. [CrossRef]

79. Annett, M. Handedness and cerebral dominance: The right shift theory. *J. Neuropsychiatry Clin. Neurosci.* **1998**, *10*, 459–469. [CrossRef]

80. Al Lawati, I.; Al Maskari, H.; Ma, S. "I am a lefty in a right-handed world": Qualitative analysis of clinical learning experience of left-handed undergraduate dental students. *Eur. J. Dent. Educ.* **2019**, *23*, 316–322. [CrossRef] [PubMed]
81. Marcori, A.J.; Monteiro, P.H.M.; Okazaki, V.H.A. Changing handedness: What can we learn from preference shift studies? *Neurosci. Biobehav. Rev.* **2019**, *107*, 313–319. [CrossRef] [PubMed]
82. Innes, E. Handgrip strength testing: A review of the literature. *Aust. Occup. Ther. J.* **1999**, *46*, 120–140. [CrossRef]
83. Nicolay, C.W.; Walker, A.L. Grip strength and endurance: Influences of anthropometric variation, hand dominance, and gender. *Int. J. Ind. Ergon.* **2005**, *35*, 605–618. [CrossRef]
84. Werle, S.; Goldhahn, J.; Drerup, S.; Sprott, H.; Simmen, B.; Herren, D.B. Age- and gender-specific normative data of grip and pinch strength in a healthy adult Swiss Population. *J. Hand Surg. Eur. Vol.* **2009**, *34*, 76–84. [CrossRef]
85. Koley, S.; Singh, A.P. Effect of hand dominance in grip strength in collegiate population of Amritsar, Punjab, India. *Anthropologist* **2010**, *12*, 13–16. [CrossRef]
86. McGrath, R.; Hackney, K.J.; Ratamess, N.A.; Vincent, B.M.; Clark, B.C.; Kraemer, W.J. Absolute and body mass index normalized handgrip strength percentiles by gender, ethnicity, and hand dominance in Americans. *Adv. Geriatr. Med. Res.* **2019**, *2*. [CrossRef]
87. Dopsaj, M.; Ivanović, J.; Blagojević, M.; Vučković, G. Descriptive, functional and sexual dimorphism of explosive isometric hand grip force in healthy university students in Serbia. *Facta Univ. Ser. Phys. Educ. Sport.* **2009**, *7*, 125–139.
88. Günther, C.M.; Burger, A.; Rickert, M.; Crispin, A.; Schulz, C.U. Grip strength in healthy caucasian adults: Reference values. *J. Hand Surg.* **2008**, *33*, 558–565. [CrossRef]
89. Chun, S.-W.; Kim, W.; Choi, K.H. Comparison between grip strength and grip strength divided by body weight in their relationship with metabolic syndrome and quality of life in the elderly. *PLoS ONE* **2019**, *14*. [CrossRef]
90. Schmidt, R.T.; Toews, J.V. Grip strength as measured by the Jamar dynamometer. *Arch. Phys. Med. Rehabil.* **1970**, *51*, 321–327. [PubMed]
91. Crosby, C.A.; Wehbé, M.A. Hand strength: Normative values. *J. Hand Surg.* **1994**, *19*, 665–670. [CrossRef]
92. Lee, K.-S.; Hwang, J. Investigation of grip strength by various body postures and gender in Korean adults. *Work* **2019**, *62*, 117–123. [CrossRef] [PubMed]
93. Adam, A.; De Luca, C.J.; Erim, Z. Hand dominance and motor unit firing behavior. *J. Neurophysiol.* **1998**, *80*, 1373–1382. [CrossRef]
94. Deora, H.; Tripathi, M.; Yagnick, N.S.; Deora, S.; Mohindra, S.; Batish, A. Changing hands: Why being ambidextrous is a trait that needs to be acquired and nurtured in neurosurgery. *World Neurosurg.* **2019**, *122*, 487–490. [CrossRef]
95. Özcan, A.; Tulum, Z.; Pınar, L.; Başkurt, F. Comparison of pressure pain threshold, grip strength, dexterity and touch pressure of dominant and non-dominant hands within and between right- and left-handed subjects. *J. Korean Med. Sci.* **2004**, *19*, 874–878. [CrossRef]
96. Chandrasekaran, B.; Ghosh, A.; Prasad, C.; Krishnan, K.; Chandrasharma, B. Age and anthropometric traits predict handgrip strength in healthy normals. *J. Hand Microsurg.* **2010**, *2*, 58–61. [CrossRef]
97. Otten, L.; Bosy-Westphal, A.; Ordemann, J.; Rothkegel, E.; Stobäus, N.; Elbelt, U.; Norman, K. Abdominal fat distribution differently affects muscle strength of the upper and lower extremities in women. *Eur. J. Clin. Nutr.* **2016**, *71*, 372–376. [CrossRef]
98. Ploegmakers, J.J.; Hepping, A.M.; Geertzen, J.H.; Bulstra, S.K.; Stevens, M. Grip strength is strongly associated with height, weight and gender in childhood: A cross sectional study of 2241 children and adolescents providing reference values. *J. Physiother.* **2013**, *59*, 255–261. [CrossRef]
99. Keevil, V.L.; Luben, R.N.; Dalzell, N.; Hayat, S.; Sayer, A.A.; Wareham, N.J.; Khaw, K.-T. Cross-sectional associations between different measures of obesity and muscle strength in men and women in a British cohort study. *J. Nutr. Health Aging* **2015**, *19*, 3–11. [CrossRef]
100. Detanico, D.; Arins, F.B.; Pupo, J.D.; Dos Santos, S.G. Strength parameters in judo athletes: An approach using hand dominance and weight categories. *Hum. Mov.* **2012**, *13*, 330–336. [CrossRef]
101. Malina, R.M.; Katzmarzyk, P.T. Validity of the body mass index as an indicator of the risk and presence of overweight in adolescents. *Am. J. Clin. Nutr.* **1999**, *70*, 131S–136S. [CrossRef] [PubMed]

102. Pizzigalli, L.; Cremasco, M.M.; La Torre, A.; Rainoldi, A.; Benis, R. Hand grip strength and anthropometric characteristics in Italian female national basketball teams. *J. Sports Med. Phys. Fit.* **2016**, *57*, 521–528.

103. Haynes, E.; DeBeliso, M. The relationship between CrossFit performance and grip strength. *Turk. J. Kinesiol.* **2019**, *5*, 15–21. [CrossRef]

104. Jakobsen, L.H.; Rask, I.K.; Kondrup, J. Validation of handgrip strength and endurance as a measure of physical function and quality of life in healthy subjects and patients. *Nutrition* **2010**, *26*, 542–550. [CrossRef]

105. Angst, F.; Drerup, S.; Werle, S.; Herren, D.; Simmen, B.R.; Goldhahn, J. Prediction of grip and key pinch strength in 978 healthy subjects. *BMC Musculoskelet. Disord.* **2010**, *11*, 94. [CrossRef]

106. Gale, C.R.; Martyn, C.N.; Cooper, C.; Sayer, A.A. Grip strength, body composition, and mortality. *Int. J. Epidemiol.* **2006**, *36*, 228–235. [CrossRef]

107. Norman, K.; Stobäus, N.; Gonzalez, M.C.; Schulzke, J.-D.; Pirlich, M. Hand grip strength: Outcome predictor and marker of nutritional status. *Clin. Nutr.* **2011**, *30*, 135–142. [CrossRef]

108. Norman, K.; Kirchner, H.; Freudenreich, M.; Ockenga, J.; Lochs, H.; Pirlich, M. Three month intervention with protein and energy rich supplements improve muscle function and quality of life in malnourished patients with non-neoplastic gastrointestinal disease—A randomized controlled trial. *Clin. Nutr.* **2008**, *27*, 48–56. [CrossRef]

International Journal of
*Environmental Research
and Public Health*

Article

Predicting Cardiovascular Risk in Athletes: Resampling Improves Classification Performance

Davide Barbieri [1,†], Nitesh Chawla [2], Luciana Zaccagni [1,3,*], Tonći Grgurinović [4], Jelena Šarac [5], Miran Čoklo [5] and Saša Missoni [6,7]

[1] Department of Biomedical and Specialty Surgical Sciences, Faculty of Medicine, Pharmacy and Prevention, University of Ferrara, 44121 Ferrara, Italy; davide.barbieri@unife.it

[2] Interdisciplinary Center for Network Science and Applications, University of Notre Dame, Notre Dame, IN 46556, USA; nchawla@nd.edu

[3] Biomedical Sport Studies Center, University of Ferrara, 44123 Ferrara, Italy

[4] Polyclinic for Occupational Health and Sports of Zagreb Sports Association with Laboratory of Medical Biochemistry, 10000 Zagreb, Croatia; toncimd@gmail.com

[5] Centre for Applied Bioanthropology, Institute for Anthropological Research, 10000 Zagreb, Croatia; jelena.sarac@inantro.hr (J.Š.); miran.coklo@inantro.hr (M.Č.)

[6] Institute for Anthropological Research, 10000 Zagreb, Croatia; sasa.missoni@inantro.hr

[7] School of Medicine, Josip Juraj Strossmayer University of Osijek, 31000 Osijek, Croatia

* Correspondence: luciana.zaccagni@unife.it

† From 1 November 2020 the name of Department will be Neuroscience and Rehabilitation.

Received: 21 August 2020; Accepted: 25 October 2020; Published: 28 October 2020

Abstract: Cardiovascular diseases are the main cause of death worldwide. The aim of the present study is to verify the performances of a data mining methodology in the evaluation of cardiovascular risk in athletes, and whether the results may be used to support clinical decision making. Anthropometric (height and weight), demographic (age and sex) and biomedical (blood pressure and pulse rate) data of 26,002 athletes were collected in 2012 during routine sport medical examinations, which included electrocardiography at rest. Subjects were involved in competitive sport practice, for which medical clearance was needed. Outcomes were negative for the largest majority, as expected in an active population. Resampling was applied to balance positive/negative class ratio. A decision tree and logistic regression were used to classify individuals as either at risk or not. The receiver operating characteristic curve was used to assess classification performances. Data mining and resampling improved cardiovascular risk assessment in terms of increased area under the curve. The proposed methodology can be effectively applied to biomedical data in order to optimize clinical decision making, and—at the same time—minimize the amount of unnecessary examinations.

Keywords: medical diagnostic; decision tree; logistic regression; machine learning

1. Introduction

Cardiovascular diseases (CVDs) are reportedly the major cause of death worldwide, taking an estimated 17.9 million lives each year, according to the World Health Organization [1]. Obesity, smoking, physical inactivity and high blood pressure are among the most important risk factors [2]. Even if they may be mitigated by consistent sport practice [3,4], CVDs can still be considered an actual danger for individuals engaged in competitions or strenuous training [5–8] because of intense and repeated efforts. Therefore, athletes are routinely monitored by sport physicians, who collect some biomedical and personal data and screen them by means of electrocardiography (ECG).

According to the outcome of the ECG examination, individuals are diagnosed as either at risk (positive, P) or not (negative, N). Subjects at risk are denied medical clearance for sport practice and eventually undergo further examination. The two classes are usually imbalanced, since the N class contains the majority of individuals, while the more interesting P class is under-represented. In general, a missing P (false negative, FN) may have a very high cost—in some cases the loss of a human life—while a false alarm (false positive, FP) usually has the cost of some further clinical investigations and temporary suspension of sport activities.

Classification is a *machine learning* technique which can be applied to predict categorical binary values, like P or N, and for such reason it may be of great value in the medical field, and in diagnostics in particular. Machine learning, a branch of artificial intelligence, consists in the application of computer algorithms in order to (semi) automatically extract knowledge from collected data. When classification is applied to large datasets, we usually speak of *data mining*, which is defined as "the process of discovering patterns in data (...) The patterns discovered must be meaningful in that they lead to some advantage, usually an economic advantage" [9]. As the amount of collected data has increased, researchers and physicians are interested in evaluating their diagnostic value by means of data mining, and eventually suggest that the observed variables may be changed or increased in order to support medical decisions. Several data mining methods have already been used as decision support systems for medical diagnosis [10]. These methods may be applied to large datasets in order to estimate health risk. Wong et al. [11] used Bayesian networks for early disease outbreak detection and obtained good performances on real data from an emergency department database, containing 7 years of medical data. The aim of the study was not diagnostic, since the data were collected from hospital patients who were actually ill, but rather epidemiological: to verify an incipient influenza outbreak (this approach could be adopted also in counter-terrorism, to detect a biological attack).

Campbell and Bennet [12] adopted a kernel-based method, which performed well on a medical dataset in identifying a rare disease. Still, the dataset size was limited and the proportion of interesting instances in the test set was very high (27 normal observations and 67 anomalies), compared to the prevalence of the disease in the general population.

Marinić et al. [13] adopted WEKA as a data mining tool. They applied a random forest classifier on a relatively small sample ($n = 102$) of psychiatric patients in order to diagnose Post Traumatic Stress Disorder (PTSD) and achieved significant results. Class distribution though was perfectly balanced (51 P and 51 N) and therefore different from that of the general population, where PTSD has a much lower prevalence. In addition, Fontaine et al. [14] explored data mining techniques in order to improve clinical evaluation of patients with neuropsychiatric disorders.

A data mining approach was proposed by Salam and McGrath [15] in dermatology. A multi-disease classifier improved medical diagnosis of skin disorders. Sacchi et al. [16] adopted a Naïve Bayes classification algorithm for the prediction of glaucoma. Having a small and imbalanced dataset, they applied both bootstrapping and resampling to train the model. Chan et al. [17] showed that machine learning classifiers outperformed traditional statistical approaches in the diagnosis of the same medical condition.

An increasing interest in the adoption of data mining for classification and prediction in cardiology was reported by Kadi et al. [18] in a recent systematic literature review. For example, Karaolis et al. [19] adopted a decision tree (DT) for the assessment of coronary risk.

Comparing different classification methods in medical statistics has been suggested by several authors [20–22] in order to assess the real advantages of one technique over the other. Still, there is a lack of studies applied to large datasets in domains where diseases have a low prevalence (like CVDs in sport medicine) but individuals may be at greater risk because of increased stress or pressure. Further, there is an on-going debate on the necessity of a sustainable and cost-effective health care [23], also by means of a more sensible use of medical tests [24].

The aims of this study were to assess the performances of a data mining method in the prediction of ECG outcomes in an imbalanced dataset, when a resampling technique is applied, and to verify whether the results may be used to support clinical decision making. The underlying hypothesis was that, given a limited set of predictive biomedical variables, data mining could achieve good predictive accuracy if the proper algorithms were trained with a large amount of data.

2. Materials and Methods

2.1. Sample

A dataset including medical examinations of 26,002 athletes, both sexes, was collected at the Polyclinic for Occupational Health and Sports in Zagreb (Croatia) by medical staff in 2012. All individuals were involved in competitive sport practice, for which medical clearance was needed. The following data were collected for all subjects: sex, age, height, weight, resting pulse rate, diastolic and systolic pressure, and ECG at rest (P or N). The largest majority (91.2%) of outcomes was N, while a minority (8.8%) was P.

This study is the result of the collaborative research project "Health status and life quality of athletes", involving the Institute for Anthropological Research in Zagreb, the Department of Biomedical Sciences and Surgical Specialties of the University of Ferrara, the Polyclinic for Occupational Health and Sports of Zagreb Sports Association with Laboratory of Medical Biochemistry in Zagreb, and the Interdisciplinary Center for Network Science and Applications of the University of Notre Dame. The research was approved by the Ethical Committee of the Institute for Anthropological Research in Zagreb (registration number: 1.14-1169/13).

2.2. Machine Learning Background

Two classification techniques were trained and tested in order to predict the class (P or N) of the athletes. DT was chosen because it allows us to describe the extracted knowledge (patterns) in a simple and intuitive way, which can be easily understood by domain experts, like medical doctors. It is commonly preferred when explanation (understanding) is as important as prediction (knowing). Further, DT is a well-established support tool in medical decision making [25–29].

Logistic regression (LR) is a technique commonly applied to medical datasets, half-way between classic statistics and machine learning. The main difference between regression and classification is that in the former the predicted variable is numeric, while in the latter it is categorical. Since in logistic regression, the predicted variable can have only two numeric values (1 or 0), it can be used for binary classification [30], where 1 stands for P and 0 for N.

The assessment of classification performance is a major issue in imbalanced datasets. Usually, the basic evaluation index is accuracy, the rate of correct guesses (i.e., true positives, TP, and true negatives, TN) on total instances (TP + TN)/(P + N). It is an acceptable choice when class distribution is symmetric or close to it.

In case the distribution is imbalanced, accuracy can be misleading, unless a trivial solution is acceptable [31]. In fact, given a low prevalence, a high accuracy can be easily achieved classifying all instances as N, but it would imply missing all P individuals. Therefore, classification algorithms may have a high specificity or true negative rate (TNR = TN/N), as the majority class (N) is well represented, while sensitivity, or true positive rate (TPR = TP/P) may be significantly lower, as the minority class (P) is under-represented. Therefore, performance indexes other than accuracy should be taken into consideration.

A trade-off between TPR and TNR can be represented in receiver operating characteristic (ROC) space (for an introduction see [32]). Different cut-off values for the same classifier correspond to points in ROC space. The interpolation of such points draws a curve. The area under the ROC curve (AUC) is the accepted standard in the assessment of classification performances in imbalanced datasets [33], in particular in diagnostic systems [34–36].

Youden index J = TPR + TNR − 1 [37] is a summary measure corresponding to the distance between a point (corresponding to a cut-off value) on the ROC curve and the underlying 45° (random-guessing) line. J represents the probability of an informed decision. It has been used to assess the ability of biomarkers to correctly classify healthy and non-healthy individuals when equal weight is given to both sensitivity and specificity [38,39].

Different resampling methods [40] can be applied in order to balance the class distribution and thus improve performances. Resampling can either undersample the majority class and/or oversample the minority class. Synthetic Minority Oversampling Technique (SMOTE) [41] has proved to be reliable in different domains, including the prediction of type 2 diabetes [42], SMOTE does not simply duplicate existing instances, which would easily lead to overfitting. It creates a new instance in feature (i.e., variable, in data mining jargon) space, between an existing instance and one, randomly-chosen, of its k nearest neighbors. Euclidean distance between two neighboring instances is calculated and then it is multiplied by a random number between 0 and 1. The distance is used to calculate the position in feature space of the new instance.

2.3. Data Cleaning and Resampling

The raw dataset was cleaned, removing outliers. Body weight and height were replaced by body mass index (BMI = weight/height^2), a common proxy for obesity and cardiovascular risk in the general population [43–47]. Only systolic pressure was used, since systolic and diastolic blood pressures are correlated, and according to recent findings, the former is a better predictor of risk [48,49].

SMOTE was applied to the minority class with 100% oversampling (P instances were doubled). Higher percentages were tried but they did not improve classification or led to overfitting. Then random undersampling was applied to the majority class with an even distribution spread, in order to balance the two classes, so that class ratio P/N was set to 1.

2.4. Data Mining and Statistical Analyses

The dataset was divided into training (66% of total instances) and test (remaining 34%). DT was first applied without resampling (1st run), then with resampling (2nd run, k = 5). In the latter case, only training set was resampled, while test set kept the same distribution as the original subset (which is supposed to be similar to the population, giving the high cardinality of the sample). This prevented performances from being artificially high. Pulse rate was the best predictor in both cases. Two thresholds were identified: low (L) and high (H). Risk was low in between, moderate for pulse rate <L and high for pulse rate >H.

LR was applied twice. In the 1st run it was applied to all the collected variables directly. In the 2nd run, a discrete variable was created, with the following values: 0 (low risk) for pulse rate values between L and H, 1 (moderate risk) for values <L and 2 (high risk) for values >H. This variable replaced pulse rate in a data-driven way, instead of using a-priori values from the literature. Cross-validation (10-fold) was used to assess the model's performances and test whether this statistical regression technique could improve the performances of a standard machine learning algorithm like DT.

TPR, TNR, J and AUC were used to assess the performances of the adopted method, which is described in Figure 1.

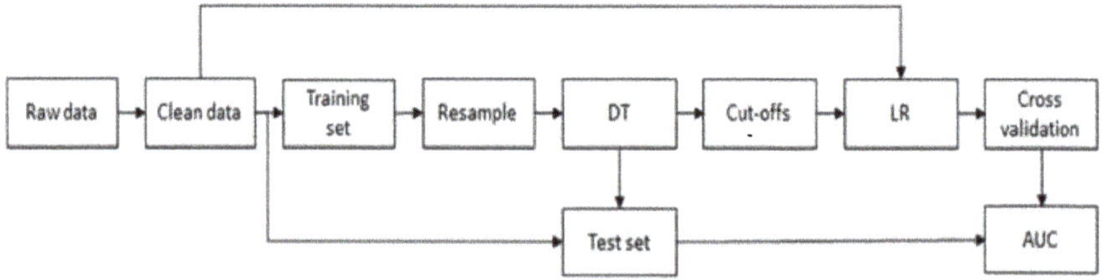

Figure 1. Data mining process. DT: decision tree; LR: logistic regression; AUC: area under curve.

MS Excel 2016 (Microsoft Corporation, Redmond, WA, USA) was used for data collection and data cleaning. WEKA 3.6 (University of Waikato, Waikato, New Zealand) was used for data mining.

Stata IC 13.1 (StataCorp LLC, College Station, TX, USA) was used to perform LR, cross validation and ROC analysis.

3. Results

Descriptive statistics by sex and age classes are shown in Table 1. Values are reported as mean and standard deviation, with the exception of ECG positives, which are reported as count and percentage.

Table 1. Descriptive statistics by sex and age classes.

Variables	6–10 Years	11–14 Years	15–18 Years	≥19 Years	Total
Females	$n = 1372$	$n = 1884$	$n = 970$	$n = 757$	$n = 4983$
Weight (kg)	34.2 ± 8.8	51.7 ± 10.8	61.8 ± 8.8	65.1 ± 10.5	50.9 ± 15.1
Height (cm)	138.3 ± 10.0	160.6 ± 8.7	168.5 ± 7.0	168.9 ± 7.52	157.3 ± 14.9
BMI (kg/m^2)	17.7 ± 2.8	19.9 ± 3.1	21.8 ± 2.6	22.8 ± 3.2	20.1 ± 3.5
Pulse rate (n)	83.1 ± 13.1	77.1 ± 13.0	68.3 ± 11.1	65.4 ± 11.7	75.2 ± 14.1
Systolic pressure (mm Hg)	97.3 ± 9.7	105.8 ± 9.9	108.7 ± 10.4	113.2 ± 11.8	105.2 ± 11.6
Diastolic pressure (mm Hg)	62.8 ± 7.47	66.4 ± 7.8	68.6 ± 7.8	73.3 ± 8.4	66.9 ± 8.6
ECG Ps (n (%))	140 (10.2)	160 (8.5)	63 (6.5)	65 (8.6)	428 (8.6)
Males	$n = 4787$	$n = 5776$	$n = 4253$	$n = 6203$	$n = 21{,}019$
Weight (kg)	33.6 ± 8.6	52.4 ± 13.4	71.4 ± 11.7	83.9 ± 12.7	61.3 ± 22.6
Height (cm)	136.9 ± 9.2	161.4 ± 11.6	178.6 ± 7.6	180.0 ± 7.3	164.8 ± 19.2
BMI (kg/m^2)	17.7 ± 2.8	19.9 ± 3.3	22.3 ± 3.0	25.9 ± 3.4	21.6 ± 4.4
Pulse rate (n)	79.0 ± 12.6	73.1 ± 12.6	66.8 ± 12.4	63.8 ± 11.9	70.4 ± 13.7
Systolic pressure (mm Hg)	97.1 ± 9.1	106.8 ± 11.1	118.2 ± 10.9	126.4 ± 11.8	112.7 ± 15.6
Diastolic pressure (mm Hg)	61.5 ± 7.8	65.2 ± 7.8	69.6 ± 8.2	77.9 ± 9.4	69.0 ± 10.5
ECG Ps (n (%))	379 (7.9)	405 (7.0)	397 (9.3)	699 (11.3)	1879 (8.9)

BMI: body mass index; ECG Ps: electrocardiography positives.

Classification performances of DT and LR are shown in Table 2.

Table 2. Algorithm classification performances.

Algorithm	TPR	TNR	J	AUC
DT (1st run)	0.29	0.97	0.26	0.68
LR (1st run)	0	1	0.00	0.56
DT (2nd run)	0.68	0.82	0.50	0.76
LR (2nd run)	0.65	0.82	0.47	0.78

TPR: True positive rate; TNR: True negative rate; J: Youden index; AUC: Area under the ROC curve; DT: Decision tree; LR: Logistic regression.

DT performances improved considerably by means of resampling, particularly in terms of greater sensitivity, from 0.29 to 0.68.

LR was highly significant in both runs ($p < 0.001$). Still, in the first run, with the default cut-off = 0.5, performance was so low (AUC = 0.56) that no further attempt with other cut-off values was made. DT performed better, even without resampling.

After a discrete variable was introduced in place of pulse rate, LR increased sensitivity (with cut-off = 0.12) from 0 to 0.65, without considerably diminishing specificity. This result was achieved because risk was high in both the lower and upper range of the continuous variable, and not monotonously increasing. It became evident after DT was applied, since it classified as P instances with either low or high pulse rate values. L and H were found inductively in the data (L = 60 bpm, H = 99 bpm), but they were close to those which can be found in domain-specific literature.

4. Discussion

In the present study, we tested the effectiveness and accuracy of a data mining methodology in the prediction of cardiovascular risk in athletes. The application of resampling and two classification techniques on an imbalanced dataset was evaluated. Resampling improved classification by means of

DT. LR has proved to be not an accurate diagnostic tool if applied directly to continuous variables, since in the medical domain high risk is often associated with feature values which are either too low or too high. This is particularly true in the assessment of cardiovascular risk, since pulse rate and blood pressure have upper and lower thresholds.

Feature cut-off values, which were acquired inductively from the collected data, can be immediately converted into a small set of simple decision rules to assist the diagnostic process. Additionally, they can be used to create categorical or discrete variables for LR, in order to improve its performances, without the need for a-priori assumptions. The chosen predictors can be easily measured by a family practitioner at almost no cost, which would make cardiovascular risk evaluation very efficient.

Classification performances were not assessed by means of accuracy because of the asymmetric distribution of the two classes. AUC and J have been suggested, because they represent a trade-off between sensitivity and specificity, in case they are both given the same importance. Even if this assumption can be reasonable from a statistical point of view, in medical diagnostics it poses a serious ethical concern. In fact, a FP comes at a cost of some unnecessary examinations and temporary suspension of sport activities, while a FN may imply the loss of a human life, in case a serious condition is present. Since a FN has a higher cost than a FP, a cost matrix may be adopted, where correct predictions (TP and TN) have no cost, and errors (FN and FP) have different weights [50]. Still, in the medical domain, it is not always possible to compare the cost of errors. Since there is no general agreement on the amount of resources which can be allocated to reduce the risk related to FN, it is not easy to give the proper weights to the two kinds of errors.

Even if resampling improved DT performances, in terms of both TPR and J, sensitivity remained lower than specificity, implying a high false negative rate. This may not be acceptable in a medical domain where a high risk may imply the loss of a human life. Increased oversampling may lead to overfitting, even if a minimum TPR may be required or imposed by medical authorities. Therefore, LR cut-off values may be lowered to improve sensitivity, even if it implies a large reduction in both specificity and J. Nonetheless, it is important to note that DT and LR performances were similar, and therefore DT alone can be adopted as a predictive algorithm, with the advantage over LR of improved understanding. DT confirmed to be a good means to capture and represent domain-specific (medical, in this case) knowledge, in an intuitive and easy-to-understand way, something which cannot be achieved by means of LR. These facts may even question the application to medical datasets of statistical techniques instead of machine learning, when a binary classification is required, as in diagnostics.

The main limitation of this study lies in the lack of comparison with a standard control method (beside ECG, which is used as a gold-standard) for cardiovascular risk assessment. In order to overcome this limitation, cardiologists and general practitioners as well should be involved in a future study from the design phase.

5. Conclusions

This study tested the effectiveness of a simple resampling technique in improving the assessment of cardiovascular risk, when data are imbalanced, as it is often the case in real-world situations. SMOTE improved the performances of DT in terms of greater AUC and sensitivity. In addition, DT produced actionable knowledge, which can be applied in the prediction of CVDs, diminishing the need for assumptions.

Further research is required to test the performances of the proposed approach in the general (i.e., non-athletic and older) population, to determine an acceptable and agreed classification performance index—which will highlight the best trade-off between medical risk and sustainable welfare—and to verify whether this data mining methodology can be applied to improve diagnosis and optimize healthcare policies, thanks to a reduction of unnecessary examinations.

Author Contributions: Conceptualization, D.B., N.C., and L.Z.; methodology, D.B., N.C., L.Z., and S.M.; formal analysis, D.B. and L.Z.; investigation, T.G.; data curation, D.B. and T.G.; writing—original draft preparation

D.B., N.C., L.Z., T.G., J.Š., M.Č., and S.M.; writing—review and editing D.B., N.C., L.Z., T.G., J.Š., M.Č., and S.M. All authors have read and agreed to the published version of the manuscript.

Funding: This research received no external funding.

Acknowledgments: The authors would like to thank the Polyclinic for Occupational Health and Sports of Zagreb Sports Association with Laboratory of Medical Biochemistry in Zagreb for providing the data. The authors would also like to acknowledge their late colleague Joško Sindik, from the Institute for Anthropological Research in Zagreb, who contributed substantially to the content of this paper as well as the whole project.

Conflicts of Interest: The authors declare no conflict of interest.

References

1. World Health Organization. Cardiovascular Diseases. Available online: https://www.who.int/health-topics/cardiovascular-diseases/#tab=tab_1 (accessed on 20 September 2020).
2. Mendis, S.; Puska, P.; Norrving, B. (Eds.) *Global Atlas on Cardiovascular Disease Prevention and Control*; World Health Organization: Geneva, Switzerland, 2011.
3. Güçlü, M. Comparing Women Doing Regular Exercise with Sedentary Women in Terms of Certain Blood Parameters, Leptin Level and Body Fat Percentage. *Coll. Antropol.* **2014**, *38*, 453–458.
4. Wronka, I.; Suliga, E.; Pawliñska-Chmara, R. Evaluation of Lifestyle of Underweight, Normal Weight and Overweight Young Women. *Coll. Antropol.* **2013**, *37*, 359–365.
5. Duraković, Z.; Duraković, M.M.; Skavić, J. Arrhythmogenic right ventricular dysplasia and sudden cardiac death in Croatians' young athletes in 25 years. *Coll. Antropol.* **2011**, *35*, 793–796.
6. Duraković, Z.; Duraković, M.M.; Skavić, J. Hypertrophic cardiomyopathy and sudden cardiac death due to physical exercise in Croatia in a 27-year period. *Coll. Antropol.* **2011**, *35*, 1051–1054. [PubMed]
7. Duraković, Z.; Misigoj Duraković, M.; Skavić, J.; Tomljenović, A. Myopericarditis and sudden cardiac death due to physical exercise in male athletes. *Coll. Antropol.* **2008**, *32*, 399–401. [PubMed]
8. Chatard, J.C.; Mujika, I.; Goiriena, J.J.; Carré, F. Screening young athletes for prevention of sudden cardiac death: Practical recommendations for sports physicians. *Scand. J. Med. Sci. Sports* **2016**, *26*, 362–374. [CrossRef] [PubMed]
9. Witten, I.H.; Frank, E. *Data Mining: Practical Machine Learning Tools and Techniques*, 2nd ed.; Morgan Kaufmann Publishers (Elsevier): San Francisco, CA, USA, 2005.
10. Bellazzi, R.; Zupan, B. Predictive data mining in clinical medicine: Current issues and guidelines. *Int. J. Med. Inform.* **2008**, *77*, 81–97. [CrossRef] [PubMed]
11. Wong, W.K.; Moore, A.; Cooper, G.; Wagner, M. Bayesian Network Anomaly Pattern Detection for Disease Outbreaks. In Proceedings of the 20th International Conference on Machine Learning, Washington, DC, USA, 21–24 August 2003; Fawcett, T., Mishra, N., Eds.; The AAAI Press: Manlo Park, CA, USA, 2003.
12. Campbell, C.; Bennett, K. A linear programming approach to novelty detection. In Proceedings of the Conference on Advances in Neural Information Processing Systems, Vancouver, BC, Canada, 3–8 December 2001; MIT Press: Cambridge, MA, USA, 2001; Volume 14.
13. Marinić, I.; Supek, F.; Kovačić, Z.; Rukavina, L.; Jendričko, T.; Kozarić-Kovačić, D. Posttraumatic Stress Disorder: Diagnostic Data Analysis by Data Mining Methodology. *Croat. Med. J.* **2007**, *48*, 185–197. [PubMed]
14. Fontaine, J.F.; Priller, J.; Spruth, E.; Perez-Iratxeta, C.; Andrade-Navarro, M.A. Assessment of curated phenotype mining in neuro, psychiatric disorder literature. *Methods* **2015**, *74*, 90–96. [CrossRef]
15. Salam, A.; McGrath, J.A. Diagnosis by numbers: Defining skin disease pathogenesis through collated gene signatures. *J. Investig. Dermatol.* **2015**, *135*, 17–19. [CrossRef]
16. Sacchi, L.; Tucker, A.; Counsell, S.; Garway-Heath, D.; Swift, S. Improving predictive models of glaucoma severity by incorporating quality indicators. *Artif. Intell. Med.* **2014**, *6*, 103–112. [CrossRef] [PubMed]
17. Chan, K.; Lee, T.W.; Sample, P.A.; Goldbaum, M.H.; Weinreb, R.N.; Sejnowski, T.J. Comparison of machine learning and traditional classifiers in glaucoma diagnosis. *IEEE Trans. Biomed. Eng.* **2002**, *49*, 963–974. [CrossRef] [PubMed]
18. Kadi, I.; Idri, A.; Fernandez-Aleman, J.L. Knowledge discovery in cardiology: A systematic literature review. *Int. J. Med. Inform.* **2017**, *97*, 12–32. [CrossRef] [PubMed]

19. Karaolis, M.A.; Moutiris, J.A.; Hadjipanayi, D.; Pattichis, C.S. Assessment of the risk factors of coronary heart events based on data mining with decision trees. *IEEE Trans. Inf. Technol. Biomed.* **2010**, *14*, 559–566. [CrossRef]

20. Schwarzer, G.; Vach, W.; Schumacher, M. On the misuses of artificial neural networks for prognostic and diagnostic classification in oncology. *Stat. Med.* **2000**, *19*, 541–561. [CrossRef]

21. Zhang, S.; Tjortjis, C.; Zeng, X.; Qiao, H.; Buchan, I.; Keane, J. Comparing data mining methods with logistic regression in childhood obesity prediction. *Inf. Syst. Front.* **2009**, *11*, 449–460. [CrossRef]

22. Maroco, J.; Silva, D.; Rodrigues, A.; Guerreiro, M.; Santana, I.; de Mendonça, A. Data mining methods in the prediction of Dementia: A real-data comparison of the accuracy, sensitivity and specificity of linear discriminant analysis, logistic regression, neural networks, support vector machines, classification trees and random forests. *BMC Res. Notes* **2011**, *4*, 299. [CrossRef]

23. Hood, V.L.; Weinberger, S.E. High value, cost-conscious care: An international imperative. *Eur. J. Intern. Med.* **2018**, *23*, 495–498. [CrossRef]

24. Qaseem, A.; Alguire, P.; Dallas, P.; Feinberg, L.E.; Fitzgerald, F.T.; Horwitch, C.; Humphrey, L.; LeBlond, R.; Moyer, D.; Wiese, J.G.; et al. Appropriate use of screening and diagnostic tests to foster high-value, cost-conscious care. *Ann. Intern. Med.* **2012**, *156*, 147–149. [CrossRef]

25. Murphy, C.K. Identifying diagnostic errors with induced decision trees. *Med. Decis. Mak.* **2001**, *21*, 368–375. [CrossRef]

26. Tanner, L.; Schreiber, M.; Low, J.G.; Ong, A.; Tolfvenstam, T.; Lai, Y.L.; Ng, L.C.; Leo, Y.S.; Thi Puong, L.; Vasudevan, S.G.; et al. Decision tree algorithms predict the diagnosis and outcome of dengue fever in the early phase of illness. *PLoS Negl. Trop. Dis.* **2008**, *2*, e196. [CrossRef]

27. Azar, A.T.; El-Metwally, S.M. Decision tree classifiers for automated medical diagnosis. *Neural Comput. Applic.* **2013**, *23*, 2387–2403. [CrossRef]

28. Christopher, J.J.; Nehemiah, H.K.; Kannan, A. A Swarm Optimization approach for clinical knowledge mining. *Comput. Methods Programs Biomed.* **2015**, *121*, 137–148. [CrossRef] [PubMed]

29. Gopinath, B.; Shanthi, N. Development of an Automated Medical Diagnosis System for Classifying Thyroid Tumor Cells using Multiple Classifier Fusion. *Technol. Cancer Res. Treat.* **2015**, *14*, 653–662. [CrossRef] [PubMed]

30. James, G.; Witten, D.; Hastie, T.; Tibshirani, R. *An Introduction to Statistical Learning with Applications in R*; Springer: New York, NY, USA, 2013.

31. Provost, F.; Fawcett, T. Analysis and Visualization of Classifier Performance: Comparison under Imprecise Cost and Classifier Distribution. In Proceedings of the Third International Conference on Knowledge Discovery and Data Mining (KDD), Huntington Beach, CA, USA, 14–17 August 1997.

32. Fawcett, T. An introduction to ROC analysis. *Pattern Recognit. Lett.* **2006**, *27*, 861–874. [CrossRef]

33. Chawla, N.V.; Bowyer, K.W.; Hall, L.O.; Kegelmeyer, W.P. SMOTE: Synthetic Minority Over-sampling Technique. *J. Artif. Intell. Res.* **2002**, *16*, 321–357. [CrossRef]

34. Swets, J.A. Measuring the accuracy of diagnostic systems. *Science* **1988**, *240*, 1285–1293. [CrossRef]

35. Ichikawa, D.; Saito, T.; Oyama, H. Impact of predicting health-guidance candidates using massive health check-up data: A data-driven analysis. *Int. J. Med. Inform.* **2017**, *106*, 32–36. [CrossRef]

36. Shimoda, A.; Ichikawa, D.; Oyama, H. Prediction models to identify individuals at risk of metabolic syndrome who are unlikely to participate in a health intervention program. *Int. J. Med. Inform.* **2018**, *111*, 90–99. [CrossRef]

37. Youden, W.J. Index for rating diagnostic tests. *Cancer* **1950**, *3*, 32–35. [CrossRef]

38. Ruopp, M.D.; Perkins, N.J.; Whitcomb, B.W.; Schisterman, E.F. Youden Index and optimal cut-point estimated from observations affected by a lower limit of detection. *Biom. J.* **2008**, *50*, 419–430. [CrossRef] [PubMed]

39. Perkins, N.J.; Schisterman, E.F. The Youden Index and the optimal cut-point corrected for measurement error. *Biom. J.* **2005**, *47*, 428–441. [CrossRef]

40. Lee, P.H. Resampling methods improve the predictive power of modeling in class-imbalanced datasets. *Int J. Environ. Res. Public Health* **2014**, *11*, 9776–9789. [CrossRef] [PubMed]

41. Chawla, N.V. Data Mining for Imbalanced Datasets: An Overview. In *Data Mining and Knowledge Discovery Handbook*; Maimon, O., Rokach, L., Eds.; Springer: New York, NY, USA, 2005; pp. 853–867.

42. Ramezankhani, A.; Pournik, O.; Shahrabi, J.; Azizi, F.; Hadaegh, F.; Khalili, D. The Impact of Oversampling with SMOTE on the Performance of 3 Classifiers in Prediction of Type 2 Diabetes. *Med. Decis. Mak.* **2016**, *36*, 137–144. [CrossRef]

43. Flegal, K.M.; Shepherd, J.A.; Looker, A.C.; Graubard, B.I.; Borrud, L.G.; Ogden, C.L.; Harris, T.B.; Everhart, J.E.; Schenker, N. Comparisons of percentage body fat, body mass index, waist circumference, and waist-stature ratio in adults. *Am. J. Clin. Nutr.* **2009**, *89*, 500–508. [CrossRef] [PubMed]

44. Freedman, D.S.; Kahn, H.S.; Mei, Z.; Grummer-Strawn, L.M.; Dietz, W.H.; Srinivasan, S.R.; Berenson, G.S. Relation of body mass index and waist-to-height ratio to cardiovascular disease risk factors in children and adolescents: The Bogalusa Heart Study. *Am. J. Clin. Nutr.* **2007**, *86*, 33–40. [CrossRef]

45. Zaccagni, L.; Lunghi, B.; Barbieri, D.; Rinaldo, N.; Missoni, S.; Šaric, T.; Šarac, J.; Babic, V.; Rakovac, M.; Bernardi, F.; et al. Performance prediction models based on anthropometric, genetic and psychological traits of Croatian sprinters. *Biol. Sport* **2019**, *36*, 17–23. [CrossRef]

46. Lam, B.C.; Koh, G.C.; Chen, C.; Wong, M.T.; Fallows, S.J. Comparison of Body Mass Index (BMI), Body Adiposity Index (BAI), Waist Circumference (WC), Waist-To-Hip Ratio (WHR) and Waist-To-Height Ratio (WHtR) as predictors of cardiovascular disease risk factors in an adult population in Singapore. *PLoS ONE* **2015**, *10*, e0122985. [CrossRef]

47. Suchanek, P.; Kralova Lesna, I.; Mengerova, O.; Mrazkova, J.; Lanska, V.; Stavek, P. Which index best correlates with body fat mass: BAI, BMI, waist or WHR? *Neuro Endocrinol. Lett.* **2012**, *33*, 78–82.

48. Borghi, C.; Dormi, A.; L'Italien, G.; Lapuerta, P.; Franklin, S.S.; Collatina, S.; Gaddi, A. The relationship between systolic blood pressure and cardiovascular risk–results of the Brisighella Heart Study. *J. Clin. Hypertens. (Greenwich)* **2003**, *5*, 47–52. [CrossRef]

49. Strandberg, T.E.; Pitkala, K. What is the most important component of blood pressure: Systolic, diastolic or pulse pressure? *Curr. Opin. Nephrol. Hypertens.* **2003**, *12*, 293–297. [CrossRef] [PubMed]

50. Weng Cheng, G.; Poon, J. A New Evaluation Measure for Imbalanced Datasets. In Proceedings of the 7th Australasian Data Mining Conference (AusDM '08), Glenelg/Adelaide, SA, Australia, 27–28 November 2008; Roddick, J.F., Li, J., Christen, P., Kennedy, P.J., Eds.; Australian Computer Society, Inc.: Darlinghurst, NSW, Australia, 2008; Volume 87, pp. 27–32.

Publisher's Note: MDPI stays neutral with regard to jurisdictional claims in published maps and institutional affiliations.

International Journal of
***Environmental Research
and Public Health***

Article

Influence of Size and Maturity on Injury in Young Elite Soccer Players

Natascia Rinaldo, Emanuela Gualdi-Russo * and Luciana Zaccagni

Department of Neuroscience and Rehabilitation, Faculty of Medicine, Pharmacy and Prevention, University of Ferrara, 44121 Ferrara, Italy; rnlnsc@unife.it (N.R.); luciana.zaccagni@unife.it (L.Z.)
* Correspondence: emanuela.gualdi@unife.it

Abstract: The involvement of pre-adolescents in soccer is becoming more and more frequent, and this growing participation generates some concerns about the potential factors for sports injuries. The purpose of this study was to investigate sports injuries in younger (U9–U11) and older (U12–U13) children playing soccer at an elite level, analyzing potential anthropometric and maturity risk factors. A total of 88 elite soccer players aged 9–13 years were investigated. Weight, stature, and sitting height were measured at the start and at the end of the competitive season, computing the relative growth velocities. Additional body composition parameters were taken during a second survey. Maturity offset was calculated using predictive equations based on anthropometric traits such as years from age at peak height velocity (YPHV). Injuries suffered during the competitive season were recorded. Maturity and some anthropometric characteristics were significantly different according to the presence or absence of injuries among the players. Multiple logistic regression revealed that YPHV, body mass index (BMI), and calf muscle area were the factors most significantly correlated with injuries. Players with increased BMI, with decreased calf muscle area, and who were closer to their peak height velocity, were at a higher risk of injury. Findings showed that a monitoring program of anthropometric characteristics taking into account the maturational stage needs to be developed to prevent injuries.

Keywords: youth athletes; soccer; injury; overuse; maturation; anthropometry

Citation: Rinaldo, N.; Gualdi-Russo, E.; Zaccagni, L. Influence of Size and Maturity on Injury in Young Elite Soccer Players. *Int. J. Environ. Res. Public Health* **2021**, *18*, 3120. https://doi.org/10.3390/ijerph18063120

Academic Editor: David Berrigan

Received: 7 February 2021
Accepted: 16 March 2021
Published: 18 March 2021

Publisher's Note: MDPI stays neutral with regard to jurisdictional claims in published maps and institutional affiliations.

1. Introduction

Soccer is one of the most popular sports in the world, and it is especially enjoyed by children and teenagers. It is estimated that 3.9 million children and adolescents participate in soccer annually [1,2]. The psychological and physical health benefits of physical activity and sports practice are well known and documented [3–6]. However, given its nature as a high-intensity contact sport, soccer is associated with a great injury risk, which makes it a target sport for injury prevention [7,8].

For this reason, it is fundamental to analyze the factors related to injury occurrence in soccer players, especially in elite and sub-elite players, as they usually result in a considerable amount of time loss from training and matches and represent a significant economic burden for the health care system [1,9]. Previous studies analyzed the risk factors and the rate of injuries in adult elite soccer players, both males [7] and females [10], but there is a paucity of studies focusing on injury rate and modalities in children and adolescents [1,11]. Moreover, the studies focusing on this topic have several limitations [12] as they are usually based only on adolescents [9] or high school and college players [13]. Sometimes, they also report combined data from adolescents and adults [7], or they combine several sports [14].

Injury incidence among soccer players differs across levels of participation, age, type of exposure, and sex [1,8,15–17] with soccer players under the age of 12 years exhibiting a lower injury rate (1.0–1.6 injuries per 1000 h) compared to older players [1,16,17]. However, this rate varies considerably among different studies. It is estimated that 2,995,765 children and adolescents aged 7–17 had an emergency department visit in the United States during

45

a 25-year epidemiological study, with an annual average of about 120,000 and an increased rate of 111% from 1990 to 2014 [12].

The majority of injuries in young soccer players are caused by acute events resulting from player-to-player contact, especially during competitions [18,19], and by overuse, especially in the lower extremities (ankles and knees) [20,21]. It must be considered that the type and location of injuries differ by age, with 5–14 year old players more likely to be exposed to upper-extremity injuries [1,12,22] in comparison to their older peers.

Only a few studies were aimed at identifying risk factors for injuries in youth soccer, such as neuromuscular imbalances, fatigue, biomechanical factors, and previous injuries [15,21,23,24]. However, in addition to these, other factors which may increase the potential for injury in sport should be analyzed. Although there is an increasing interest in the role played by body composition, and anthropometric and growth characteristics as injury risk factors in youth soccer, the research on this topic is still scarce and controversial [21,25–27].

Previous studies suggested that the player's age, evaluated as growth and maturation, is an important risk factor, especially in the life period around the peak height velocity (PHV) [25,26,28,29]. PHV is considered the moment with the largest increase in stature, implying a high growth velocity [30]. In particular, an increased likelihood of overuse injuries was observed during this adolescent growth spurt. During rapid growth, skeletally immature athletes do not have the same resistance to tensile, shear, and compressive forces as skeletally mature athletes or athletes in a prepubertal stage of greater immaturity [31]. Moreover, the anthropometric changes associated with adolescent growth spurts affect coordination skills and, consequently, the injury risk [32]. Moreover, a decrease in flexibility and bone density occurs during the maturation process, increasing the vulnerability of the skeletal system [31,33]. In a recent study, Rommers et al. [11] reported that age, total growth, and lower-limb growth were strongly associated with increased injury risk in both preadolescent and adolescent elite soccer players. According to this research, body weight and lower-limb length were also correlated with injuries [11]. From this perspective, an increased body mass index (BMI) and a low fat percentage were identified as anthropometric injury risk factors in elite-standard young soccer players [25,34]. However, the results concerning the association between body composition parameters and risk of injuries are controversial. It is possible that there could be an optimal range in fat and fat-free mass percentage in terms of injury risk, whereby every deviation could represent a potential injury risk [35–37]. Further studies are needed to understand the effect of growth rate, maturation, and anthropometric characteristics on the increase in injury rate risk.

Therefore, the main aims of this study were to: (i) describe the prevalence of injuries in a sample of pre-adolescent elite soccer players of different ages, and (ii) evaluate whether anthropometric characteristics, growth, and maturation are possible predictors of injury risk. We hypothesized that great changes in anthropometric characteristics and being close to the growth spurt would have a negative influence on the injury risk of young players.

2. Materials and Methods

2.1. Participants and Procedures

This research was carried out during the 2018–2019 competitive season from September to May. Figure 1 shows the experimental design of the study.

Anthropometric data were collected during the first two weeks of September in an appropriate indoor space at the soccer field. Moreover, the tracking of these anthropometric traits was investigated using a prospective design (second survey in the first two weeks of May), supplementing the anthropometric profile of soccer players with some other useful measurements for the assessment of body composition. The occurrence of injuries per season was also recorded using questionnaires distributed at the beginning of the season and collected at the end of the season. The coaches of each team were asked to check that all injuries were recorded. The injuries associated with the anthropometric characteristics and biological maturity of the players were assessed.

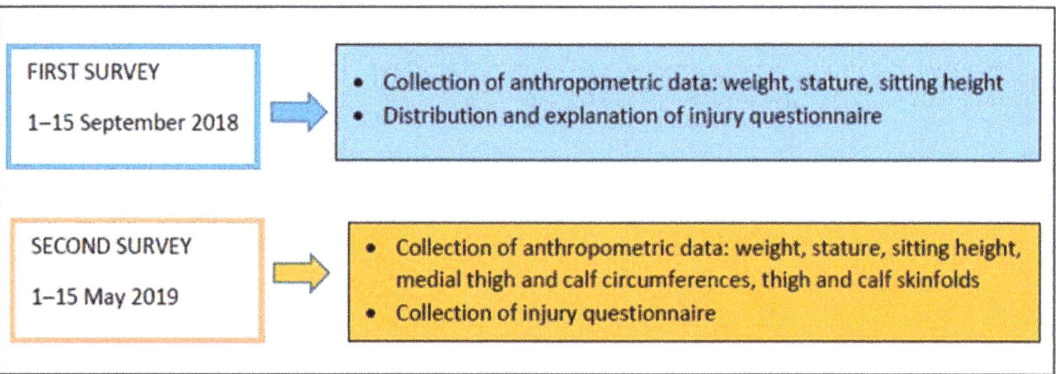

Figure 1. Schematic figure showing the experimental design of the study.

The study protocol was approved by the Ethical Committee of the Province of Ferrara (N. 140797). Children could only participate in the study if their parents/guardians provided informed consent. Participants were also informed that participation was voluntary and that they could withdraw from the study at any time.

This research monitored 88 male players aged 9–13. This convenience sample was recruited from teams under 13 (U13) (26 players), U12 (22 players), U11 (17 players), U10 (12 players), and U9 (11 players) of the youth academy of the professional soccer club S.P.A.L. (Società Polisportiva *Ars et Labor*, Ferrara, Italy), playing in the first division of the Italian soccer league. The sample (aged 11.5 ± 1.4 years, on average) was divided into two subgroups U9–U11 (aged 10.2 ± 0.8 years) and U12–U13 (aged 12.5 ± 0.5 years). The training program of this elite academy included at least three training sessions and one match per week, usually on Saturday. In particular, U9–U11 participants trained three days a week with a total weekly soccer training duration of 5 h and 30 min. U12–U13 participants trained four days a week with a total weekly soccer training duration of 7 h.

2.2. Anthropometric Measurements

At the start of the competitive season (September), we measured the following anthropometric traits: weight, stature, and sitting height. The length of the lower limb was determined using an indirect method. The repetition of these anthropometric measurements, collected during the first survey and particularly relevant for the assessment of the child's growth, and the measurement of other anthropometric characteristics for the assessment of body composition, were included in the second survey at the end of competitive season (May): stature, weight, sitting height, lower-limb circumference (medial thigh and calf) and skinfold thickness (thigh and calf). In selecting the anthropometric characteristics, we took into account the "fundamental" anthropometric measurements (height and weight) [38] and those measurements (length of the lower limb, obtained from stature minus sitting height, and body composition parameters) considered crucial during the prepubertal and pubertal periods [39].The changes for the season in stature, weight, lower limb length, and BMI were computed and then transformed into yearly growth rates by dividing the change by the time between the two measurements. Anthropometric characteristics of children were collected according to standardized methodologies [40–42].

Stature and sitting height were measured to the nearest 0.1 cm using an anthropometer (Raven Equipment Ltd., UK). During the stature measurement, the child was upright without shoes, with heels together, and with the head on the horizontal plane of Frankfurt. During the sitting height measurement, the child was sitting on a box of known height with his head oriented on the Frankfurt plane. By subtracting the height of this box from the measured height, sitting height was determined. Weight was measured to the nearest 0.1 kg using a Seca weighing scale (Seca Deutschland Medical Measuring Systems and Scale, Hamburg, Germany) for children dressed in light clothing. The lower-limb length (subischial) was indirectly obtained as stature minus sitting height. Thigh and calf circumferences were measured to the nearest 0.1 cm using an anelastic tape (GPM measuring tape, DKSH, Swiss).

The same trained operator (N.R.) performed a double measurement of thigh and calf skinfold thicknesses to the nearest 0.5 mm using a Lange skinfold caliper (Beta Technology Inc., Houston, TX, USA). In the analysis, we used the average of two plyometric values. The technical error of measurement (assessed prior to project start) was <5% for skinfolds and <1% for other measurements.

All bilateral measurements were conventionally taken on the left side.

Among the anthropometric indices, we calculated the body mass index (BMI) as weight (kg)/stature squared (m^2) and the Skelic index as (lower limb length/sitting height) $\times$ 100. Young players were classified as underweight, normal weight, overweight or obese according to international age- and sex-specific BMI thresholds [43,44]. Body composition and cross-sectional areas of the thigh and leg were also estimated from thigh or calf circumference (C) in cm, and thigh or calf skinfold thickness (ST) in cm [45,46] as follows:

Total area (cm^2) = C^2/4π,
Muscle area (cm^2) = (C $-$ ($\pi \times$ ST))^2/4π,
Fat area (cm^2) = Total area $-$ Muscle area,
Fat Index (%) = (Fat area/Total area) $\times$ 100.

2.3. Peak Height Velocity

To determine the age at PHV of the players, we applied a sex-specific equation developed in Canadian and Belgian boys [30] on the basis of some anthropometric traits described above. The predictive equation [47] for boys is as follows: $-9.236 + (0.0002708 \times$ (lower-limb length $\times$ sitting height)) $+ (-0.001663 \times$ (age $\times$ lower-limb length)) $+ (0.007216 \times$ (age $\times$ sitting height)) $+ (0.02292 \times$ (weight/stature $\times$ 100)). This equation predicts maturity offset (time before or after PHV). According to the estimated year of PHV, four maturity categories (YPHV = decimal age–PHV) were identified: category 0, YPHV >-1 year; category 1, -2 years < YPHV ≤ -1 year; category 2, -3 years < YPHV ≤ -2 years; category 3, YPHV ≤ -3 years.

2.4. Injuries

The injuries suffered by players during the 2018–2019 competitive season were collected through questionnaires distributed and explained at the beginning of the season. Players reported type, severity, and mechanism of injury (acute or overuse) according to the Fuller et al. [48] questionnaire. In compliance with these authors, injuries were defined as any physical complaint suffered by a player resulting from a soccer match or training, regardless of the necessity of medical care or time lost from soccer activities. Injuries resulting from a specific and identifiable event were denoted as acute, while injuries caused by repeated microtrauma without a single identifiable event were denoted as overuse. Injury number was reported separately by each player. Their severity was reported as time lost (number of days that the player was not able to take full part in competition or training): slight for 0 days; minimal for 1–3 days; mild for 4–7 days; moderate for 8–28 days; severe for >28 days [17,28,48]. Data on injury location and type were also collected. In terms of anatomic location, injuries were grouped for simplicity into three main body regions: trunk, upper limbs, and lower limbs [27].

We calculated injury incidence as the number of injuries per 1000 exposure player-hours in training and matches, as follows: (Σ injuries/Σ exposure hours) $\times$ 1000 [17,49].

2.5. Statistical Analysis

Descriptive statistics (means and standard deviations) were separately computed for the younger (U9–U11) and older (U12–U13) teams. We tested normality using the

Kolmogorov-Smirnov test. The log-transformed values of thigh and calf skinfolds were carried out before statistical comparisons. Percentage frequency was computed for qualitative variables (weight status and YPHV categories).

Anthropometric variables were compared between injured and uninjured subgroups using the Student t-test for independent samples, and Cohen's d was calculated to estimate the effect size. By convention, t-test effect size (ES) values of 0.2, 0.5, and 0.8 are considered small, medium, and large, respectively [50]. Differences between observed frequencies of qualitative traits (weight status; YPHV categories) and their frequencies expected under the null hypothesis were tested using the chi-squared test.

Multivariate logistic regression analysis was executed to investigate the injury risk factors. The dependent variable (selected criterion) was the injury occurrence (no injury = 0, injury = 1), while the independent predictor variables for multivariate analysis were identified using automated backward stepwise selection. The general fit of the model was tested using the Hosmer and Lemeshow goodness-of-fit test (LS). The power of the model was determined by Nagelkerke's R-squared. The odds ratios (ORs) and 95% confidence intervals (CI) were computed referring to the injury occurrence.

Analyses were done with STATISTICA software, version 11 (StatSoft, Tulsa, OK, USA). Statistical significance was set at $p < 0.05$.

3. Results

3.1. Anthropometric, Maturation and Injury Characteristics

In the 2018–2019 competitive season, we surveyed a sample of 88 elite soccer players. Table 1 shows the anthropometric characteristics of the total sample and of the two age subgroups (U9–U11 and U12–U13) according to the survey. As expected, the older group had higher mean values of anthropometric variables in both surveys, except for skinfold thicknesses. With regard to the weight status, there were 11 underweight (12.5%), 72 normal weight (81.8%), and five overweight (5.7%) players during the first survey (i.e., in September), whereas, at the end of the competitive season (i.e., in May), there were five underweight (5.7%), 80 normal weight (90.9%), and three overweight (3.4%) players, with non-significant changes in frequency. The growth rate in weight, stature, lower-limb length, and BMI were higher in older than in younger players; the former players were, on average, closer to maturity offset, as highlighted by the lesser distance from peak height velocity (PHV).

Table 2 displays the descriptive statistics of injuries. During the 8-month period, a total of 57 injuries were registered in the total sample: 34 injuries (equal to 60%) were a result of overuse, whereas the remaining 23 (equal to 40%) occurred traumatically. Of the injuries, 40 occurred during matches, and 17 occurred during training sessions. A total of 37 players (equal to 42% of entire sample) suffered no injury, 47 players (53.4%) suffered one injury and four players (4.5%) suffered multiple injuries. Half of the younger group (i.e., U9–U11) was injured, sustaining a total of 22 injuries (equal to 38.6% of all injuries), while 64.6% of the older group (i.e., U12–U13) was injured, sustaining a total of 35 injuries (the remaining 61.4% of all injuries). The number of injuries and the number of injured players increased with age, albeit not significantly, as highlighted by the higher injury incidence in the older group (0.73 vs. 0.55). The injury incidence for 1000 h of exposure was 2.71 in the younger and 2.82 in the older subgroup. The majority of injuries were to the lower limb ($n = 44$; 77.2%), followed by injuries to the upper limb ($n = 7$; 12.3%) and trunk ($n = 6$; 10.5%). The most frequent injuries among the players were strains, sprains, and contusions. As regards injury severity, the half (50.8%) of the injuries resulted in absence from sport fewer than 8 days, 28.1% in absence 8 to 28 days, and 21.1% in absence more than 28 days. In both age groups, there were similar proportions with regard to the type of injury (~60% for overuse and ~40% for acute); however, in the younger group, the incidence of both overuse and acute injuries was lower, especially with respect to overuse.

Table 1. Anthropometric characteristics (mean and SD) for the total sample and age subgroups of elite soccer players during the first (above) and second (below) surveys.

Variable	Total Sample (n = 88)		U9–U10–U11 (n = 40)		U12–U13 (n = 48)	
	Mean	SD	Mean	SD	Mean	SD
First Survey						
Age (years)	11.45	1.37	10.15	0.83	12.54	0.50
Weight (kg)	38.47	7.48	34.07	4.95	42.05	7.31
Stature (cm)	146.67	9.80	139.38	6.05	152.58	8.15
Sitting height (cm)	75.72	4.17	73.06	2.85	77.88	3.83
Lower-limb length (cm)	70.95	6.52	66.32	4.69	74.71	5.28
BMI (kg/m^2)	17.75	1.95	17.50	2.02	17.95	1.88
Skelic index	93.67	6.51	90.84	6.57	95.97	5.53
YPHV (years)	−2.33	0.98	−3.21	0.53	−1.63	0.61
Second Survey						
Weight (kg)	41.33	8.83	35.64	4.79	46.08	8.64
Stature (cm)	150.61	10.79	142.45	6.21	157.42	8.92
Sitting height (cm)	77.02	4.86	73.79	2.97	79.72	4.48
Lower-limb length (cm)	73.59	6.86	68.66	4.68	77.70	5.57
BMI (kg/m^2)	18.03	1.90	17.52	1.69	18.45	1.98
Skelic index	95.54	6.56	93.13	6.64	97.54	5.99
YPHV (years)	−2.18	1.09	−3.13	0.54	−1.38	0.72
Weight gain rate (kg/year)	4.40	3.90	2.43	1.43	6.00	4.50
Stature gain rate (cm/year)	6.11	3.49	4.75	3.04	7.21	3.46
BMI gain rate (kg/m^2/year)	0.43	1.38	0.03	1.09	0.75	1.52
Lower-limb length gain rate (cm/year)	4.08	3.32	3.61	3.44	4.46	3.20
Thigh circumference (cm)	39.96	3.93	37.77	3.11	41.78	3.62
Thigh skinfold (mm)	14.32	4.35	14.35	4.52	14.29	4.25
Calf circumference (cm)	30.99	2.51	29.58	1.81	32.17	2.41
Calf skinfold (mm)	8.30	3.11	8.15	3.25	8.42	3.02
Total thigh area (cm^2)	128.30	25.74	114.29	19.41	139.98	24.65
Thigh muscle area (cm^2)	101.03	19.59	88.50	12.30	111.47	18.44
Thigh fat area (cm^2)	27.28	9.93	25.79	9.73	28.51	10.02
Thigh fat index (%)	21.00	5.07	22.09	5.24	20.10	4.78
Total calf area (cm^2)	76.95	12.66	69.90	8.66	82.83	12.53
Calf muscle area (cm^2)	64.57	10.39	58.31	6.38	69.78	10.24
Calf fat area (cm^2)	12.38	5.08	11.59	5.06	13.04	5.05
Calf fat index (%)	15.90	5.22	16.30	5.51	15.57	5.00

Note: U, under; YPHV, years from age at peak height velocity; BMI, body mass index.

3.2. Injury Risk Factors

Table 3 shows the comparison between injured and uninjured players at the end of the competitive season. The group of injured players was significantly closer to maturity offset, with higher BMI and growth rates in stature and weight. According to the benchmarks suggested by Cohen [50], the effect size was large for BMI and rate of weight gain, whereas it was medium for rate stature gain and distance from peak height velocity. Repeating the analysis separately for overuse injuries and acute injuries, the BMI, and the growth rates in weight and lower-limb length, were found to be significantly different only when comparing overuse injured and uninjured (BMI: p = 0.025, ES = 0.48; weight rate: p = 0.025, ES = 0.46; limb length: p = 0.047, ES = 0.42). A further comparison of injured and uninjured players within the two age subgroups confirmed the general trend with significant differences in weight (p = 0.011, ES = 0.85) and BMI (p = 0.008, ES = 0.88) in the U9-U11 subgroup, and in weight gain (p = 0.032, ES = 0.71) in the U12–U13 subgroup.

Table 2. Injury characteristics for the total sample and both subgroups of elite soccer players.

All Injuries	Total Sample $n = 88$	U9–U10–U11 $n = 40$	U12–U13 $n = 48$
N (%)	57 (100)	22 (100)	35 (100)
Injured players	51 (58.0)	20 (50.0)	31 (64.6)
Incidence (n/player season)	0.65	0.55	0.73
Injury risk per injured athlete	1.12	1.10	1.13
Injury severity			
slight	9 (15.8)	4 (18.2)	5 (14.3)
minimal	10 (17.5)	4 (18.2)	6 (17.1)
mild	10 (17.5)	3 (13.6)	7 (20.0)
moderate	16 (28.1)	6 (27.3)	10 (28.6)
severe	12 (21.1)	5 (22.7)	7 (20.0)
Overuse injuries			
N (%)	34 (59.6)	13 (59.1)	21 (60.0)
Injured players	29 (56.9)	12 (60.0)	17 (54.8)
Incidence (n/player season)	0.39	0.33	0.44
Injury risk per injured athlete	1.18	1.08	1.23
Acute injuries			
N (%)	23 (40.4)	9 (40.9)	14 (40.0)
Injured players	22 (43.1)	8 (40)	14 (45.2)
Incidence (n/player season)	0.26	0.23	0.29
Injury risk per injured athlete	1.05	1.13	1.00

Note: Numbers in brackets are percentages.

Table 3. Comparison between injured and uninjured youth soccer players by risk factors. The anthropometric characteristics and growth rate during the second survey are presented above, whereas the distribution of injured/uninjured players according to their biological maturity (YPHV categories) is presented below.

Variable	Injured $n = 51$ Mean	SD	Uninjured $n = 37$ Mean	SD	p^a	Effect Size
Weight (kg)	42.89	9.20	39.18	7.92	0.0513	0.43
Stature (cm)	151.65	10.29	149.19	11.44	0.2943	0.23
Sitting height (cm)	77.67	4.81	76.13	4.86	0.1427	0.32
Lower-limb length (cm)	73.97	6.54	73.06	7.33	0.5399	0.13
BMI (kg/m^2)	18.48	2.16	17.41	1.26	**0.0086**	0.61
Skelic index	95.28	6.59	95.89	6.58	0.6634	0.09
YPHV (years)	−1.98	1.04	−2.44	1.11	**0.0478**	0.43
Rate of weight gain (kg/year)	5.34	4.45	3.08	2.43	**0.0070**	0.63
Rate stature gain (cm/year)	6.76	3.71	5.18	2.95	**0.0368**	0.47
Rate of BMI gain (kg/m^2/year)	0.62	1.58	0.15	1.00	0.1179	0.36
Rate of gain in lower limb length (cm/year)	4.38	3.64	3.66	2.78	0.3184	0.22
Thigh circumference (cm)	40.60	4.26	39.08	3.28	0.0723	0.40
Thigh skinfold (mm)	14.94	4.66	13.46	3.77	0.1315	0.35
Calf circumference (cm)	31.32	2.57	30.55	2.39	0.1579	0.31
Calf skinfold (mm)	8.64	3.29	7.81	2.83	0.1810	0.27
Total thigh area (cm^2)	132.61	28.33	122.37	20.58	0.0650	0.41
Thigh muscle area (cm^2)	103.65	21.01	97.41	17.08	0.1413	0.33
Thigh fat area (cm^2)	28.96	11.08	24.95	7.64	0.0612	0.42
Thigh fat index (%)	21.50	5.11	20.32	4.99	0.2862	0.23
Total calf area (cm^2)	78.57	13.12	74.72	11.81	0.1602	0.31
Calf muscle area (cm^2)	65.51	10.29	63.27	10.53	0.3198	0.22
Calf fat area (cm^2)	13.06	5.52	11.45	4.30	0.1442	0.33
Calf fat index (%)	16.34	5.24	15.29	5.20	0.3521	0.20

YPHV category	N	%	N	%	p^b	Effect size
0	10	77	3	23	0.0625	0.41
1	16	64	9	36		
2	15	52	14	48		
3	10	48	11	52		

Note: YPHV, years from age at peak height velocity; BMI, body mass index; category 0, YPHV > −1 year; category 1, −2 years < YPHV ≤ −1 year; category 2, −3 years < YPHV ≤ −2 years; category 3, YPHV ≤ −3 years; p^a, probability associated with the *t*-test; p^b, probability associated with the chi-squared test; bold values indicate statistical significance.

The lower part of Table 3 also shows the distribution of injuries in the four categories of YPHV. Upon approaching peak height velocity, the occurrence of injuries increased from the minimum of 48% registered in category 3, i.e., the furthest from maturity offset, to the maximum of 77% in the closest category (i.e., YPHV category 0).

Table 4 displays the odds ratios of the multivariate model with 95% confidence intervals and *p*-values. Logistic regression analysis revealed that three variables (YPHV, BMI, and calf muscle area) were significant risk factors for injuries: increased YPHV with 2.6-fold increased risk of injury, increased BMI with almost twofold increased risk of injury, increased calf muscle area with 13% decrease in risk of injury. The other anthropometric variables provided no further contribution to the explanation of the model in the multivariate analysis. The Hosmer-Lemeshow statistic adequately fit the data ($p > 0.05$). Nagelkerke's statistic indicated that the model's explained variance was 22%.

Table 4. Multivariate logistic regression analysis of potential injury risk factors in young soccer players.

Risk Factors	OR (95% CI)	Wald	*p*
YPHV (years)	2.633 (1.170–5.928)	5.471	0.019
BMI (kg/m^2)	1.923 (1.225–3.018)	8.079	0.005
Calf muscular area (cm^2)	0.874 (0.786–0.972)	6.139	0.013
Hosmer-Lemeshow	9.593		0.295
Nagelkerke R^2	0.216		

Note: YPHV, years from age at peak height velocity; OR, odds ratio; BMI, body mass index.

4. Discussion

In this study, we analyzed the prevalence of injuries in teams of young elite soccer players by age, evaluating the main anthropometric and growth factors associated with an increased risk of injuries. Our findings demonstrate a trend in the incidence of injuries with age and biological maturity (PHV), as well as an association of some anthropometric characteristics with the risk of injuries in young elite soccer players.

A higher overall number of injuries per 1000 exposure player-hours was found in this study (2.77) than in other previous studies on soccer players (1.20 in children and 1.30 in adolescent Brazilian soccer players in [27]; 0.61 in 712-year-old Czechoslovak and Swiss players in [17]), but similar to other findings from the literature (2.81 in high-school soccer players, according to [51]). Considering the impact of different sports on the level of exposure in male adolescents, the literature shows that this rate per 1000 exposures achieves its highest values in football, hockey, and basketball, and its lowest values in track, swimming, tennis, and badminton athletes [9,51,52].

The results of this study highlighted that the number of injuries varies among different age groups (U9–U11 vs. U12–U13), regardless of the type of injury (acute or overuse). In particular, older groups reported a greater number of injured players and a greater incidence in comparison to younger players, especially with regard to overuse injuries. This could be due to the increase in training load in the older groups, as reported in the study by Vanderlei et al. [27], showing that injured adolescent players trained for longer weekly hours. Although there is no concordance in the literature regarding the relationship between athlete age and injury risk, lighter and smaller soccer players are generally deemed to be less prone to injury [11]. Furthermore, it is recognized that, as the level of competition and contact in sport increases with age, so does the risk of injury in children and adolescents [9]. In soccer, as in other sports, the peak of injuries lies consistently in the age group of older adolescents [9,53]. However, a major role could be played by the maturity status, rather than by the chronological age.

Analyzing the frequencies of injuries according to the distance from the age at PHV, we found an increasing trend of injury rate with the approach of the peak, as confirmed by the ES. This was further confirmed by the comparison between injured and uninjured players, where the mean YPHV appeared significantly lower in the first group, and by the multivariate logistic regression, where the YPHV increased the injury risk more than twofold. Previous studies have consistently underlined that maturity status influences the structural body changes of young soccer players, in addition to functional growth capacity [11,21,25,26,28,32,54].

The findings of the current study support the relevance of specific anthropometric variables from the perspective of injury prevention. BMI, calf muscle area, and growth rates in weight, stature, and lower-limb length were identified as risk factors for injuries. This finding is in line with previous studies showing that numerous anthropometric changes linked to the growth spurt of adolescence [55] led to a period of increased risk of sports injuries [56], so much so that it was suggested that earlier and later players follow different training programs [28]. Therefore, regular anthropometric measurements of players (every 3 months according to [54]) should be used to monitor their somatic maturity [57] and identify specific growth phases involving different injury risk [54].

In particular, with reference to the results of our prospective study, higher growth rates in weight and stature were found to be significant risk factors; players were more likely to be injured as the growth rates in weight and stature increased. In our study the gain rate in lower-limb length, in addition to BMI and weight gain rate, was also significantly different between overuse injured and uninjured players. These growth parameters, particularly the increase in leg length and stature [55], define the tempo of maturation and are strictly connected to the PHV, whereby the growth rate increases as the YPHV tends to zero, approaching the peak. The tendency for the risk of injury to increase with increasing maturation is probably due to the high training loads [27] that overlap with the rapid yearly growth changes. A high tempo of maturation could be one of the causes of the increase in injury rate around the PHV, as underlined by other studies performed in young soccer players [11,24,28] and in young athletes of other sports [9,27]. This could be connected to a temporary increase in skeletal fragility, which increases the rate of acute injuries [58,59], and to a temporary decline in essential motor performance during years of maximal growth caused by an imbalance between lower limb and trunk growth and muscular size and strength [55,60]. Among injuries, however, it is also well known that growth-related factors predispose players to overuse injuries [31,56], as observed in our study. The increase in weight could also be a consequence of the growth [32], but this anthropometric characteristic may be influenced by several other factors; thus, it is also important to analyze the changes in BMI among injured and uninjured players. However, the results in the literature are not always consistent. A study by Rommers et al. [11] found that an increase in leg length in younger players (U10–U12) and having long legs in older players (U13-U15) were associated with overuse injuries, while a higher weight and a slower growth rate were associated with acute injuries.

Although we found no significant difference between injured and uninjured players in body composition parameters, the relevance of calf muscle area in the occurrence of injuries emerged through multivariate logistic regression analysis: injuries decreased as the calf muscle area increased. This result underlines the importance of fat free mass in the lower extremities, which strongly contributes to strength and power performance. Players with greater muscle development in the lower limbs, due to training and/or structural characteristics, have greater dynamic stability in their sporting actions with lower risk of injury. An association of muscle weakness or alterations in muscle recruitment with injury was previously reported [61]. From this point of view, it should be noted that the stage between the ages of 12 and 15 represents a very sensitive period of neuromuscular maturation, which has to be maintained through systematic training [62].

The increase in BMI, an index commonly used as a proxy for adiposity, was another significant risk factor for injuries in our sample, as evidenced by the multivariate logistic regression analysis, as well as by the statistical comparison between injured and uninjured players. Kemper et al. [25] suggested that every change in BMI, both increases and decreases, should be monitored by coaches to reduce the risk of injuries, especially if the increase is higher than the normal growth and maturation rate. Vanderlei et al. [27], analyzing different sports, reported a significant association of weight and stature with injury risk in children over 12 years, but not in younger children. In our study, the stature did not seem to be associated with injured risk, probably because the majority of elite players examined were still far from their growth spurt (i.e., PHV), growing at an average rate of 6.1 cm/year compared to the maximum growth rate reported for soccer players of 8–9.7 cm/year [63,64]. The fact that a BMI increase of at least 0.3 kg/m^2 per month also led to an increase in the relative risk of injury [25] is supported by our results: injured players had a yearly BMI rate that was four times higher than that of uninjured players.

The major strength of this study is its prospective research design with the repetition of some anthropometric measurements relevant to the assessment of the velocity of growth over the competitive season of elite youth soccer players. Some body composition parameters were also assessed during the second survey. All anthropometric measurements were taken directly, whereas previous studies often employed self-reported measures or were restricted to BMI values in non-elite samples, using a cross-sectional research design. Our study also has two main limitations. The first is the small size of the sample, even if it comprised only a selected group of elite soccer players. Anyway, as we analyzed an elite sample, the results of this study can provide new indications useful for future in-depth studies. The second limitation concerns the questionnaire on injury data collection [48]; despite it being the result of the Injury Consensus Group activities under the aegis of the Fédération Internationale de Football Association Medical Assessment and Research Centre (F-MARC), it was completed without medical supervision. Lastly, we could not control for every other possible injury risk factor, and we could not exclude the possibility of residual confounding. Future studies need to also consider injury risk factors for players that have already reached or surpassed their PHV, when growth rates cannot provide satisfactory explanations.

5. Conclusions

This study highlights the remarkable effect of biological maturity, assessed in terms of PHV and anthropometric variables, on the injury occurrence of elite young soccer players. Players who are closer to their biological maturity (YPHV) are at a higher risk of injury. On the basis of our findings, we propose that anthropometric characteristics and growth rate should be monitored because they could be connected with an increased risk of injuries in young soccer players.

With regard to the practical implications of our study, the large individual variation in adolescent growth spurts needs to be taken into consideration by coaches not only during talent identification, but also during training. Indeed, training programs must be tailored according to both the chronological age and the maturity status of the players. Moreover, some body composition and anthropometric characteristics need to be monitored as part of a prevention program in order to decrease injury risk, with particular attention paid to rapid changes in weight status during early adolescence.

Author Contributions: Conceptualization, N.R.; methodology, N.R., L.Z., and E.G.-R.; formal analysis, N.R. and L.Z.; investigation, N.R.; data curation, N.R. and L.Z.; writing–original draft preparation, N.R., E.G.-R. and L.Z.; writing–review and editing, E.G.-R. and L.Z.; supervision, N.R.; project administration, N.R. All authors have read and agreed to the published version of the manuscript.

Funding: This research received no external funding.

Institutional Review Board Statement: The study was conducted according to the guidelines of the Declaration of Helsinki, and approved by the Ethics Committee of the Province of Ferrara (protocol code 140797, 16/10/14).

Informed Consent Statement: Informed consent was obtained from all subjects involved in the study.

Data Availability Statement: The data presented in this study are available on request from the authors (N.R., L.Z.). The data are not publicly available due to privacy restrictions.

Acknowledgments: The authors would like to thank the players and coaches for their participation in the research, with special thanks to Alessio Timperanza for his help during data collection.

Conflicts of Interest: The authors declare no conflict of interest.

References

1. Watson, A.; Mjaanes, J.M.; Council on Sports Medicine and Fitness. Soccer Injuries in Children and Adolescents. *Pediatrics* **2019**, *144*, e20192759. [CrossRef] [PubMed]
2. FIFA Big Count 2006: 270 Million People Active in Football. Available online: https://resources.fifa.com/image/upload/big-count-estadisticas-520058.pdf?cloudid=mzid0qmguixkcmruvema (accessed on 12 January 2021).
3. Blair, S.N.; Kohl, H.W.; Barlow, C.E.; Paffenbarger, R.S., Jr.; Gibbons, L.W.; Macera, C.A. Changes in physical fitness and all-cause mortality: A prospective study of healthy and unhealthy men. *JAMA* **1995**, *273*, 1093–1098. [CrossRef]
4. Rinaldo, N.; Zaccagni, L.; Gualdi-Russo, E. Soccer training programme improved the body composition of pre-adolescent boys and increased their satisfaction with their body image. *Acta Paediatr.* **2016**, *105*, e492–e495. [CrossRef] [PubMed]
5. Rinaldo, N.; Toselli, S.; Gualdi-Russo, E.; Zedda, N.; Zaccagni, L. Effects of Anthropometric Growth and Basketball Experience on Physical Performance in Pre-Adolescent Male Players. *Int. J. Environ. Res. Public Health* **2020**, *17*, 2196. [CrossRef]
6. Zaccagni, L.; Rinaldo, N.; Gualdi-Russo, E.G. Anthropometric Indicators of Body Image Dissatisfaction and Perception Inconsistency in Young Rhythmic Gymnastics. *Asian J. Sports Med.* **2019**, *10*, e87871. [CrossRef]
7. Schmikli, S.L.; De Vries, W.R.; Inklaar, H.; Backx, F.J. Injury prevention target groups in soccer: Injury characteristics and incidence rates in male junior and senior players. *J. Sci. Med. Sport* **2011**, *14*, 199–203. [CrossRef]
8. Owoeye, O.B.A.; Vanderwey, M.J.; Pike, I. Reducing Injuries in Soccer (Football): An Umbrella Review of Best Evidence Across the Epidemiological Framework for Prevention. *Sports Med. Open* **2020**, *6*, 46. [CrossRef]
9. Emery, C.A. Risk Factors for Injury in Child and Adolescent Sport: A Systematic Review of the Literature. *Clin. J. Sport Med.* **2003**, *13*, 256–268. [CrossRef]
10. Le Gall, F.; Carling, C.; Reilly, T. Injuries in Young Elite Female Soccer Players: An 8-Season Prospective Study. *Am. J. Sports Med.* **2008**, *36*, 276–284. [CrossRef]
11. Rommers, N.; Rössler, R.; Goossens, L.; Vaeyens, R.; Lenoir, M.; Witvrouw, E.; D'Hondt, E. Risk of acute and overuse injuries in youth elite soccer players: Body size and growth matter. *J. Sci. Med. Sport* **2020**, *23*, 246–251. [CrossRef]
12. Smith, N.A.; Chounthirath, T.; Xiang, H. Soccer-Related Injuries Treated in Emergency Departments: 1990–2014. *Pediatrics* **2016**, *138*, 20160346. [CrossRef] [PubMed]
13. Hall, E.C.; Larruskain, J.; Gil, S.M.; Lekue, J.A.; Baumert, P.; Rienzi, E.; Moreno, S.; Tannure, M.; Murtagh, C.F.; Ade, J.D.; et al. An injury audit in high-level male youth soccer players from English, Spanish, Uruguayan and Brazilian academies. *Phys. Ther. Sport* **2020**, *44*, 53–60. [CrossRef]
14. Rechel, J.A.; Yard, E.E.; Comstock, R.D. An Epidemiologic Comparison of High School Sports Injuries Sustained in Practice and Competition. *J. Athl. Train.* **2008**, *43*, 197–204. [CrossRef] [PubMed]
15. Zarei, M.; Namazi, P.; Abbasi, H.; Noruzyan, M.; Mahmoodzade, S.; Seifbarghi, T.; Zareei, M. The Effect of Ten-Week FIFA 11+ Injury Prevention Program for Kids on Performance and Fitness of Adolescent Soccer Players. *Asian J. Sports Med.* **2018**, *9*, e61013. [CrossRef]
16. Froholdt, A.; Olsen, O.E.; Bahr, R. Low risk of injuries among children playing organized soccer: A prospective cohort study. *Am. J. Sports Med.* **2009**, *37*, 1155–1160. [CrossRef]
17. Rössler, R.; Junge, A.; Chomiak, J.; Dvorak, J.; Faude, O. Soccer Injuries in Players Aged 7 to 12 Years: A Descriptive Epidemiological Study Over 2 Seasons. *Am. J. Sports Med.* **2015**, *44*, 309–317. [CrossRef] [PubMed]
18. Kerr, Z.Y.; Marshall, S.W.; Simon, J.E.; Hayden, R.; Snook, E.M.; Dodge, T.; Gallo, J.A.; McLeod, T.C.V.; Mensch, J.; Murphy, J.M.; et al. Injury Rates in Age-Only Versus Age-and-Weight Playing Standard Conditions in American Youth Football. *Orthop. J. Sports Med.* **2015**, *3*, 325967115603979. [CrossRef]
19. Tirabassi, J.; Brou, L.; Khodaee, M.; Lefort, R.; Fields, S.K.; Comstock, R.D. Epidemiology of High School Sports-Related Injuries Resulting in Medical Disqualification: 2005–2006 through 2013–2014 Academic Years. *Am. J. Sports Med.* **2016**, *44*, 2925–2932. [CrossRef]
20. Leppänen, M.; Pasanen, K.; Clarsen, B.; Kannus, P.; Bahr, R.; Parkkari, J.; Haapasalo, H.; Vasankari, T. Overuse injuries are prevalent in children's competitive football: A prospective study using the OSTRC Overuse Injury Questionnaire. *Br. J. Sports Med.* **2019**, *53*, 165–171. [CrossRef] [PubMed]
21. Read, P.J.; Jimenez, P.; Oliver, J.L.; Lloyd, R.S. Injury prevention in male youth soccer: Current practices and perceptions of practitioners working at elite English academies. *J. Sports Sci.* **2018**, *36*, 1423–1431. [CrossRef] [PubMed]
22. Leininger, R.E.; Knox, C.L.; Comstock, R.D. Epidemiology of 1.6 Million Pediatric Soccer-Related Injuries Presenting to US Emergency Departments from 1990 to 2003. *Am. J. Sports Med.* **2007**, *35*, 288–293. [CrossRef] [PubMed]
23. O'Kane, J.W.; Tencer, A.; Neradilek, M.; Polissar, N.; Sabado, L.; Schiff, M.A. Is Knee Separation During a Drop Jump Associated with Lower Extremity Injury in Adolescent Female Soccer Players? *Am. J. Sports Med.* **2016**, *44*, 318–323. [CrossRef] [PubMed]
24. O'Kane, J.W.; Neradilek, M.; Polissar, N.; Sabado, L.; Tencer, A.; Schiff, M.A. Risk Factors for Lower Extremity Overuse Injuries in Female Youth Soccer Players. *Orthop. J. Sports Med.* **2017**, *5*, 2325967117733963. [CrossRef] [PubMed]
25. Kemper, G.L.J.; Van Der Sluis, A.; Brink, M.S.; Visscher, C.; Frencken, W.G.P.; Elferink-Gemser, M.T. Anthropometric Injury Risk Factors in Elite-standard Youth Soccer. *Int. J. Sports Med.* **2015**, *36*, 1112–1117. [CrossRef] [PubMed]

26. Bult, H.J.; Barendrecht, M.; Tak, I.J.R. Injury Risk and Injury Burden Are Related to Age Group and Peak Height Velocity Among Talented Male Youth Soccer Players. *Orthop. J. Sports Med.* **2018**, *6*. [CrossRef]

27. Vanderlei, F.M.; Vanderlei, L.C.M.; Bastos, F.N.; Netto, J., Jr.; Pastre, C.M. Characteristics and associated factors with sports injuries among children and adolescents. *Braz. J. Phys. Ther.* **2014**, *18*, 530–537. [CrossRef]

28. Van Der Sluis, A.; Elferink-Gemser, M.; Coelho-E-Silva, M.; Nijboer, J.; Brink, M.; Visscher, C. Sport Injuries Aligned to Peak Height Velocity in Talented Pubertal Soccer Players. *Int. J. Sports Med.* **2013**, *35*, 351–355. [CrossRef]

29. Malina, R.M.; Reyes, M.E.P.; Figueiredo, A.J.; Coelho e Silva, M.J.; Horta, L.; Miller, R.; Chamorro, M.; Serratosa, L.; Morate, F. Skeletal Age in Youth Soccer Players: Implication for Age Verification. *Clin. J. Sport Med.* **2010**, *20*, 469–474. [CrossRef]

30. Mirwald, R.L.; Baxter-Jones, A.D.; Bailey, D.A.; Beunen, G.P. An assessment of maturity from anthropometric measurements. *Med. Sci. Sports Exerc.* **2002**, *34*, 689–694.

31. DiFiori, J.P.; Benjamin, H.J.; Brenner, J.S.; Gregory, A.; Jayanthi, N.; Landry, G.L.; Luke, A. Overuse injuries and burnout in youth sports: A position statement from the American Medical Society for Sports Medicine. *Br. J. Sports Med.* **2014**, *48*, 287–288. [CrossRef]

32. Malina, R.M.; Eisenmann, J.C.; Cumming, S.P.; Ribeiro, B.; Aroso, J. Maturity-associated variation in the growth and functional capacities of youth football (soccer) players 13–15 years. *Eur. J. Appl. Physiol.* **2004**, *91*, 555–562. [CrossRef]

33. Swain, M.; Kamper, S.J.; Maher, C.G.; Broderick, C.; McKay, D.; Henschke, N. Relationship between growth, maturation and musculoskeletal conditions in adolescents: A systematic review. *Br. J. Sports Med.* **2018**, *52*, 1246–1252. [CrossRef]

34. Meltzer, S.; Fuller, C. *Sports Nutrition, a Practical Guide to Eating for Sport*; New Holland Publishers: London, UK, 2005; pp. 125–146.

35. Arnason, A.; Sigurdsson, S.B.; Gudmundsson, A.; Holme, I.; Engebretsen, L.; Bahr, R. Risk Factors for Injuries in Football. *Am. J. Sports Med.* **2004**, *32*, 4–16. [CrossRef]

36. Jackowski, S.A.; Faulkner, R.A.; Farthing, J.P.; Kontulainen, S.A.; Beck, T.J.; Baxter-Jones, A.D. Peak lean tissue mass accrual precedes changes in bone strength indices at the proximal femur during the pubertal growth spurt. *Bone* **2009**, *44*, 1186–1190. [CrossRef]

37. Oppliger, R.A.; Case, H.S.; Horswill, C.A.; Landry, G.L.; Shelter, A.C. American College of Sports Medicine position stand. Weight loss in wrestlers. *Med. Sci. Sports Exerc.* **1996**, *28*, 9–12. [CrossRef]

38. Facchini, F. *Antropologia. Evoluzione Uomo Ambiente*, 2nd ed.; UTET Libreria: Torino, Italy, 1995.

39. Malina, R.M.; Meleski, B.W.; Shoup, R.F. Anthropometric, Body Composition, and Maturity Characteristics of Selected School-Age Athletes. *Pediatr. Clin. N. Am.* **1982**, *29*, 1305–1323. [CrossRef]

40. Weiner, J.S.; Lourie, J.A. *Practical Human Biology*; Academic Press: Cambridge, MA, USA, 1981.

41. Lohman, T.G.; Roche, A.F.; Martorell, R. *Anthropometric Standardization Reference Manual*; Human Kinetics: Champaign, IL, USA, 1988.

42. Rinaldo, N.; Gualdi-Russo, E. Anthropometric Techniques. *Ann. Online dell'Università Ferrara Sez. Didatt. Form. Docente* **2015**, *10*, 275–289.

43. Cole, T.J.; Bellizzi, M.C.; Flegal, K.M.; Dietz, W.H. Establishing a standard definition for child overweight and obesity worldwide: International survey. *BMJ* **2000**, *320*, 1240. [CrossRef]

44. Cole, T.J.; Flegal, K.M.; Nicholls, D.; Jackson, A.A. Body mass index cut offs to define thinness in children and adolescents: International survey. *BMJ* **2007**, *335*, 194. [CrossRef] [PubMed]

45. Frisancho, A.R. *Anthropometric Standards: An Interactive Nutritional Reference of Body Size and Body Composition for Children and Adults*, 2nd ed.; University of Michigan Press: Ann Arbor, MI, USA, 2008.

46. Gibson, R. *Principles of Nutritional Assessment*, 2nd ed.; Oxford University Press: New York, NY, USA, 2005; pp. 285–286.

47. Sherar, L.B.; Mirwald, R.L.; Baxter-Jones, A.D.; Thomis, M. Prediction of adult height using maturity-based cumulative height velocity curves. *J. Pediatr.* **2005**, *147*, 508–514. [CrossRef] [PubMed]

48. Fuller, C.W.; Ekstrand, J.; Junge, A.; Andersen, T.E.; Bahr, R.; Dvorak, J.; Hägglund, M.; McCrory, P.; Meeuwisse, W.H. Consensus statement on injury definitions and data collection procedures in studies of football (soccer) injuries. *Br. J. Sports Med.* **2006**, *40*, 193–201. [CrossRef]

49. Hagglund, M.; Waldén, M.; Bahr, R.; Ekstrand, J. Methods for epidemiological study of injuries to professional football players: Developing the UEFA model. *Br. J. Sports Med.* **2005**, *39*, 340–346. [CrossRef]

50. Cohen, J. *Statistical Power Analysis for the Behavioral Sciences*, 2nd ed.; Lawrence Erlbaum: Hillsdale, NJ, USA, 1988; p. 567.

51. Knowles, S.B.; Marshall, S.W.; Bowling, J.M.; Loomis, D.; Millikan, R.; Yang, J.; Weaver, N.L.; Kalsbeek, W.; Mueller, F.O. A Prospective Study of Injury Incidence among North Carolina High School Athletes. *Am. J. Epidemiol.* **2006**, *164*, 1209–1221. [CrossRef] [PubMed]

52. Powell, J.W.; Barber-Foss, K.D. Injury patterns in selected high school sports: A review of the 1995–1997 seasons. *J. Athl. Train.* **1999**, *34*, 277–284.

53. Yde, J.; Nielsen, A.B. Sports injuries in adolescents' ball games: Soccer, handball and basketball. *Br. J. Sports Med.* **1990**, *24*, 51–54. [CrossRef] [PubMed]

54. Lloyd, R.S.; Oliver, J.L.; Faigenbaum, A.D.; Myer, G.D.; De Ste Croix, M.B.A. Chronological Age vs. Biological Maturation: Implications for Exercise Programming in Youth. *J. Strength Cond. Res.* **2014**, *28*, 1454–1464. [CrossRef] [PubMed]

55. Malina, R.M.; Bouchard, C.; OBar-Or, O. *Growth, Maturation, and Physical Activity*; Human Kinetics: Champaign, IL, USA, 2004.

56. Steidl-Müller, L.; Hildebrandt, C.; Müller, E.; Raschner, C. Relationship of Changes in Physical Fitness and Anthropometric Characteristics over One Season, Biological Maturity Status and Injury Risk in Elite Youth Ski Racers: A Prospective Study. *Int. J. Environ. Res. Public Health* **2020**, *17*, 364. [CrossRef] [PubMed]
57. Beunen, G.P.; Rogol, A.D.; Malina, R.M. Indicators of Biological Maturation and Secular Changes in Biological Maturation. *Food Nutr. Bull.* **2006**, *27*, S244–S256. [CrossRef] [PubMed]
58. Ford, K.R.; Myer, G.D.; Hewett, T.E. Longitudinal effects of maturation on lower extremity joint stiff ness in adolescent athletes. *Am. J. Sports Med.* **2010**, *38*, 1829–1837. [CrossRef]
59. Iuliano-Burns, S.; Mirwald, R.L.; Bailey, D.A. Timing and magnitude of peak height velocity and peak tissue velocity for early, average, and late maturing boys and girls. *Am. J. Hum. Biol.* **2001**, *13*, 1–8. [CrossRef]
60. Davies, P.L.; Rose, J.L. Motor skills of typically developing adolescents: Awkwardness or improvement? *Phys. Occup. Ther. Pediatr.* **2000**, *20*, 19–42. [CrossRef]
61. Bliven, K.C.H.; Anderson, B.E. Core Stability Training for Injury Prevention. *Sports Health* **2013**, *5*, 514–522. [CrossRef]
62. Mala, L.; Maly, T.; Cabell, L.; Hank, M.; Bujnovsky, D.; Zahalka, F. Anthropometric, Body Composition, and Morphological Lower Limb Asymmetries in Elite Soccer Players: A Prospective Cohort Study. *Int. J. Environ. Res. Public Health* **2020**, *17*, 1140. [CrossRef] [PubMed]
63. Malina, R.M.; Morano, P.J.; Barron, M.; Miller, S.J.; Cumming, S.P. Growth Status and Estimated Growth Rate of Youth Football Players. *Clin. J. Sport Med.* **2005**, *15*, 125–132. [CrossRef] [PubMed]
64. Philippaerts, R.M.; Vaeyens, R.; Janssens, M.; Van Renterghem, B.; Matthys, D.; Craen, R.; Bourgois, J.; Vrijens, J.; Beunen, G.; Malina, R.M. The relationship between peak height velocity and physical performance in youth soccer players. *J. Sports Sci.* **2006**, *24*, 221–230. [CrossRef] [PubMed]

International Journal of
Environmental Research and Public Health

Article

Association of Health Utility Score with Physical Activity Outcomes in Stroke Survivors

Masashi Kanai [1,2,3], **Kazuhiro P. Izawa** [2,3,*], **Hiroki Kubo** [4], **Masafumi Nozoe** [1], **Kyoshi Mase** [1] and **Shinichi Shimada** [2,3,5]

1   Department of Physical Therapy, Faculty of Nursing and Rehabilitation, Konan Women's University, Kobe 658-0001, Japan; kanaimasa07@gmail.com (M.K.); nozoe@konan-wu.ac.jp (M.N.); tkjwg268@yahoo.co.jp (K.M.)
2   Department of Public Health, Kobe University Graduate School of Health Sciences, Kobe 654-0142, Japan; itamikousei1@yahoo.co.jp
3   Cardiovascular Stroke Renal Project (CRP), Kobe 654-0142, Japan
4   Department of Rehabilitation, Itami Kousei Neurosurgical Hospital, Itami 664-0028, Japan; hiro.k16862@gmail.com
5   Department of Neurosurgery, Itami Kousei Neurosurgical Hospital, Itami 664-0028, Japan
*   Correspondence: izawapk@harbor.kobe-u.ac.jp; Tel.: +81-78-796-4566

Abstract: Health-related quality of life (HRQoL) after stroke tends to vary across studies or across stages of stroke. It is useful to use the health utility score to compare HRQoL across studies. Physical activity after stroke also tends to vary similarly. The purpose of the present study was to determine associations between the health utility score and physical activity outcomes in stroke survivors. This cross-sectional study recruited stroke survivors who could ambulate outside, free of assistance. We assessed the health utility score with the EuroQoL 5-Dimension 3-Level questionnaire. The physical activity outcomes were the number of steps taken and duration of moderate-to-vigorous physical activity (MVPA) as measured with an accelerometer. Multiple linear regression analyses were used to determine whether the physical activity outcomes were independently associated with the health utility score. Fifty patients (age: 68.0 years; 40 men, 10 women) were included. Multiple linear regression analysis showed the health utility score to be significantly associated with the number of steps taken ($\beta = 0.304$, $p = 0.035$) but not with MVPA. This is the first study to examine the association between the health utility score and objectively measured physical activity in stroke survivors. Promoting physical activity especially by increasing the number of steps taken might be a priority goal in improving a patient's health utility score after stroke.

Keywords: health utility; physical activity; quality of life; rehabilitation; stroke

Citation: Kanai, M.; Izawa, K.P.; Kubo, H.; Nozoe, M.; Mase, K.; Shimada, S. Association of Health Utility Score with Physical Activity Outcomes in Stroke Survivors. *Int. J. Environ. Res. Public Health* **2021**, *18*, 251. https://doi.org/10.3390/ijerph18010251

Received: 3 December 2020
Accepted: 28 December 2020
Published: 31 December 2020

Publisher's Note: MDPI stays neutral with regard to jurisdictional claims in published maps and institutional affiliations.

1. Introduction

Stroke has a direct impact on overall health. The interval between the onset of symptoms and arrival at the hospital can greatly influence the effectiveness of treatment and patient prognosis [1,2]. Stroke survivors are additionally affected by long-term physical and psychosocial well-being [3]. Several studies reported that health-related quality of life (HRQoL) is also altered in stroke survivors. Most stroke survivors have a lower HRQoL than healthy subjects [4–6], even when adjusting for confounding factors [6]. De Wit et al. indicated that HRQoL of stroke survivors was more than 1/2 standard deviations below that of a healthy population [4]. However, there are strong associations between levels of HRQoL assessed by the EuroQoL 5-Dimension 3-level (EQ-5D-3L) questionnaire at three months and survival within the first year [7]. Thus, assessing HRQoL may contribute to predicting prognosis after stroke.

HRQoL after stroke tends to vary across studies [7,8] or stages of stroke [9]. Thus, it is useful to use the health utility score to compare HRQoL across different studies. The EQ-5D-3L is particularly useful for calculating the health utility score to compare health

59

over time between different populations [10]. The EQ-5D-3L is utilized for the analysis of cost-effectiveness as a means to use the outcome for quality-adjusted life years (QALY) [11].

Physical activity after stroke also tends to vary across studies or stages of stroke because of factors such as stroke severity, physical function, and environmental factors [12]. Physical activity is just as important as HRQoL as a predictor of prognosis in terms of mortality [13] and recurrent stroke [14,15]. Therefore, promoting physical activity may be one of the most important strategies in improving stroke survival and preventing recurrence. There are several reports on the relationship between HRQoL and physical activity or exercise after stroke. Adaptive physical activity improves mobility function and HRQoL in chronic stroke [16]. Hou et al. suggested thar long-term regular mild exercise such as walking could improve HRQoL after stroke [17]. Based on the above reports, exercise intervention might have a positive effect on HRQoL. In a cross-sectional study, Rand et al. reported that HRQoL correlated significantly with the amount of daily physical activity as measured with an accelerometer and self-reported questionnaire [8]. However, their study did not calculate a health utility score based on HRQoL, so the association of the health utility score with physical activity outcomes was not clear. Previously, we tried to clarify these relationships in a preliminary cross-sectional study [18]. As a result, the health utility score correlated significantly with the number of steps taken in community-dwelling ambulatory patients with stroke. Thus, we concluded that the more the patients with stroke walked, the higher their health utility score would be. However, because the study sample was quite small, we could not conduct a multivariate analysis. Additionally, we could not investigate the association between the health utility score and physical activity intensity. If these associations are clarified even when considering confounding factors, we may be able to develop appropriate methods to increase health utility by promoting physical activity. In addition, the results of the present study might also serve as a steppingstone for examining the relationship between QALY and physical activity in future trials.

We thus hypothesized that the health utility score would be associated with the amount of physical activity or intensity of physical activity in stroke survivors. Therefore, the purpose of the present study was to determine associations between the health utility score and physical activity outcomes in stroke survivors.

2. Materials and Methods

2.1. Study Design and Participant Recruitment

This cross-sectional study was approved by the research ethics committee of Kobe University Graduate School of Health Sciences (approval no. 690, 20 April 2018). Informed consent was obtained from all patients. Participants were selected from June 2017 to November 2018 at Itami Kousei Neurosurgical Hospital by a medical doctor or physical therapist based on the inclusion criteria. The sample size used in the present study was determined based on the total number of patients seen over the selection period.

The inclusion criteria were a previous history of stroke, ability to ambulate outside free of assistance, and consent to measurement of physical activity. Exclusion criteria were those younger than 18 years of age, patients with dementia or aphasia as evaluated by their primary care physician, those with a modified Rankin Scale score [19] > 3 (moderate to severe disability conditions that require assistance with walking and physical demands) due to musculoskeletal disease, and those with severe cardiopulmonary disease or psychiatric disease such as schizophrenia based on evaluation of the patient's medical records by a physical therapist.

2.2. Clinical Characteristics

Patient characteristics, including age, sex, body mass index (BMI), subtypes of stroke (ischemic or hemorrhage stroke), neurological deficit by the National Institutes of Stroke Scale (NIHSS) [20], time from stroke onset, comorbidities (hypertension, diabetes mellitus, dyslipidemia), handgrip strength, and comfortable walking speed [21] were collected from electronic medical records. The NIHSS is one of the measures of stroke severity and assesses

11 items related to cognition, vision, motor and sensory function, speech and language, ataxia, and inattention. It consists of a score of 0–42, with lower scores indicating milder neurological symptoms and higher scores indicating more severe symptoms. Comfortable walking speed was determined from a 10 m walking test as 10 m/time required in seconds [21]. We used a stopwatch to time the patient's walking time over a 10 m length of a 14 m walkway. Grip strength was measured using a digital grip strength meter. During the measurement, the patient was instructed to sit in a chair without a backrest and to keep the upper limbs off the side of the body [22]. The grip strength was measured twice on each side, and the maximum value was adopted.

2.3. Assessment of Health Utility Score

To assess the health utility score, we used the EQ-5D-3L questionnaire [23]. The EuroQoL-5D-3L was introduced by the EuroQoL group and has been used in Japanese populations. The EQ-5D-3L has an index score as the first component, and patients select outcomes from choices of no problems, some problems, and severe problems (scored 1–3) for the following five dimensions: mobility, self-care, usual activities, pain/discomfort, and anxiety/depression. The responses obtained from the EQ-5D-3L were converted by a physical therapist to a health utility score, which was considered the primary outcome in the present study based on a set of Japanese values [23]. The health utility score assesses HRQoL quantitatively as a fraction of ideal health, with a score of 1 representing perfect health, a score of 0 representing death, and a negative score representing health states worse than death [23].

2.4. Physical Activity Measurement

The physical activity outcomes were the number of steps taken and duration of moderate-to-vigorous physical activity (MVPA). We used a Fitbit One 3-dimensional accelerometer (Fitbit, Inc., San Francisco, CA, USA) to measure the physical activity values. The Fitbit One has been used in previous studies of stroke patients [18–25]. After patient enrollment, the device was worn on the waist belt of all patients 24/h day for more than one week, except when bathing or changing clothes. We used the first 7 days (1 week) of continuous data to determine physical activity outcomes in the present study. We confirmed the number of steps, exercise energy expenditure, and duration of activity time after downloading the data files to Fitbit online dashboard software [26]. We calculated the average number of steps (/day) and duration of MVPA (min/day). Duration of MVPA was calculated by the sum of MVPA time at greater than three metabolic equivalents.

2.5. Statistical Analysis

The results are shown as median (interquartile range) or as ordinal variables and counts (%) for categorical variables. The Shapiro-Wilk test was used to access normality of the values. Nonparametric analyses were used. Multiple linear regression analyses were used to determine whether the physical activity outcomes were independently associated with the health utility score. The health utility score was the dependent variable, and the independent variables were the physical activity outcome and the following relevant confounding variables that correlated with the health utility score by Spearman correlation coefficient ($p < 0.05$): age, sex, BMI, handgrip strength, and walking speed. To account for the effects of multicollinearity, physical activity outcomes were selected as the independent variables in two separate models (Model 1: number of steps, Model 2: MVPA). A p-value of < 0.05 was considered to indicate statistical significance. Statistical analyses were performed with IBM SPSS 25.0 statistical software (IBM SPSS Japan, Inc., Tokyo, Japan).

3. Results

3.1. Participants and Clinical Characteristics

Participant flow through the present study is shown in Figure 1. Of the original 98 patients, 55 patients met the inclusion criteria, but five patients later dropped out because they did not wear the accelerometer (*n* = 2), did not respond to the questionnaire correctly (*n* = 2), or declined to participate (*n* = 1). Therefore, the study sample comprised 50 patients. Clinical characteristics of the patients are shown in Table 1.

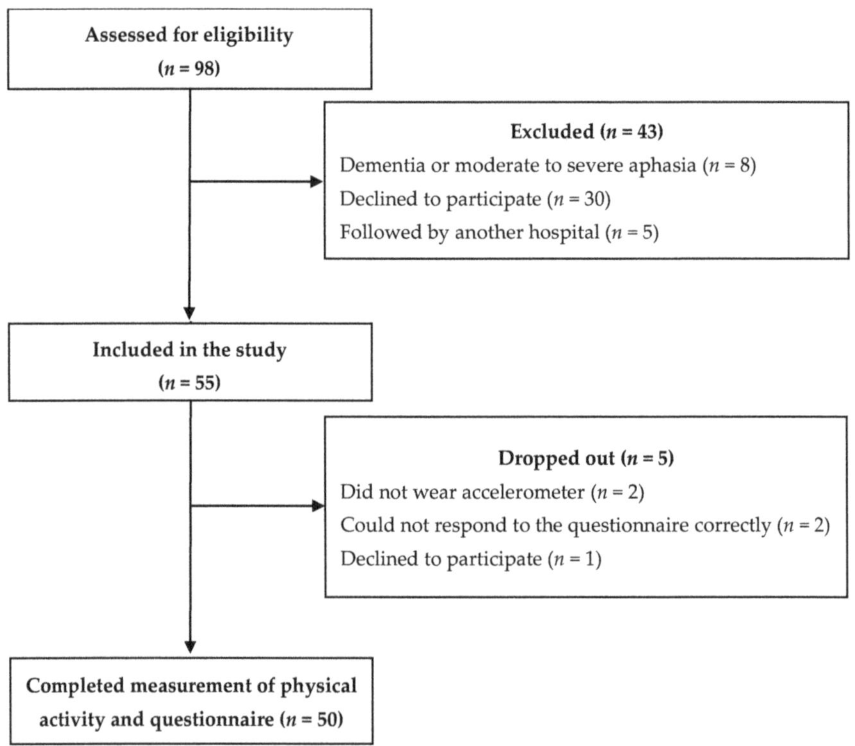

Figure 1. Participant flow in the present study.

Table 1. Clinical characteristics.

Characteristic	All Participants (*n* = 50)
Age (years)	68.0 (53.8–77.0)
Sex (male), *n* (%)	40 (80.0)
Body mass index (kg/m^2)	23.1 (21.8–24.8)
Subtypes, *n* (%)	
Ischemic	32 (64.0)
Hemorrhage	18 (36.0)
NIHSS (score)	1.0 (1.0–2.0)
Time since stroke (months)	4.2 (1.7–6.4)
Comorbidity, *n* (%)	
Hypertension	44 (88.8)
Diabetes mellitus	18 (36.0)
Dyslipidemia	28 (56.0)
Handgrip strength (kgf)	31.9 (24.3–37.6)
Walking speed (m/s)	1.1 (0.9–1.3)

Abbreviation: NIHSS, National Institutes of Health Stroke Scale. Values are shown as median (interquartile range) or as ordinal variables and counts (%) for categorical variables.

3.2. Health Utility and Physical Activity

The health utility score was 0.77 (0.71–0.85). The physical activity outcomes as indicated by the number of steps taken and the duration of MVPA were 5472.9 (3445.2–7399.9) steps/day and 10.2 (1.7–33.0) min/day, respectively. The health utility score showed a significant positive correlation with BMI, handgrip strength, walking speed, number of steps, and MVPA and a significant negative correlation with sex (female) and NIHSS (Table 2).

Table 2. Relation between the health utility score and other characteristics.

	Age	Sex (0, Male; 1, Female)	BMI	NIHSS	Handgrip Strength	Walking Speed	Number of Steps	MVPA
HU score	−0.204 (0.154)	−0.438 (0.001)	0.465 (0.001)	−0.326 (0.021)	0.396 (0.004)	0.429 (0.002)	0.454 (0.001)	0.497 (<0.001)

Abbreviations: HU, health utility; BMI, body mass index; MVPA, moderate-to-vigorous physical activity; NIHSS, National Institutes of Health Stroke Scale. Values are shown as ϱ value (p value).

The results of the multiple linear regression analyses are shown in Table 3. The health utility score in Model 1 was significantly associated with sex ($\beta = -0.366$, $p = 0.026$) and the number of steps taken ($\beta = 0.304$, $p = 0.035$). The health utility score in Model 2 was significantly associated only with sex ($\beta = -0.354$, $p = 0.035$) and not with MVPA ($\beta = 0.231$, $p = 0.102$). In both Model 1 and Model 2, women had lower health utility scores.

Table 3. Multivariate regression analysis for the health utility score.

	Model 1 (Adjusted $R^2 = 0.383$)		Model 2 (Adjusted $R^2 = 0.357$)	
	β	p Value	β	p Value
Sex	−0.366	0.026	−0.354	0.035
BMI	0.210	0.109	0.218	0.106
NIHSS	−0.124	0.365	−0.142	0.309
Handgrip strength	−0.158	0.382	−0.082	0.641
Walking speed	0.188	0.182	0.171	0.257
Number of steps	0.304	0.035		
MVPA			0.231	0.102

Abbreviations: BMI, body mass index; MVPA, moderate-to-vigorous physical activity; NIHSS, National Institutes of Health Stroke Scale.

4. Discussion

4.1. Key Findings

To our knowledge, this is the first study to examine the associations between health utility score and objectively measured physical activity in stroke survivors using multivariate analysis. The results indicated that there was an association between the health utility score and the number of steps taken in stroke survivors.

In addition to the EQ-5D-3L, there are other methods for calculating the health utility score from other scales, such as the Short Form 6-Dimension questionnaire [27] and the Health Utility Index 2 (HUI 2) and Health Utility Index 3 (HUI 3) [28]. The HUI 2 overall score includes sensation, mobility, emotion, cognition, self-care, and pain, whereas the HUI 3 includes vision, hearing, speech, ambulation, dexterity, emotion, cognition, and pain. The health utility score was 0.77 in stroke survivors in the present study. Haacke et al. indicated that the health utility scores in long-term survivors of stroke as calculated by the HUI 2, HUI 3, and EQ-5D-3L were 0.67, 0.47, and 0.73, respectively [28]. Post et al. reported that health utility score was 0.64 for patients with minor stroke from a systematic review of the literature [29]. Carod-Artal and Egido also presented several stroke sequels in which the health utility score as assessed by the EQ-5D ranged from 0.60 to 0.70 for minor stroke (NIHSS < 6) [3]. Most of the stroke survivors in the present study suffered strokes of minor severity (median NIHSS score of 1.0), so the health utility score was generally consistent with those of previous studies.

4.2. Association of Health Utility Score with Physical Activity Outcomes

Rand et al. reported on the association between HRQoL and physical activity [8], although not the health utility score. We previously reported that the health utility score showed a positive correlation with the number of steps taken in stroke survivors [18]. In terms of the positive correlation between health utility score or HRQoL and physical activity outcomes, the results of the present study support the findings of these two previous studies. The present study added further evidence that there appears to be an association between the health utility score and the number of steps taken, but not with the duration of MVPA after adjusting for confounding factors in stroke survivors.

Macko et al. reported that a structured adaptive physical activity program comprised of mobility, balance, and stretching exercises improves mobility function and QoL in patients with chronic stroke [16]. Another study found that regular mild exercise such as walking or Tai Chi improved QoL after stroke [17]. Grau-Pellicer et al. reported that walking speed was a predictor of community mobility and HRQoL as assessed by the EQ-5D-5L in a population of stroke survivors that included some patients who required supervision to ambulate [30]. Although the health utility score in the present study showed a significant correlation with walking speed, the correlation was not significant after multivariate analysis. Because we only included stroke survivors who could ambulate outside free of assistance, our result might differ from that of the Grau-Pellicer et al. study. Walking activity represented by the number of steps taken is supposed to categorize a community ambulatory level [31], so it is possible that the number of steps may have been altered by engagement with others, and this social interaction might have affected health utility.

Several reports investigated predictors of the health utility score including HRQoL. White et al. suggested that potentially modifiable risk factors such as community participation and stroke-related disability affected HRQoL [9]. Tse et al. reported that the ability to re-engage in work and social activities positively influenced HRQoL in the domains of physical function, participation, and perceived recovery in stroke survivors [32]. Although we could not investigate return to work or return to social activities after stroke in the present patients, the health utility score showed some variability for each stroke survivor, such that the health utility score might have been influenced by work and social activities. Additionally, because these HRQoL or health utility-related factors such as community participation and social activities do not necessarily require high-intensity physical activity, the health utility score was not significantly associated with the duration of MVPA in the present study. Therefore, the health utility score might be more strongly associated with amount of physical activity rather than its intensity in stroke survivors.

The results of this study could be used to suggest that encouragement and support for stroke survivors in order to increase the number of steps for community participation and maintain independence could be a major strategy to increase health utility after stroke. Further study should be conducted to examine QALY and number of steps in stroke survivors.

4.3. Limitations

The present study has several limitations. First, we included only stroke survivors who could ambulate outside free of assistance, so generalizability of the results of this study requires caution. The reason we included only minor stroke survivors was because of the accuracy of measuring the number of steps on the instrument. Although health utility after stroke depends on stroke severity [29], physical activity after stroke also depends on stroke severity [33]. Thus, the results of the present study might be partly generalizable, even if the severity and level of mobility of stroke have changed. Second, we could not know the exact wearing time of the accelerometer. The wearing time may have had some impact on the degree of physical activity. Third, because we used the EQ-5D-3L to evaluate the health utility score, there may have been a ceiling effect. In addition, due to the small sample size, it was not possible to compare health utility score by sex difference. Because female stroke survivors tend to have a lower quality of life [34,35], the relationship between the health utility score and physical activity outcome may have needed to be confirmed by sex-specific models. Fourth, we did not evaluate the association of the health utility score with physical activity outcomes considering confounding factors such as depression or anxiety [32–36], which are quite common in stroke survivors and might have been present in the participants of the present study. Finally, in the multivariate regression analysis, we were not able to add a third model that includes number of steps and MVPA due to the small sample size. We need to address these deficiencies in future studies.

5. Conclusions

The present study showed that the health utility score was significantly associated with the number of steps taken by stroke survivors but not with the duration of MVPA. The more stroke survivors walked, the higher their health utility score was. Promoting physical activity especially through increasing the number of steps might be a priority goal in improving the health utility score of patients after stroke. Additional study is needed to clarify the association between QALY and number of steps.

Author Contributions: Conceptualization, M.K. and K.P.I.; formal analysis, M.K., H.K. and M.N.; investigation, M.K. and H.K.; resources, M.N. and K.M.; data curation, M.K., H.K. and M.N.; writing—original draft preparation, M.K.; writing—review and editing, K.P.I., and K.M.; supervision, K.M. and S.S.; project administration, K.P.I. and S.S. All authors approved the manuscript for submission. All authors have read and agreed to the published version of the manuscript.

Funding: This research received no external funding.

Institutional Review Board Statement: This cross-sectional study was approved by the research ethics committee of Kobe University Graduate School of Health Sciences (approval no. 690, 20 April 2018).

Informed Consent Statement: Informed consent was obtained from all subjects involved in the study.

Acknowledgments: The authors thank all of the staff and participants at Itami Kousei Neurosurgical Hospital.

Conflicts of Interest: The authors declare no conflict of interest.

References

1. Powers, W.; Rabinstein, A.; Ackerson, T.; Adevoe, O.; Bambakidis, N.; Becker, K. 2018 Guidelines for the Early Management of Patients with Acute Ischemic Stroke: A Guideline for Healthcare Professionals from the American Heart Association/American Stroke Association. *J. Vasc. Surg.* **2018**, *67*, 1934. [CrossRef]
2. Soto-Cámara, R.; González-Santos, J.; González-Bernal, J.; Martín-Santidrian, A.; Cubo, E.; Trejo-Gabriel-Galán, J.M. Factors Associated with shortening of prehospital delay among patients with acute ischemic stroke. *J. Clin. Med.* **2019**, *17*, 1712. [CrossRef] [PubMed]
3. Carod-Artal, F.J.; Egido, J.A. Quality of life after stroke: The importance of a good recovery. *Cerebrovasc. Dis.* **2009**, *27* (Suppl. 1), 204–214. [CrossRef]
4. De Wit, L.; Theuns, P.; Dejaeger, E.; Devos, S.; Gantenbein, A.R.; Kerckhofs, E.; Schuback, B.; Schupp, W.; Putman, K. Long-term impact of stroke on patients' health-related quality of life. *Disabil. Rehabil.* **2017**, *39*, 1435–1440. [CrossRef] [PubMed]
5. Jeon, N.E.; Kwon, K.M.; Kim, Y.H.; Lee, J.S. The factors associated with health-related quality of life in stroke survivors age 40 and older. *Ann. Rehabil. Med.* **2017**, *41*, 743–752. [CrossRef] [PubMed]
6. Kwon, S.; Park, J.H.; Kim, W.S.; Han, K.; Lee, Y.; Paik, N.J. Health-related quality of life and related factors in stroke survivors: Data from Korea National Health and Nutrition Examination Survey (KNHANES) 2008 to 2014. *PLoS ONE* **2018**, *13*, e0195713. [CrossRef]
7. Ayis, S.; Wellwood, I.; Rudd, A.G.; McKevitt, C.; Parkin, D.; Wolfe, C.D. Variations in health-related quality of life (HRQoL) and survival 1 year after stroke: Five European population-based registers. *BMJ Open* **2015**, *5*, e007101. [CrossRef]
8. Rand, D.; Eng, J.J.; Tang, P.F.; Hung, C.; Jeng, J.S. Daily physical activity and its contribution to the health-related quality of life of ambulatory individuals with chronic stroke. *Health Qual. Life. Outcomes* **2010**, *8*, 80. [CrossRef]
9. White, J.; Magin, P.; Attia, J.; Sturm, J.; McElduff, P.; Carter, G. Predictors of health-related quality of life in community-dwelling stroke survivors: A cohort study. *Fam. Pract.* **2016**, *33*, 382–387. [CrossRef]
10. Noto, S.; Yanagi, H.; Tomura, S. Measuring utilities for various functional outcomes after stroke. Comparison of rating scale and time trade-off methods. *Jpn. J. Public Health* **2002**, *9*, 1205–1216.

11. van Eeden, M.; van Heugten, C.; van Mastrigt, G.A.; van Mierlo, M.; Visser-Meily, J.M.; Evers, S.M. The burden of stroke in The Netherlands: Estimating quality of life and costs for 1 year poststroke. *BMJ Open* **2015**, *27*, e008220. [CrossRef] [PubMed]

12. Thilarajah, S.; Mentiplay, B.F.; Bower, K.J.; Tan, D.; Pua, Y.H.; Williams, G.; Koh, G.; Clark, R.A. Factors associated with post stroke physical activity: A systematic review and meta-analysis. *Arch. Phys. Med. Rehabil.* **2018**, *99*, 1876–1889. [CrossRef] [PubMed]

13. Loprinzi, P.D.; Addoh, O. Accelerometer-determined physical activity and all-cause mortality in a national prospective cohort study of adults post-acute stroke. *Am. J. Health Promot.* **2018**, *32*, 24–27. [CrossRef]

14. Kono, Y.; Yamada, S.; Yamaguchi, J.; Hagiwara, Y.; Iritani, N.; Ishida, S.; Araki, A.; Hasegawa, Y.; Sakakibara, H.; Koike, Y. Secondary prevention of new vascular events with lifestyle intervention in patients with noncardioembolic mild ischemic stroke: A single-center randomized controlled trial. *Cerebrovasc. Dis.* **2013**, *36*, 88–97. [CrossRef]

15. Kono, Y.; Yamada, S.; Iwatsu, K.; Nitobe, S.; Tanaka, Y.; Shimizu, Y.; Shinoda, N.; Okumura, T.; Hirashiki, A.; Murohara, T. Predictive impact of daily physical activity on new vascular events in patients with mild ischemic stroke. *Int. J. Stroke* **2015**, *10*, 219–223. [CrossRef]

16. Macko, R.F.; Benvenuti, F.; Stanhope, S.; Macellari, V.; Taviani, A.; Nesi, B.; Weinrich, M.; Stuart, M. Adaptive physical activity improves mobility function and quality of life in chronic hemiparesis. *J. Rehabil. Res. Dev.* **2008**, *45*, 323–328. [CrossRef] [PubMed]

17. Hou, L.; Du, X.; Chen, L.; Li, J.; Yan, P.; Zhou, M.; Zhu, C. Exercise and quality of life after first-ever ischaemic stroke: A two-year follow-up study. *Int. J. Neurosci.* **2018**, *128*, 540–548. [CrossRef]

18. Sasaki, S.; Kanai, M.; Shinoda, T.; Morita, H.; Shimada, S.; Izawa, K.P. Relation between health utility score and physical activity in community-dwelling ambulatory patients with stroke: A preliminary cross-sectional study. *Top. Stroke Rehabil.* **2018**, *25*, 475–479. [CrossRef]

19. van Swieten, J.C.; Koudstaal, P.J.; Visser, M.C.; Schouten, H.J.; van Gijn, J. Interobserver agreement for the assessment of handicap in stroke patients. *Stroke* **1988**, *19*, 604–607. [CrossRef]

20. Lyden, P.; Brott, T.; Tilley, B.; Welch, K.M.; Mascha, E.J.; Levine, S.; Haley, E.C.; Grotta, J.; Marler, J. Improved reliability of the NIH Stroke Scale using video training. NINDS TPA Stroke Study Group. *Stroke* **1994**, *25*, 2220–2226. [CrossRef]

21. Guralnik, J.M.; Simonsick, E.M.; Ferrucci, L.; Glynn, R.J.; Berkman, L.F.; Blazer, D.G.; Scherr, P.A.; Wallace, R.B. A short physical performance battery assessing lower extremity function: Association with self-reported disability and prediction of mortality and nursing home admission. *J. Gerontol.* **1994**, *49*, M85–M94. [CrossRef] [PubMed]

22. Kamo, T.; Suzuki, R.; Ito, K.; Sugimoto, T.; Murakoshi, T.; Nishida, Y. Prevalence of sarcopenia and its relation to body composition, physiological function, and nutritional status in community-dwelling frail elderly people [Japanese with English abstract]. *J. Jpn. Phys. Ther. Assoc.* **2013**, *40*, 414–420.

23. Tsuchiya, A.; Ikeda, S.; Ikegami, N.; Nishimura, S.; Sakai, I.; Fukuda, T.; Hamashima, C.; Hisashige, A.; Tamura, M. Estimating an EQ-5D population value set: The case of Japan. *Health Econ.* **2002**, *11*, 341–353. [CrossRef] [PubMed]

24. Kanai, M.; Izawa, K.P.; Kubo, H.; Nozoe, M.; Mase, K.; Koohsari, M.J.; Oka, K.; Shimada, S. Association of perceived built environment attributes with objectively measured physical activity in community-dwelling ambulatory patients with stroke. *Int. J. Environ. Res. Public Health* **2019**, *16*, 3908. [CrossRef]

25. Kanai, M.; Izawa, K.P.; Nozoe, M.; Kubo, H.; Kobayashi, M.; Onishi, A.; Mase, K.; Shimada, S. Long-term effect of promoting in-hospital physical activity on postdischarge patients with mild ischemic stroke. *J. Stroke Cerebrovasc. Dis.* **2019**, *28*, 1048–1055. [CrossRef]

26. Fitbit Dashboard. Available online: https://accounts.fitbit.com/login (accessed on 21 December 2020).

27. Kharroubi, S.A.; Brazier, J.E.; Roberts, J.; O'Hagan, A. Modelling SF-6D health state preference data using a nonparametric Bayesian method. *J. Health Econ.* **2007**, *26*, 597–612. [CrossRef]

28. Haacke, C.; Althaus, A.; Spottke, A.; Siebert, U.; Back, T.; Dodel, R. Long-term outcome after stroke: Evaluating health-related quality of life using utility measurements. *Stroke* **2006**, *37*, 193–198. [CrossRef]

29. Post, P.N.; Stiggelbout, A.M.; Wakker, P.P. The utility of health states after stroke: A systematic review of the literature. *Stroke* **2001**, *32*, 1425–1429. [CrossRef]

30. Grau-Pellicer, M.; Chamarro-Lusar, A.; Medina-Casanovas, J.; Serdà Ferrer, B.C. Walking speed as a predictor of community mobility and quality of life after stroke. *Top. Stroke Rehabil.* **2019**, *26*, 349–358. [CrossRef]

31. Fulk, G.D.; He, Y.; Boyne, P.; Dunning, K. Predicting home and community walking activity poststroke. *Stroke* **2017**, *48*, 406–411. [CrossRef]

32. Tse, T.; Binte Yusoff, S.Z.; Churilov, L.; Ma, H.; Davis, S.; Donnan, G.A.; Carey, L.M.; START Research Team. Increased work and social engagement is associated with increased stroke specific quality of life in stroke survivors at 3 months and 12 months post-stroke: A longitudinal study of an Australian stroke cohort. *Top. Stroke Rehabil.* **2017**, *24*, 405–414. [CrossRef] [PubMed]

33. Hokstad, A.; Indredavik, B.; Bernhardt, J.; Ihle-Hansen, H.; Salvesen, Ø.; Seljeseth, Y.M.; Schüler, S.; Engstad, T.; Askim, T. Hospital differences in motor activity early after stroke: A comparison of 11 Norwegian stroke units. *J. Stroke Cerebrovasc. Dis.* **2015**, *24*, 1333–1340. [CrossRef] [PubMed]

34. Gray, L.J.; Sprigg, N.; Bath, P.M.; Boysen, G.; De Deyn, P.P.; Leys, D.; O'Neill, D.; Ringelstein, E.B.; TAIST Investigators. Sex differences in quality of life in stroke survivors: Data from the Tinzaparin in Acute Ischaemic Stroke Trial (TAIST). *Stroke* **2007**, *38*, 2960–2964. [CrossRef] [PubMed]

35. Patel, M.D.; McKevitt, C.; Lawrence, E.; Rudd, A.G.; Wolfe, C.D. Clinical determinants of long-term quality of life after stroke. *Age Ageing* **2007**, *36*, 316–322. [CrossRef]
36. Baumann, M.; Lurbe, K.; Leandro, M.E.; Chau, N. Life satisfaction of two-year post-stroke survivors: Effects of socio-economic factors, motor impairment, Newcastle stroke-specific quality of life measure and World Health Organization quality of life: Bref of informal caregivers in Luxembourg and a rural area in Portugal. *Cerebrovasc. Dis.* **2012**, *33*, 219–230.

International Journal of
Environmental Research and Public Health

Article

Does Multiple Sclerosis Differently Impact Physical Activity in Women and Man? A Quantitative Study Based on Wearable Accelerometers

Massimiliano Pau [1],*, Micaela Porta [1], Giancarlo Coghe [2], Jessica Frau [2], Lorena Lorefice [2] and Eleonora Cocco [2]

[1] Department of Mechanical, Chemical and Materials Engineering, University of Cagliari, 09123 Cagliari, Italy; m.porta@dimcm.unica.it

[2] Department of Medical Sciences and Public Health, University of Cagliari, 09123 Cagliari, Italy; gccoghe@gmail.com (G.C.); jessicafrauneuro@gmail.com (J.F.); lorena.lorefice@hotmail.it (L.L.); ecocco@unica.it (E.C.)

* Correspondence: massimiliano.pau@dimcm.unica.it; Tel.: +39-070-675-3264

Received: 19 October 2020; Accepted: 26 November 2020; Published: 28 November 2020

Abstract: In people with multiple sclerosis (pwMS), fatigue, weakness and spasticity may reduce mobility and promote sedentary behavior. However, little is known about the existence of possible differences in the way MS modifies the propensity to perform physical activity (PA) in men and women. The present study aimed to partly close this gap by means of quantitative analysis carried out using wearable sensors. Forty-five pwMS (23 F, 22 M, mean age 50.3) and 41 unaffected age- and sex-matched individuals wore a tri-axial accelerometer 24 h/day for 7 consecutive days. Raw data were processed to calculate average number of daily steps, vector magnitude (VM) counts, and percentage of time spent in sedentary behavior and in PA of different intensities (i.e., light and moderate-to-vigorous, MVPA). Women with MS spent more time in sedentary behavior and exhibited a reduced amount of light intensity activity with respect to men, while MVPA was similar across sexes. However, in comparison with unaffected individuals, the overall PA patterns appear significantly modified mostly in women who, in presence of the disease, present increased sedentary behavior, reduced MVPA, number of daily steps and VM counts. The findings of the present study highlight the urgency of including sex as variable in all studies on PA in pwMS.

Keywords: multiple sclerosis; physical activity; accelerometer; gender differences

---

1. Introduction

Multiple sclerosis (MS) is a chronic immunomediated and neurodegenerative disease of the central nervous system (CNS) which is thought to be caused by an interaction of environmental, genetic, and epigenetic factors [1,2]. This disease, which represent the most frequent cause of disability among young adults [1,3,4], presents with a variety of symptoms depending on the CNS site in which the damage is located.

A number of relevant differences in terms of impact of the disease on physical and psychological dimensions are also associated with the individual's sex, besides the well-known disproportion in terms of affected population (i.e., MS affects more women than men [5]). Men are more vulnerable to cognitive deficits [6], more susceptible to disability accumulation (i.e., characterized by faster relapse-onset, [7,8]) more at risk of poorer prognosis [9] and participation restrictions [10], while women are more likely to experience anxiety [11] but tend to cope better with the disease [12]. Focusing attention on mobility issues, recent studies also report that men and women with MS exhibit peculiar gait patterns in terms of kinematics [13] and that their functional mobility is differently influenced by muscular strength [14].

However, little is known about the existence of possible sex-related differences in amount and intensity of physical activity (PA) performed by pwMS, since few studies have explicitly considered sex as a main variable of interest. This aspect is of extreme importance because approximately 80 to 85% of pwMS complain about the presence of mild to moderate fatigue as well as some kind of gait impairment [15]. Owing to these symptoms, they tend to reduce overall daily mobility, exhibit a sedentary behavior [16] and are more reluctant to follow a scheduled program of structured PA. Such a phenomenon implies further negative consequences, as physical inactivity worsens cardiorespiratory fitness and promotes physical deconditioning. It is also a co-factor in the onset of comorbidities such as obesity, metabolic syndrome osteoporosis, etc. [17–19].

The scarce existing literature provides contrasting findings about a possible different impact of MS on amount and intensity of PA carried out by women and men with MS. Anens et al. [20] reported significantly lower levels of PA in men with MS (and thus hypothesized that they are more physically affected by the disease than women) by analyzing data obtained through the Physical Activity Disability Survey (PADS-R) administered to a cohort of 287 Swedish pwMS. In contrast, no sex-related differences were found across pwMS in terms of moderate-to-vigorous PA objectively assessed using a uniaxial accelerometer [21]. In a different study using the same device, men were found less active when considering the number of daily steps as a metric representative of PA [22]. Finally, the recent systematic review of correlates and determinants of PA in pwMS by Streber et al. [23], who analyzed data from 65 studies carried out in the period 1980–2015, concluded that sex was inconsistently associated with PA among pwMS.

Such an apparent lack of agreement might be explained, at least partly, by several factors which include: unbalanced samples composed predominantly of women (in percentages ranging from 70% to 85%), variety of data collection methods (e.g., questionnaires, accelerometers, pedometers), type of parameter selected as representative of PA (e.g., number of daily steps, daily minutes spent in moderate-to-vigorous PA, etc.) and participants' disability level. Historically, self-reported data, in the form of diaries and recall questionnaires, have been the most widespread technique used to assess amount and intensity of PA owing to their low cost, easiness of use and versatility. However, such an approach suffers from poor reliability and validity, participant recall bias, interpretation of questions [24] and overestimates the amount and intensity of movement performed [25]. To overcome such issues, starting from a decade ago objective techniques based on the use of wearable accelerometers have been successfully employed in acquiring more reliable, long-term data, from which it is possible to obtain the number of daily steps, classify the performed PA in terms of intensity and calculate energy expenditure [26]. The availability of a massive amount of continuous quantitative data on mobility of pwMS has shown itself to be quite useful, especially to elucidating the relationship between PA levels and important features associated with MS, such as risk of falls [27], cognitive performance [28], self-efficacy [29] and quality of life [30,31].

On the basis of the aforementioned considerations, the main purpose of the present study was twofold: (1) to investigate the existence of possible differences between women and men with MS in terms of amount and intensity of PA performed during a week, continuously acquired using wrist-worn wearable accelerometers and (2) to verify whether the disease has a stronger impact on men or women with MS. The latter aspect was analyzed through data comparison with a control group of unaffected individuals matched for age, sex and socioeconomic attributes. Such information should be useful in adding a sex perspective in the management of MS, especially as regards the design of training and rehabilitative programs to reduce sedentary activity and promote a healthy active lifestyle in pwMS.

2. Materials and Methods

2.1. Participants

In the period March-November 2019, a convenience sample of 45 outpatients (23 women, 22 men) presenting with relapsing–remitting MS, followed at the Regional Center for Multiple Sclerosis of Sardinia (Cagliari, Italy) were enrolled in the study following these inclusion criteria: diagnosis of MS according to the 2005 McDonald criteria [32,33], which allow MS to be diagnosed on the basis of the dissemination of lesions in the central nervous system in space and time supported by clinical and instrumental parameters (magnetic resonance imaging, evoked potential and cerebrospinal fluid analysis); age between 18 and 65 years; Expanded Disability Status Scale (EDSS, a score used to quantify the disability caused by MS based on a neurological examination of 8 functional systems [34]) score ≤ 6; being clinically stable and on treatment with disease-modifying agents for at least 6 months. Forty-one age-matched unaffected individuals (21 women, 20 men), selected among relatives and caregivers of the pwMS and hospital staff, composed the control group (HC). Both groups of pwMS and HC were thus equally balanced in terms of women to men ratio (1.05:1). After a detailed explanation of the purposes of the study and the experimental methodology, all participants agreed to participate by signing an informed consent form. The study was carried out in compliance with the ethical principles for research involving human subjects expressed in the Declaration of Helsinki and its later amendments and was approved by the local ethics committee(102/2018/CE). The anthropometric and clinical data of the participants are reported in Table 1.

Table 1. Anthropometric and clinical features of participants. Values are expressed as mean ± SD.

	Healthy Controls		Multiple Sclerosis	
	Women	Men	Women	Men
Participants	21	20	23	22
Age (years)	46.7 ± 14.6	49.6 ± 14.4	49.4 ± 9.0	51.2 ± 11.8
Height (cm)	163.2 ± 6.4	171.0 ± 6.8	159.2 ± 6.5	172.8 ± 6.5
Body Mass (kg)	61.5 ± 10.4	73.9 ± 11.7	61.5 ± 10.4	71.8 ± 9.8
EDSS score	NA		3.6 ± 1.8	3.6 ± 1.8
Type of MS	NA		16 RR/4 PP/3 SP	14 RR/4 PP/4 SP
Disease duration (years)	NA		17.6 ± 10.2	18.4 ± 13.4

EDSS: Expanded Disability Status Scale; MS: Multiple Sclerosis; PP: Primary Progressive; RR: Relapsing Remitting; SP: Secondary Progressive.

For some specific supplemental analyses, individuals with MS were also stratified into two groups according to their EDSS score as follows: mild disability (n = 29, 13 M, 16 F, EDSS in the range 0–3.5) and moderate-severe disability (n = 16, 9 M, 7 F, EDSS > 3.5). Six participants (all belonging to the moderate-severe disability group) reported the use of a walking aid on a regular basis.

2.2. Data Collection and Processing

Each participant was instructed to wear a tri-axial accelerometer (Actigraph GT3X, Acticorp Co., Pensacola, FL, USA) previously validated for use in pwMS [27,35] for 7 consecutive days 24 h/day on the non-dominant wrist. Removal of the device was allowed only for showering, bathing and when performing water-based activities (i.e., swimming). The choice of wrist as the placement site for the device was made as it results generally better tolerated by pwMS for both comfort and aesthetic reasons, thus increasing the likelihood of wear time compliance [36]. Wrist placement also makes data collection on sleep patterns easier [37]. PA data was collected in 60-s epochs at 30 Hz frequency. At the end of the acquisition period, raw data were downloaded on a PC and processed by means of the dedicated ActiLife software (v6.13.3 Acticorp Co., Pensacola, FL, USA) to extract the following variables on a weekly, daily and hourly basis:

- step counts (SC);
- vector magnitude counts (VM), which are composite measures of accelerometric counts considered a proxy for overall physical activity [38]. VM is calculated with the following equation:

$$VM = \sqrt{(x^2 + y^2 + z^2)} \tag{1}$$

in which x, y and z are the values of accelerations measured by the device in each of the three orthogonal directions;

- levels of PA intensity classified into three categories according to the associated value of metabolic equivalent (MET), namely sedentary behavior (SB, 0–1.5 MET), light intensity PA (LPA, 1.5–3 MET) and moderate-to-vigorous PA (MVPA, > 3 MET). Such a discrimination was carried out based on the cut-points for accelerometric counts per minute (cpm) proposed by Sandroff et al. [39] as reported in Table 2.

Table 2. List of accelerometric cut-points employed to classify the intensity levels of Physical Activity performed by people with Multiple Sclerosis and unaffected individuals (adapted from [39]).

Level of Physical Activity Intensity	Metabolic Equivalent (MET)	Accelerometric Counts Per Minute (cpm)	
		Healthy Controls and People with MS with EDSS Score ≤ 3.5	People with MS with EDSS Score > 3.5
Sedentary Behavior (SB)	<1.5	0–319	0–87
Light Physical Activity (LPA)	1.5–3	320–1980	88–1185
Moderate-to-Vigorous Physical Activity (MVPA)	>3	>1980	>1185

As these thresholds refer to experiments during which accelerometers were positioned at the hip, a correction (provided automatically by the ActiLife software) was applied to account for the wrist placement. In the analysis, we considered and included only acquisition days characterized by an overall wear time of at least 16 h.

2.3. Statistical Analyses

A two-way multivariate analysis of variance (MANOVA) was carried out to explore the existence of a possible differential impact of the disease on men and women with MS considering as independent variables sex (women or men) and status (pwMS or unaffected). Dependent variables were SC and VM in one case and the three PA levels in the other respectively. The level of significance was set at $p = 0.05$ and the effect sizes were assessed using the eta-squared (η^2) coefficient. Univariate ANOVA was carried out as a post-hoc test by reducing the level of significance to $p = 0.025$ (0.05/2) for SC and VM and $p = 0.016$ (0.05/3) for the PA levels, after a Bonferroni correction for multiple comparisons. Supplemental statistical tests focused on SC and VM only were performed using a two-way MANOVA by considering sex (women or men) and group (unaffected, pwMS with mild disability or pwMS with moderate-severe disability) as independent variables. All analyses were performed using the IBM SPSS Statistics v.20 software (IBM, Armonk, NY, USA).

3. Results

Results regarding the absolute values (min/day) and the percentage of time spent in PA of different intensities, as calculated from the accelerometric data, are reported in Table 3.

Table 3. Physical activity patterns for men and women with Multiple Sclerosis and those unaffected. Values are expressed as mean ± SD.

		Healthy Controls		Multiple Sclerosis	
		Women	Men	Women	Men
SB	%	59.3 ± 7.8	66.1 ± 8.0 *	68.6 ± 8.2 †	62.6 ± 9.8 *
	min	804.1 ± 132.0	899.9 ± 112.2 *	859.5 ± 149.1 †	810.1 ± 108.0 *
LPA	%	23.9 ± 3.9	23.0 ± 2.0	20.9 ± 6.5	27.1 ± 7.3 * †
	min	311.9 ± 49.2	313.3 ± 70.4	266.0 ± 77.7	347.0 ± 95.4 †
MVPA	%	16.8 ± 5.3	10.8 ± 4.6 *	10.5 ± 5.5 †	10.3 ± 5.5
	min	209.1 ± 69.8	147.7 ± 64.0 *	123.0 ± 74.2 †	134.9 ± 73.6
SC (steps/day)		12726 ± 2771	10850 ± 3532	8375 ± 3199 †	9719 ± 3071
VMC (10^6 counts/day)		2.68 ± 0.54	2.08 ± 0.63 *	1.85 ± 0.55 †	2.08 ± 0.66

LPA: Light Intensity Physical Activity; MVPA: Moderate-to-vigorous Physical Activity: SB: Sedentary Behavior; SC: Steps count; VMC: Vector Magnitude Counts. The symbol * denotes a significant difference vs. women after Bonferroni correction. The symbol † denotes a significant difference vs. unaffected individuals of the same sex after Bonferroni correction.

3.1. Steps Count (SC) and Vector Magnitude (VM) Counts

MANOVA detected a significant main effect of sex (F(3, 80) = 3.13, p = 0.03, Wilks λ = 0.89, η^2 = 0.10), status (F(3, 80) = 7.66, p < 0.001, Wilks λ = 0.78, η^2 = 0.22) and sex x status interaction (F(3, 80) = 4.52, p = 0.006, Wilks λ = 0.85, η^2 = 0.14) on SC and VM values. The post-hoc analysis (data are graphically displayed in Figure 1) found significant main effects for sex (SC and VM) and sex x status interaction (VM only) after Bonferroni correction. In particular, it was observed that in women (but not in men) the presence of the disease significantly reduced both SC (−32%, 8375 vs. 12,283, p < 0.001) and VM counts (−45%, 2.58 106 vs. 1.85 106, p < 0.001)., VM counts were also found significantly higher in unaffected women with respect to men (+24%, 2.58 106 vs. 2.08 106, p = 0.009).

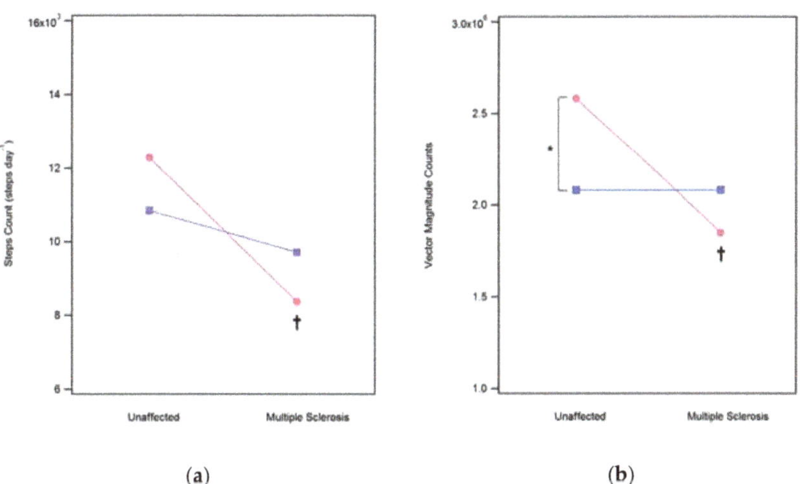

(a)   (b)

Figure 1. Comparison of Daily Steps (**a**) and Vector Magnitude Counts (**b**) for women (pink line) and men (blue line) with Multiple Sclerosis and unaffected individuals. The symbol † denotes a significant difference between unaffected individuals and people with MS after Bonferroni correction (p < 0.025). The symbol * denotes a significant difference between women and men after Bonferroni correction (p < 0.025).

Figures 2 and 3 show respectively the hourly trends and the average number of daily steps taken by unaffected individuals and pwMS of different levels of disability according to the stratification previously indicated. It appears that while the trend is quite similar for all groups, with two peaks of activity located approximately between 9–10 AM and 6–7 PM, mobility tends to decrease with increasing disability, with differences particularly evident in the women's group.

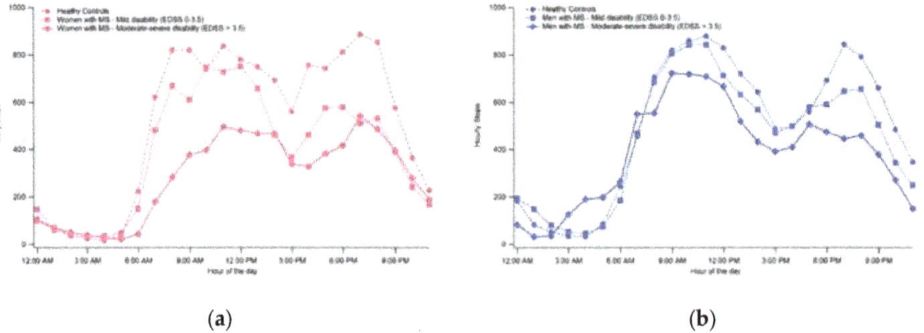

(a)   (b)

Figure 2. Comparison of hourly trends in women (**a**) and men (**b**) for step counts.

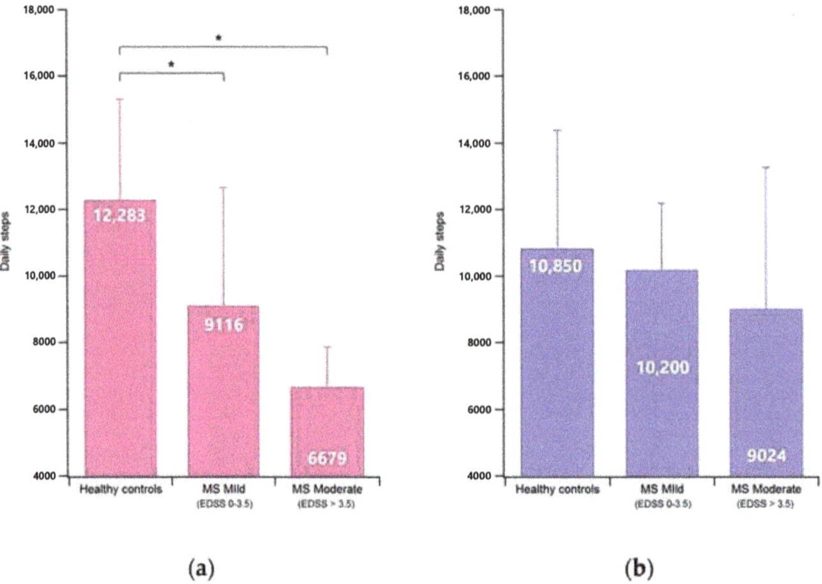

(a)   (b)

Figure 3. Average number of daily steps in women (**a**) and men (**b**) depending on their disability level. The symbol * denotes a significant difference ($p < 0.05$).

The statistical analysis carried out on daily steps considering the participants stratified into the three groups (unaffected, pwMS with mild disability and pwMS with moderate-severe disability) detected a main effect of status (F(3, 80) = 3.13, p = 0.03, Wilks λ = 0.89, η^2 = 0.10). However. the post-hoc analysis revealed that the only significant differences involved women participants. Women with MS, regardless their disability level, performed a reduced number of daily steps with respect to unaffected (mild disability 9116 vs. 12,283, p = 0.003; moderate-severe disability 6679 vs. 12,283, p = 0.003).

3.2. Physical Activity Intensity Levels

MANOVA also detected a significant effect of sex (F(3, 80) = 4.46, p = 0.006, Wilks λ = 0.86, η^2 = 0.14), status (F(3, 80) = 3.33, p = 0.024, Wilks λ = 0.89, η^2 = 0.11) and sex x status interaction (F(3, 80) = 4.03, p = 0.01, Wilks λ = 0.87, η^2 = 0.13) on PA intensity levels. The post-hoc analysis (see Figure 4) detected significant sex x status interaction for SB and LPA intensity, and a significant effect of status for MVPA intensity after Bonferroni correction. In particular, women with MS exhibited significantly higher SB than those unaffected (68.6 vs. 59.3%, $p = 0.035$) and also higher than men with MS (68.6% vs. 62.6% of men, $p = 0.021$). As regards LPA, unaffected women and men displayed similar levels, while men with MS exhibit a significant increase of LPA when compared to unaffected, passing from 23% to 27.1% ($p = 0.029$). Also, men with MS exhibit higher LPA than women (27.1 vs. 20.9%, $p < 0.001$). At last, unaffected women of our sample spent more time in MVPA than men (15.9% vs. 10.8%, $p = 0.003$) while the values become almost identical in people with MS. The decrease observed in MVPA for women (from 15.9% to 10.5%) was found statistically significant ($p < 0.001$).

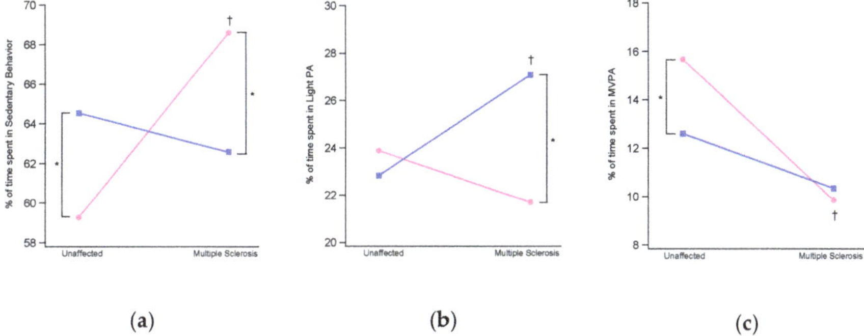

(a)	(b)	(c)

Figure 4. (a) percentages of time spent in Sedentary Behavior, (b) Light Intensity and (c) Moderate-to-vigorous intensity Physical Activity in women (pink lines) and men (blue lines) with Multiple Sclerosis and unaffected individuals. The symbol † denotes a significant difference between unaffected individuals and people with MS after Bonferroni correction ($p < 0.016$). The symbol * denotes a significant difference between women and men after Bonferroni correction ($p < 0.016$).

4. Discussion

The aim of this study was to assess the existence of possible differences in amount and intensity of PA (quantitatively assessed using accelerometers worn for a consecutive week) performed by men and women with MS and to estimate the impact of the disease across sexes through comparison with a sample of age and sex-matched unaffected individuals. The findings partly support both hypotheses since women with MS exhibited a pattern of PA characterized by a significantly higher percentage of time spent in sedentary behavior and a lower percentage of time spent in light intensity PA with respect to men with MS, while a substantial similarity between groups was detected in terms of MVPA. Nevertheless, even the trend of the remaining variables, though not fully supported by results of the statistical analysis, concur in defining the picture of an overall lower engagement in PA for women with MS. Our data also suggest that the disease has a stronger impact on women, since a significant reduction in daily steps, VM counts, and percentage of time spent in MVPA, as well as an increase in sedentary behavior, were observed in women with MS with respect to those unaffected, while this phenomenon was not detected in men.

Firstly, it appears important to discuss the results obtained for unaffected individuals, as they represent the reference basis for estimating the impact of MS on PA levels. It is commonly believed that men are more physically active than women [40], but such a statement is not fully supported by experimental findings. In fact, even though some epidemiologic studies observed higher activity levels in men in terms of moderate-intensity, vigorous-intensity and total leisure-time physical activity practice [41], others pointed out that sex-related differences closely depend on the type of activity considered and, in particular, lower levels of PA in women are expected when no distinction is made between regular sport activity and habitual physical activity in daily life. In practice, women are not actually less physically active than men but active in a different way as they prefer, for example, activity such as light intensity exercise, walking, cycling, etc. [42]. Moreover, many studies did not consider the physical effort associated with housework (in most cases still predominantly performed by women, McMunn et al. [43]) which would further increase the levels of habitual physical activity.

In this context, our results are in agreement with those of two recent large-scale studies involving more than 90,000 British [38] and 900 Finnish [44] participants, in which the same setup (i.e., wrist-worn accelerometer) was adopted. In both cases it was observed that women were characterized by a significantly higher value of daily VM counts with respect to men and, in particular, considering the study by Wennman et al. [44] which employed the same brand/type of activity tracker, agreement with our data was good even from a quantitative point of view. The sex-related differences were explained mostly by the specific position selected for placing the accelerometer, which in most previous studies was located on the hip, while more recently the wrist is becoming preferred to increase compliance and acquire data on sleep. The main differences in terms of acquired data is that hip-worn accelerometers are more suitable for describing whole-body movements, while wrist-worn accelerometers add more specific details on upper limb activity. This increases the possibility of capturing information on a range of activities that involve arm movements (such as those associated with household chores) which are more frequently carried out by women.

The pattern of PA calculated for pwMS, considered as a single group, is also generally consistent with data of previous studies, especially as regards daily steps and percentage of time spent in sedentary behavior and light intensity PA. In particular, the number of daily steps calculated in our sample, pooling both women and men regardless of disability level (9032 steps/day), or even considering it, was found quite similar to results in studies on European pwMS [45–47], but well above those found in several investigations performed in the United States. In the latter case, the value found for daily steps varied from 5800 to 7698, depending on the average level of participants' disability [48–51]. Such a discrepancy might be due, among other factors, to the intrinsic propensity to walking, which may differ significantly from country to country and which has generally been recognized as lower in the United States with respect to EU countries, as found in several studies based on accelerometric, smartphone and fitness tracker data [52–54].

In terms of PA intensity, the percentages of SB (63 to 68% for men and women respectively) and LPA (20 to 27%) here calculated, were found fully consistent with those reported in previous studies which indicated values from 60 to 70% for SB and 27 to 37% for LPA [16,42–44,55]. In contrast, the percentage of time spent in MVPA (approximately 10%), was found slightly higher than what usually observed for pwMS (1–7%). However, this result is not fully surprising, as the differences in PA classification associated with different accelerometer positioning sites (i.e., wrist vs. hip) are usually more evident in the case of intense activities [56].

The analysis of changes in PA magnitude and intensity across men and women with MS suggest that the disease has a stronger impact on the latter, who experience larger reductions in VM counts, daily steps, and percentage of time spent in MVPA, and also tend to exhibit a more marked sedentary behavior. While this phenomenon probably originates from a complex interaction between physiological, psychological and environmental factors which may differ across sexes, some explanations of this may be formulated based on evidence reported in previous studies that explicitly consider sex as the investigated variable. In the general population, it is known that women with disabilities must cope with several barriers that may limit their engagement in PA such as fatigue, lack of time, difficulty in accessing proper facilities, financial constraints and lack of knowledge about how to exercise safely [57]. In women with MS, such issues are further worsened by a greater need for assistance in performing daily activities with respect to men [58]. Moreover, of particular relevance in this context is the role potentially played by fatigue, a typical symptom lamented by most pwMS that represents a relevant barrier to PA even when performed in non-structured forms [59–61]. According to the findings of a large-scale study carried out on data extracted from the US Patient-Reported Outcome Measurement Information System (PROMIS), sex is a predictor significantly associated with fatigue (i.e., women are more likely to experience fatigue than men [62]) and this would partly explain why in women with MS, PA is affected more than in men. Furthermore, since disability associated with the disease modifies the physical capacity of women to meet both workplace and household demands [63], even the possibility of performing activities in the domestic sphere may be lessened. As a result, women with MS are more at risk of experiencing significantly reduced PA, not only that possibly associated with leisure-time, but also that associated with household activity which, as previously mentioned, represents a significant part of their daily PA balance.

Some limitations of the study are to be acknowledged. Firstly, the accuracy of the GTX3 device in terms of calculation of number of steps and classification of the PA intensity might be, in some cases, influenced by the wrist positioning adopted during our experimental tests. In particular, since the wrist acceleration on the y-axis (i.e., the one used for steps count purposes) is lower with respect to the value obtained for the typical positioning on the hip, it is possible that in some cases (i.e., in those participants with relevant gait alterations such as very low speed) some steps were not correctly detected [64,65]. Also, it is possible that women with MS with more severe disability perform PA of light intensity in which the arms are only marginally involved (for example stationary bike, yoga, pilates etc.) and thus their overall sedentary behavior may result overestimated. Similarly, previous studies on general population reported that women are more likely to perform activities of light intensity (like cycling for work/leisure time purposes [42]) that might be misclassified due to the stationary position of the arms. To overcome such discrepancies, in future studies it would probably be useful to integrate the quantitative data acquired by accelerometers with some kind of diary or questionnaire, in order to increase the accuracy of the classification of all the performed activities.

As regards the tested sample, it should be recalled that our participants were mostly individuals characterized by low-mild disability (60% of them had an EDSS score of $\leq$ 3.5) who lived in an inner-city residential area. Thus, generalization of the results presented here to different geographic and socio-economic contexts and to individuals more severely impaired should be performed cautiously. Also, most pwMS who participated in the present study were still engaged in part- or full-time working activities, but in our analysis, we did not include specific information on the type of working task (i.e., manual, office etc.). As a result, we were unable to specify whether superior levels of sedentariness were somehow associated with the performed job. At last, given the cross-sectional nature of the study, it is impossible to verify whether the sex-related differences observed here change in the course of the disease or not. Although the level of disability and duration of the disease were similar across men and women of our sample, it is to be considered that men usually take less time than women to reach the same impairment level, and thus that the trajectory of change in attitude to performing PA may be different for the two sexes. In this context, the results which refer to sedentariness should be further verified in future studies performed on larger cohorts, as there are factors (like for example fatigue, but also the presence of specific disabilities which affects mobility [55]) that are likely to change with the duration of the disease, thus also modifying the propensity to perform PA.

5. Conclusions

This study investigated the existence of possible sex-related differences in both amount and intensity of PA performed by women and men affected by MS using objective quantitative techniques (i.e., wrist-worn accelerometers). Our initial hypothesis was substantially confirmed by the results, which show a pattern of PA for women characterized by greater sedentary behavior and reduced activity of light intensity with respect to men, while similar values of MVPA were found. However, when comparing data of pwMS with those of unaffected individuals, most significant changes involved women with MS, who exhibited increased sedentary behavior, reduced MVPA, number of daily steps and VM counts with respect to unaffected ones. Taken together, such findings suggest that changes in the propensity to perform PA caused by the presence of the disease observed in previous similar studies, are mostly driven by the peculiar impact of MS on women, given their majority in the tested cohorts. Although further studies on larger cohorts also composed of individuals with higher levels of disability are needed to confirm the trend observed here, our data confirm the urgency of including sex as a variable in all studies on PA in pwMS, as well as selecting, whenever possible, quantitative objective measurements of PA. These strategies would be of great benefit in planning adequate interventions, in terms of training, physical therapy and even lifestyle guidelines, tailored to individuals' needs to maximize their effectiveness.

Author Contributions: Conceptualization, M.P. (Massimiliano Pau); Methodology, M.P. (Massimiliano Pau), M.P. (Micaela Porta); Formal analysis, M.P. (Massimiliano Pau), M.P. (Micaela Porta); recruitment and clinical assessment, E.C., J.F., G.C., L.L.; data curation, M.P. (Massimiliano Pau), M.P. (Micaela Porta); writing—original draft preparation, M.P. (Massimiliano Pau); E.C. and L.L. writing—review and editing, M.P. (Massimiliano Pau); funding acquisition, M.P. (Massimiliano Pau), E.C. All authors have read and agreed to the published version of the manuscript.

Funding: This work was partly supported by FISM—Fondazione Italiana Sclerosi Multipla (Italian Foundation for Multiple Sclerosis) grant code 2017/R/19.

Acknowledgments: The authors wish to thank the participants of the study and their families for their valuable support.

Conflicts of Interest: The authors declare no conflict of interest.

References

1. Compston, A.; Coles, A. Multiple sclerosis. *Lancet* **2008**, *372*, 1502–1517. [CrossRef]
2. Waubant, E.; Lucas, R.; Mowry, E.; Graves, J.; Olsson, T.; Alfredsson, L.; Langer-Gould, A. Environmental and genetic risk factors for MS: An integrated review. *Ann. Clin. Transl. Neurol.* **2019**, *6*, 1905–1922. [CrossRef] [PubMed]
3. Koch-Henriksen, N.; Sørensen, P.S. The changing demographic pattern of multiple sclerosis epidemiology. *Lancet Neurol.* **2010**, *9*, 520–532. [CrossRef]
4. Thompson, A.J.; Baranzini, S.E.; Geurts, J.; Hemmer, B.; Ciccarelli, O. Multiple sclerosis. *Lancet* **2018**, *391*, 1622–1636. [CrossRef]
5. Trojano, M.; Lucchese, G.; Graziano, G.; Taylor, B.V.; Simpson, S.; Lepore, V.; Grand'Maison, F.; Duquette, P.; Izquierdo, G.; Grammond, P.; et al. Geographical variations in sex ratio trends over time in multiple sclerosis. *PLoS ONE* **2012**, *7*, e48078. [CrossRef]
6. Beatty, W.W.; Aupperle, R.L. Sex differences in cognitive impairment in multiple sclerosis. *Clin. Neuropsychol.* **2002**, *16*, 472–480. [CrossRef]
7. Bergamaschi, R. Prognostic Factors in Multiple Sclerosis. *Int Rev. Neurobiol.* **2007**, *79*, 423–447.
8. Ribbons, K.A.; McElduff, P.; Boz, C.; Trojano, M.; Izquierdo, G.; Duquette, P.; Girard, M.; Grand'Maison, F.; Hupperts, R.; Grammond, P.; et al. Male Sex Is Independently Associated with Faster Disability Accumulation in Relapse-Onset MS but Not in Primary Progressive MS. *PLoS ONE.* **2015**, *10*, e0122686. [CrossRef]
9. Damasceno, A.; Von Glehn, F.; Brandão, C.O.; Damasceno, B.P.; Cendes, F. Prognostic indicators for long-term disability in multiple sclerosis patients. *J. Neurol. Sci.* **2013**, *324*, 29–33. [CrossRef]
10. Bertoni, R.; Jonsdottir, J.; Feys, P.; Lamers, I.; Cattaneo, D. Modified Functional Walking Categories and participation in people with multiple sclerosis. *Mult. Scler. Relat. Disord.* **2018**, *26*, 11–18. [CrossRef]
11. Théaudin, M.; Romero, K.; Feinstein, A. In multiple sclerosis anxiety, not depression, is related to gender. *Mult. Scler.* **2016**, *22*, 239–244. [CrossRef] [PubMed]
12. Jobin, C.; Larochelle, C.; Parpal, H.; Coyle, P.K.; Duquette, P. Gender issues in multiple sclerosis: An update. *Womens Health* **2010**, *6*, 797–820. [CrossRef] [PubMed]
13. Pau, M.; Corona, F.; Pilloni, G.; Porta, M.; Coghe, G.; Cocco, E. Do gait patterns differ in men and women with multiple sclerosis? *Mult. Scler. Relat. Disord.* **2017**, *18*, 202–208. [CrossRef] [PubMed]
14. Pau, M.; Casu, G.; Porta, M.; Pilloni, G.; Frau, J.; Coghe, G.; Cocco, E. Timed up and go in men and women with multiple sclerosis: Effect of muscular strength. *J. Bodyw. Mov. Ther.* **2020**, *24*, 124–130. [CrossRef]
15. Kister, I.; Bacon, T.E.; Chamot, E.; Salter, A.R.; Cutter, G.R.; Kalina, J.T.; Herbert, J. Natural history of multiple sclerosis symptoms. *Int. J. MS Care.* **2013**, *15*, 146–158. [CrossRef]
16. Motl, R.W.; Learmonth, Y.C.; Pilutti, L.A.; Appmaier, E.; Coote, S. Top 10 research questions related to physical activity and multiple sclerosis. *Res. Q. Exerc. Sport.* **2015**, *86*, 117–129. [CrossRef]
17. Motl, R.W.; Goldman, M. Physical inactivity, neurological disability, and cardiorespiratory fitness in multiple sclerosis. *Acta Neurol. Scand.* **2011**, *123*, 98–104. [CrossRef]

18. Sandroff, B.M.; Klaren, R.E.; Motl, R.W. Relationships among physical inactivity, deconditioning, and walking impairment in persons with multiple sclerosis. *J. Neurol Phys. Ther.* **2015**, *39*, 103–110. [CrossRef]

19. Giesser, B.S. Exercise in the management of persons with multiple sclerosis. *Ther. Adv. Neurol. Disord.* **2015**, *8*, 123–130. [CrossRef]

20. Anens, E.; Emtner, M.; Zetterberg, L.; Hellström, K. Physical activity in subjects with multiple sclerosis with focus on gender differences: A survey. *BMC Neurol.* **2014**, *14*, 47. [CrossRef]

21. Klaren, R.E.; Motl, R.W.; Dlugonski, D.; Sandroff, B.M.; Pilutti, L.A. Objectively quantified physical activity in persons with multiple sclerosis. *Arch. Phys. Med. Rehabil.* **2013**, *94*, 2342–2348. [CrossRef] [PubMed]

22. Dlugonski, D.; Pilutti, L.A.; Sandroff, B.M.; Suh, Y.; Balantrapu, S.; Motl, R.W. Steps per day among persons with multiple sclerosis: Variation by demographic, clinical, and device characteristics. *Arch. Phys. Med. Rehabil.* **2013**, *94*, 1534–1539. [CrossRef] [PubMed]

23. Streber, R.; Peters, S.; Pfeifer, K. Systematic Review of Correlates and Determinants of Physical Activity in Persons with Multiple Sclerosis. *Arch. Phys. Med. Rehabil.* **2016**, *97*, 633–645. [CrossRef] [PubMed]

24. Silfee, V.J.; Haughton, C.F.; Jake-Schoffman, D.E.; Lopez-Cepero, A.; May, C.N.; Sreedhara, M.; Rosal, M.C.; Lemon, S.C. Objective measurement of physical activity outcomes in lifestyle interventions among adults: A systematic review. *Prev. Med. Rep.* **2018**, *11*, 74–80. [CrossRef]

25. Boon, R.M.; Hamlin, M.J.; Steel, G.D. Validation of the New Zealand Physical Activity Questionnaire (NZPAQ-LF) and the International Physical Activity Questionnaire (IPAQ-LF) with accelerometry. *Br. J. Sports Med.* **2010**, *44*, 741–746. [CrossRef]

26. Casey, B.; Coote, S.; Galvin, R.; Donnelly, A. Objective physical activity levels in people with multiple sclerosis: Meta-analysis. *Scand. J. Med. Sci. Sports* **2018**, *28*, 1960–1969. [CrossRef]

27. Sebastião, E.; Learmonth, Y.C.; Motl, R.W. Lower Physical Activity in Persons with Multiple Sclerosis at Increased Fall Risk: A Cross-sectional Study. *Am. J. Phys. Med. Rehabil.* **2017**, *96*, 357–361. [CrossRef]

28. Sandroff, B.M.; Dlugonski, D.; Pilutti, L.A.; Pula, J.H.; Benedict, R.H.; Motl, R.W. Physical activity is associated with cognitive processing speed in persons with multiple sclerosis. *Mult. Scler. Relat. Disord.* **2014**, *3*, 123–128. [CrossRef]

29. Motl, R.W.; Snook, E.M.; McAuley, E.; Gliottoni, R.C. Symptoms, self-efficacy, and physical activity among individuals with multiple sclerosis. *Res. Nurs Health* **2006**, *29*, 597–606. [CrossRef]

30. Motl, R.W.; McAuley, E. Pathways between physical activity and quality of life in adults with multiple sclerosis. *Health Psychol.* **2009**, *28*, 682–689. [CrossRef]

31. Motl, R.W.; McAuley, E.; Snookm, E.M.; Gliottoni, R.C. Physical activity and quality of life in multiple sclerosis: Intermediary roles of disability, fatigue, mood, pain, self-efficacy and social support. *Psychol. Health Med.* **2009**, *14*, 111–124. [CrossRef] [PubMed]

32. Polman, C.H.; Reingold, S.C.; Edan, G.; Filippi, M.; Hartung, H.; Kappos, L.; Lublin, F.D.; Metz, L.M.; McFarland, H.F.; O'Connor, P.W.; et al. Diagnostic criteria for multiple sclerosis: 2005 revisions to the McDonald criteria. *Ann. Neurol.* **2005**, *58*, 840–846. [CrossRef] [PubMed]

33. Polman, C.H.; Reingold, S.C.; Banwell, B.; Clanet, M.; Cohen, J.A.; Filippi, M.; Fujihara, K.; Havrdova, E.; Hutchinson, M.; Kappos, L.; et al. Diagnostic criteria for multiple sclerosis: 2010 revisions to the McDonald criteria. *Ann. Neurol.* **2011**, *69*, 292–302. [CrossRef] [PubMed]

34. Kurtze, J.F. Rating neurologic impairment in multiple sclerosis: An expanded disability status scale (EDSS). *Neurology* **1983**, *33*, 1444–1452. [CrossRef] [PubMed]

35. Sandroff, B.M.; Motl, R.W.; Suh, Y. Accelerometer output and its association with energy expenditure in persons with multiple sclerosis. *J. Rehabil Res. Dev.* **2012**, *49*, 467–475. [CrossRef]

36. Kos, D.; Nagels, G.; D'Hooghe, M.B.; Ilsbrouks, S.; Delbeke, S.; Kerckhofs, E. Measuring activity patterns using actigraphy in multiple sclerosis. *Chronobiol. Int.* **2007**, *24*, 345–356. [CrossRef]

37. Buchan, D.S.; McSeveney, F.; McLellan, G. A comparison of physical activity from Actigraph GT3X+ accelerometers worn on the dominant and non-dominant wrist. *Clin. Physiol. Funct. Imaging* **2019**, *39*, 51–56. [CrossRef]

38. Doherty, A.; Jackson, D.; Hammerla, N.; Plotz, T.; Olivier, P.; Granat, M.H.; van Hees, V.T.; Trenell, M.I.; Owen, C.G. Large-Scale Population Assessment of Physical Activity Using Wrist-Worn Accelerometers: The UK Biobank Study. *PLoS ONE* **2017**, *12*, e0169649. [CrossRef]

39. Sandroff, B.M.; Riskin, B.J.; Agiovlasitis, S.; Motl, R.W. Accelerometer cut-points derived during over-ground walking in persons with mild, moderate, and severe multiple sclerosis. *J. Neurol. Sci.* **2014**, *340*, 50–57. [CrossRef]

40. The Lancet Public Health. Time to tackle the physical activity gender gap. *Lancet Pub. Health* **2019**, *4*, e360. [CrossRef]

41. Azevedo, M.R.; Araújo, C.L.; Reichert, F.F.; Wernek, G.L.; Griep, R.H.; Ponce de Leon, A.C.; Faerestin, E. Gender differences in leisure-time physical activity. *Int. J. Pub. Health* **2007**, *52*, 8–15. [CrossRef] [PubMed]

42. Meyer, K.; Niemann, S.; Abel, T. Gender differences in physical activity and fitness—Association with self-reported health and health-relevant attitudes in a middle-aged Swiss urban population. *J. Pub. Health* **2004**, *12*, 283–290. [CrossRef]

43. McMunn, A.; Bird, L.; Webb, E.; Saker, A. Gender Divisions of Paid and Unpaid Work in Contemporary UK Couples. *Work Employ. Soc.* **2020**, *34*, 155–173. [CrossRef]

44. Wennman, H.; Pietilä, A.; Rissanen, H.; Valkeinen, H.; Partonen, T.; Maki-Opas, T.; Borodulin, B. Gender, age and socioeconomic variation in 24-hour physical activity by wrist-worn accelerometers: The Fin Health 2017 Survey. *Sci. Rep.* **2019**, *9*, 6534. [CrossRef]

45. Shammas, L.; Zentek, T.; von Haaren, B.; Schlesinger, S.; Hey, S.; Rashid, A. Home-based system for physical activity monitoring in patients with multiple sclerosis (Pilot study). *Biomed. Eng. Online* **2014**, *13*, 10. [CrossRef]

46. Romberg, A.; Ruutianen, J.; Daumer, M. Physical Activity in Finnish Persons with Multiple Sclerosis. *J. Nov. Physiother.* **2013**, *3*, 3. [CrossRef]

47. Norris, M.; Anderson, R.; Motl, R.W.; Heyes, S.; Coote, S. Minimum number of days required for a reliable estimate of daily step count and energy expenditure, in people with MS who walk unaided. *Gait Posture* **2017**, *53*, 201–206. [CrossRef]

48. Motl, R.W.; McAuley, E.; Snook, E.M.; Scott, J.A. Validity of physical activity measures in ambulatory individuals with multiple sclerosis. *Disabil. Rehabil.* **2006**, *28*, 1151–1156. [CrossRef]

49. Sandroff, B.M.; Dlugonski, D.; Weikert, M.; Suh, S.; Motl, R.W. Physical activity and multiple sclerosis: New insights regarding inactivity. *Acta Neurol. Scand.* **2012**, *126*, 256–262. [CrossRef]

50. Fjeldstad, C.; Fjeldstad, A.S.; Pardo, G. Use of Accelerometers to Measure Real-Life Physical Activity in Ambulatory Individuals with Multiple Sclerosis: A Pilot Study. *Int. J. MS Care* **2015**, *17*, 215–220. [CrossRef]

51. Engelhard, M.M.; Patek, S.D.; Lach, J.C.; Goldman, M.D. Real-world walking in multiple sclerosis: Separating capacity from behavior. *Gait Posture* **2018**, *59*, 211–216. [CrossRef] [PubMed]

52. Bassett, D.R.; Wyatt, H.R.; Thompson, H.; Peters, J.C.; Hill, J.O. Pedometer-Measured Physical Activity and Health Behaviors in U.S. Adults. *Med. Sci Sports Exerc.* **2010**, *42*, 1819–1825. [CrossRef] [PubMed]

53. Althoff, T.; Sosi, R.; Hicks, J.L.; King, A.C.; Delp, S.L.; Leskovec, J. Large-scale physical activity data reveal worldwide activity inequality. *Nature* **2017**, *547*, 336–339. [CrossRef] [PubMed]

54. Kosecki, D. Fitbit's Healthiest Cities, Countries, and Days for 2017. **2017**. Available online: https://blog.fitbit.com/fitbit-year-in-review/ (accessed on 24 June 2020).

55. Neal, W.N.; Cederberg, K.L.; Jeng, B.; Sasaki, J.E.; Motl, R.W. Is Symptomatic Fatigue Associated with Physical Activity and Sedentary Behaviors Among Persons with Multiple Sclerosis? *Neurorehabil. Neural. Repair* **2020**, *34*, 505–511. [CrossRef]

56. Hildebrand, M.; Van Hees, V.T.; Hansen, B.H.; Ekelund, U. Age group comparability of raw accelerometer output from wrist- and hip-worn monitors. *Med. Sci Sports Exerc.* **2014**, *46*, 1816–1824. [CrossRef]

57. Stuifbergen, A.K.; Roberts, G.J. Health promotion practices of women with multiple sclerosis. *Arch. Phys. Med. Rehabil.* **1997**, *78*, S3–S9. [CrossRef]

58. Stuifbergen, A.K.; Becker, H.; Blozis, S.; Timmerman, G.; Kullberg, V. A randomized clinical trial of a wellness intervention for women with multiple sclerosis. *Arch. Phys. Med. Rehabil.* **2003**, *84*, 467–476. [CrossRef]

59. Ploughman, M. Breaking down the barriers to physical activity among people with multiple sclerosis—A narrative review. *Phys. Ther. Rev.* **2017**, *22*, 124–132. [CrossRef]

60. Kalron, A.; Menascu, S.; Frid, L.; Aloni, R.; Achiron, A. Physical activity in mild multiple sclerosis: Contribution of perceived fatigue, energy cost, and speed of walking. *Disabil. Rehabil.* **2020**, *42*, 1240–1246. [CrossRef]

61. Mezini, S.; Soundy, A. A Thematic Synthesis Considering the Factors which Influence Multiple Sclerosis Related Fatigue during Physical Activity. *Behav. Sci.* **2019**, *1*, 70. [CrossRef]

62. Salter, A.; Fox, R.J.; Tyry, T.; Cutter, G.; Marrie, R.A. The association of fatigue and social participation in multiple sclerosis as assessed using two different instruments. *Mult. Scler. Relat. Disord.* **2019**, *31*, 165–172. [CrossRef] [PubMed]
63. Dyck, I. Hidden geographies: The changing lifeworlds of women with multiple sclerosis. *Soc. Sci. Med.* **1995**, *40*, 307–320. [CrossRef]
64. John, D.; Morton, A.; Arguello, D.; Lyden, K.; Bassett, D. "What is a step?" differences in how a step is detected among three popular activity monitors that have impacted physical activity research. *Sensors* **2018**, *18*, 1206. [CrossRef] [PubMed]
65. Santos-Lozano, A.; Torres-Luque, G.; Garatachea, N. Inter-trial variability of GT3X accelerometer. *Sci. Sports* **2014**, *29*, e7. [CrossRef]

Publisher's Note: MDPI stays neutral with regard to jurisdictional claims in published maps and institutional affiliations.

International Journal of
*Environmental Research
and Public Health*

Article

Mediating Effect of Perceived Stress on the Association between Physical Activity and Sleep Quality among Chinese College Students

Xiangyu Zhai [1,†], Na Wu [2,3,†], Sakura Koriyama [1], Can Wang [4], Mengyao Shi [4], Tao Huang [4], Kun Wang [4], Susumu S. Sawada [5] and Xiang Fan [4,6,*]

[1] Graduate School of Sport Sciences, Waseda University, Saitama 359-1192, Japan; xiangyu.zhai@akane.waseda.jp (X.Z.); gnymskr221@akane.waseda.jp (S.K.)

[2] Key Laboratory for the Genetics of Developmental and Neuropsychiatric Disorders, Bio-X Institutes, Shanghai Jiao Tong University, Shanghai 200030, China; Lorna.STJU@sjtu.edu.cn

[3] Shanghai Key Laboratory of Psychotic Disorders, Brain Science and Technology Research Center, Shanghai Jiao Tong University, Shanghai 200030, China

[4] Department of Physical Education, Shanghai Jiao Tong University, Shanghai 200240, China; wang_c@sjtu.edu.cn (C.W.); smy0214@sjtu.edu.cn (M.S.); taohuang@sjtu.edu.cn (T.H.); wangkunz@sjtu.edu.cn (K.W.)

[5] Faculty of Sport Sciences, Waseda University, Saitama 359-1192, Japan; s-sawada@waseda.jp

[6] Shanghai Research Center for Physical Fitness and Health of Health of Children and Adolescents, Shanghai University of Sport, Shanghai 200438, China

* Correspondence: fansheva@sjtu.edu.cn

† These authors contributed equally to this work.

Citation: Zhai, X.; Wu, N.; Koriyama, S.; Wang, C.; Shi, M.; Huang, T.; Wang, K.; Sawada, S.S.; Fan, X. Mediating Effect of Perceived Stress on the Association between Physical Activity and Sleep Quality among Chinese College Students. *Int. J. Environ. Res. Public Health* **2021**, *18*, 289. https://doi.org/10.3390/ijerph18010289

Received: 23 November 2020
Accepted: 28 December 2020
Published: 2 January 2021

Publisher's Note: MDPI stays neutral with regard to jurisdictional claims in published maps and institutional affiliations.

Abstract: Background: While physical activity has been reported to positively affect stress and sleep quality, less is known about the potential relationships among them. The present study aimed to investigate the mediating effect of stress on the association between physical activity and sleep quality in Chinese college students, after controlling for age, nationality, and tobacco and alcohol use. Participants: The sample comprised 6973 college students representing three Chinese universities. Methods: Physical activity, perceived stress, and sleep quality were respectively measured using the International Physical Activity Questionnaire—Short Form (IPAQ-SF), Perceived Stress Scale—10 Items (PSS-10), and Pittsburgh Sleep Quality Index (PSQI). Results: Mediating effects of perceived stress on the association between physical activity and sleep quality were observed in males and females, with 42.4% (partial mediating effect) and 306.3% (complete mediating effect) as percentages of mediation, respectively. Conclusion: The results of this study may provide some suggestions that physical activity could improve sleep by aiding individuals in coping with stress and indicate that stress management might be an effective non-pharmaceutical therapy for sleep improvement.

Keywords: physical activity; sleep quality; perceived stress; mediating effect; Chinese college students

1. Introduction

Poor sleep quality (SQ) is a crucial public health problem increasing the risk of premature morbidity and mortality [1]. There is evidence that poor SQ is associated with impaired attention and memory, physical and mental disorders, and increased healthcare costs [2,3]. Notably, insomnia and other sleep problems are quite common in young adults, especially college students [4,5]. Evidence emphasized that insomnia prevalence in university is higher than that in the general population [6]. It is reported that the prevalence rate of poor SQ is as high as 25.7% in Chinese college students [7].

Subjective SQ is defined as an individual's general level of satisfaction with the sleep experience [8], which can be influenced by environmental factors and lifestyles [9]. There has been increasing research focusing on the association between physical activity (PA) and sleep situation [10], as it may imply and prevent health consequences in later life [11,12].

Regular PA has been suggested as a non-pharmaceutical cure to improve SQ [13] which is also easily accessible and less costly for treating insomnia [14]. Although overwhelming evidence shows that light (e.g., walking) [15,16], moderate (e.g., yoga and tai chi) [17,18], and vigorous (e.g., aerobic and endurance exercise) PA [19,20] have positive effects on SQ, some cross-sectional research has indicated that no correlation between PA and SQ has been observed [21–23]. Nevertheless, we suspected that this may be caused by a lack of several key variables acting as mediators.

Previous studies indicate that stress was one of the main predictors of SQ [24]. Stress is a normal reaction to everyday pressure, and excessive stress has numerous deleterious effects on physical and mental health outcomes. According to the 2019 survey from the American Psychological Association, more than three quarters of adults reported symptoms of stress, including changes in sleeping habits [25]. Stress also commonly exists among college students [26], as they frequently encounter it, possibly caused by experiences of academic failure, high expectations from parents, and changes in friendships under this unique developmental period of transition from adolescence into young adulthood [27]. Doolin et al. reported that stress was negatively associated with sleep in American and Bolivian university students [28]. The role of stress on the structure of neuroplasticity, such as the release of endocannabinoids and brain-derived neurotrophic factor (BDNF), could lead to restored sleep and improvement of insomnia but also lead to sleep deprivation [29–31], which may have an effect on sleep.

Additionally, PA was also widely recommended as a strategy to cope with stress in view of its protective effects against stress, including increased stress tolerance and lower subsequent stress [32,33]. A cross-sectional survey of 36,984 Canadians reported that 40% used exercise to cope with stress [34]. College students with vigorous PA were less likely to experience stress [35] and even lower light PA was associated with a higher level of stress [36], and thus a lack of PA was regarded as a predictor of stress.

Tobacco and alcohol abuse are critical problems that young adults face. According to the data from the Substance Abuse and Mental Health Services Administration in 2019, 27.9% and 69.5% persons aged 18–25 reported tobacco and alcohol use in the past year [37]. Substance use may be seen by students as a way to cope with stress [38]; however, this unhealthy behavior may cause even more stress and other health problems such as insomnia [39–44]. In fact, PA has been shown to be beneficial for decreasing substance use, relieving stress, and improving SQ [13,35,45]. Therefore, tobacco and alcohol use were also included as covariables to reduce bias.

In this study, we attempted to consider stress as a mediating variable to investigate the relationship between PA and SQ in college students, after controlling for age, nationality, and tobacco and alcohol use.

2. Materials and Methods

2.1. Participants and Procedures

This study, named Physical Activity and Sleep Quality in Chinese College Students, was conducted by researchers at Shanghai Jiao Tong University. Participants who were physically healthy were recruited from three public universities in Shanghai, China. Students who were interested in this study filled out the electronic questionnaire via scanning the Quick Response code (a type of two-dimensional barcode containing the link for the online questionnaire) to complete the survey.

After excluding the participants with missing information ($n = 2$, 0.2%), outliers for age ($n = 166$, 2.3%), and total time (the sum of sleep duration, vigorous PA time, moderate PA time, low PA time, and sitting time) per day more than 24 h ($n = 21$, 0.3%), a total of 6973 students (age: 19.0 ± 0.9 years old) participated in this study, with a higher number of male students ($n = 4752$, 68.2%) than female students ($n = 2221$, 31.9%).

2.2. Ethical Considerations

The procedures were reviewed and approved by the Ethics Committee of Shanghai Jiao Tong University (No. 20170100). Each participant was asked to indicate his or her willingness to participate in this study before filling out the survey. The data in this study were collected and analyzed anonymously.

2.3. Physical Activity Measurement

PA over the last seven days was measured using the International Physical Activity Questionnaire—Short Form (IPAQ-SF) [46]. The total physical activity recorded by this questionnaire was acceptably reliable (single measure intraclass correlation coefficient: 0.79; the coefficient of variation as a percentage of the mean score: 26%) in a Chinese population [47]. Three intensity levels of PA, including low-intensity activities (3.3 metabolic equivalents, METs), moderate-intensity activities (4.0 METs), and vigorous-intensity activities (8.0 METs), were evaluated and calculated via this questionnaire. Participants were required to report the frequency and duration that they engaged in each level of PA for at least 10 min. The total PA per week for each participant was calculated by following formula:

Total MET-minutes/week = Low PA (METs × min × days) + Moderate PA (METs × min × days) + Vigorous PA (METs × min × days).

2.4. Perceived Stress Measurement

Perceived stress, the degree that participants viewed their daily lives as unpredictable, uncontrollable, and overwhelming, was measured using the Perceived Stress Scale—10 Items (PSS-10) [48], which is a reliable (Cronbach's α: 0.85) and valid (goodness-of-fit index of two-factor model: 0.940) instrument in a Chinese population [49]. This scale comprises 10 items to indicate how often participants felt or thought a certain way during the last month; each item uses a 5-point Likert scale ranging from 0 (never) to 4 (very often). Scores range from 0 to 40, with higher composite scores indicating greater levels of perceived stress.

2.5. Sleep Quality Measurement

SQ was evaluated by the Pittsburgh Sleep Quality Index (PSQI) including subjective SQ, sleep latency, sleep duration, sleep efficiency, sleep disturbances, use of sleep medications, and daytime dysfunction [50]. The Chinese version of PSQI has been verified with good reliability (Cronbach's α: 0.84) and validity (factor loading of each component: >0.5) in Chinese students [51]. Each component is scored from 0 to 3; scores range from 0 to 21, with higher composite scores indicating poorer SQ.

2.6. Covariates

Several confounding factors including age, nationality, and tobacco and alcohol use were considered as covariates in the present study. The age of participants was calculated using the birthdate. Nationality was classified as Han Chinese or others. Tobacco and alcohol use were divided into three categories (never, rarely, or always use) by the following questions, respectively: "Have you ever used tobacco?" and "Have you ever used alcohol?".

2.7. Statistical Analysis

Participants' characteristics were examined using means, standard deviations (SD), and percentages. Moreover, gender differences in age, total PA MET-minutes, PSS-10 scores, and PSQI scores were investigated using t-tests. Gender differences in nationality and tobacco and alcohol use were analyzed using chi-square tests.

In the causal steps approach of mediation, described by Baron & Kenny (1986) [52], the starting point is to establish first that there is a significant zero-order effect of independent variable (X) on dependent variable (Y). In other words, they consider that there is no point in further investigating whether the effect of X on Y is in fact mediated by Mediator (M) if the X-Y test fails. However, some authors argue for waiving the X-Y test because if c and $a*b$ are of opposite signs (competitive mediation), then c can be close to zero and the X-Y test may fail [53–56].

Therefore, a new typology of the mediation model being developed by Zhao, Lynch, and Chen was used to estimate the mediating effect of perceived stress on the association between PA and SQ [56]. In contrast to this traditional mediation analyses, the method in this study indicates that the indirect path $a*b$ test is the first step to estimate the mediating effect. As shown in Figure 1, three regression models were established to verify the mediating effect. Regression coefficient of PA to SQ was calculated in the first regression model (path a). The regression coefficients for both PA and stress were calculated in the second regression model (path b and c'). The effect of PA, excluding stress, as a predictor of SQ was shown in the third regression model (path c). Continuous variables including age, stress, PA, and SQ in regression models were standardized. The mediated effect was examined with 95% bootstrapped confidence intervals (CIs), using 5000 bootstrapped samples. Effects with CIs not including zero were interpreted as statistically significant. Percentage of mediation was calculated by dividing the indirect effect by the total effect to examine how much of the total effect was explained by the mediation. These analyses were controlled for the effects of several confounding factors mentioned above.

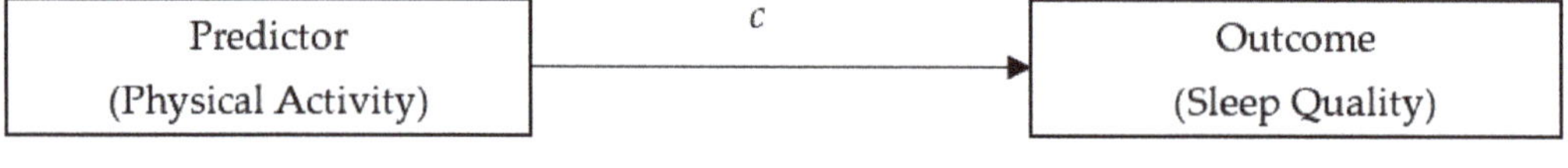

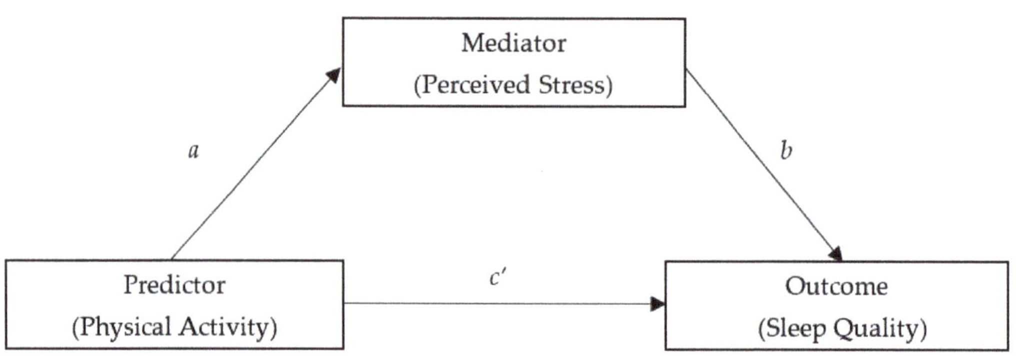

Figure 1. Conceptual model: how perceived stress mediates the association between physical activity and sleep quality. Note: a, b, c, and c' refer to the path of models (more details in the method documentation).

All statistical analyses were conducted using R program (4.0 version). *Lavaan* package in R was used to estimate the mediation analyses [57]. The acceptable threshold of statistical significance was specified as 0.05 (two-tailed).

3. Results

3.1. Participant Characteristics

In total, 6973 students participated in this study. The characteristics of participants are shown in Table 1. Compared to the female students, male students reported more PA, less perceived stress, and better SQ. Significant differences in nationality and tobacco and alcohol use were also observed between genders.

Table 1. Demographic and characteristics of participating students in the survey.

Participant Characteristics	Male (*n* = 4752)		Female (*n* = 2221)		Total (*n* = 6973)		
	Mean or *n*	SD or %	Mean or *n*	SD or %	Mean or *n*	SD or %	*p*
Age (years)	19.0	0.7	19.0	0.7	19.0	0.7	0.31
Nationality							
Han Chinese	4315	90.8%	1930	86.9%	6245	89.6%	<0.001
Others	437	9.2%	291	13.1%	728	10.4%	
Tobacco use							
Never	4624	97.3%	2196	98.9%	6820	97.8%	<0.001
Rarely	95	2.0%	19	0.8%	114	1.6%	
Always	33	0.7%	6	0.3%	39	0.6%	
Alcohol use							
Never	2537	53.4%	1527	68.8%	4064	58.3%	<0.001
Rarely	2162	45.5%	676	30.4%	2838	40.7%	
Always	53	1.1%	18	0.8%	71	1.0%	
Physical activity							
Total MET-minutes/week	3049.1	1908.8	2553.7	1667.2	2891.3	1849.6	<0.001
Perceived stress							
PSS-10 (scores)	17.5	6.8	19.3	7.1	18.1	6.9	<0.001
Sleep quality							
PSQI (scores)	4.7	2.6	5.2	2.8	4.8	2.7	<0.001

Note: SD, standard deviation; PSS, Perceived Stress Scale; PSQI, Pittsburgh Sleep Quality Index.

3.2. Mediation Models

The results of bootstrapped mediation models in male and female students after controlling for age, nationality, and tobacco and alcohol use are presented in Table 2. In the path *a*, PA was negatively associated with PSS in both male and female students. The total effect (path *c*) and the direct effect (path *c'*) of PA on PSQI in the model were significant only in male students. In female students, PSS was positively associated with PSQI (path *b*), although there was no significant association between PA and PSQI (path *c'*).

As shown in Table 3, bootstrapped CIs of total, direct, and indirect effects in males were all statistically significant, with 42.4% as percentage of mediation (partial mediating effect). Only bootstrapped CI of indirect effects in females was statistically significant, with 306.3% as percentage of mediation (complete mediating effect).

Table 2. Mediation analyses: association between physical activity and sleep quality via perceived stress.

	Variables	Total Effect Model (PSQI)						PSS						Direct Effect Model (PSQI)					
		β	Boot SE	t	p	Bootstrap 95%CI		β	Boot SE	t	p	Bootstrap 95%CI		β	Boot SE	t	p	Bootstrap 95%CI	
						Lower	Upper					Lower	Upper					Lower	Upper
Male	PA	−0.074	0.014	−5.421	<0.001	−0.101	−0.047	−0.082	0.013	−6.140	<0.001	−0.109	−0.057	−0.043	0.013	−3.363	<0.001	−0.068	−0.018
	PSS													0.381	0.013	28.500	<0.001	0.355	0.407
	Age	0.026	0.014	1.877	0.06	−0.001	0.053	0.010	0.014	0.694	0.49	−0.019	0.038	0.022	0.013	1.752	0.08	−0.003	0.048
	Nationality	0.127	0.052	2.424	0.02	0.025	0.229	0.107	0.045	2.374	0.02	0.020	0.194	0.086	0.049	1.771	0.08	−0.009	0.182
	Tobacco use	0.361	0.088	4.082	<0.001	0.194	0.545	0.182	0.055	3.318	<0.001	0.076	0.294	0.291	0.084	3.476	<0.001	0.131	0.461
	Alcohol use	0.160	0.028	5.670	<0.001	0.104	0.216	0.128	0.028	4.581	<0.001	0.072	0.183	0.112	0.026	4.280	<0.001	0.060	0.162
Female	PA	−0.012	0.028	−0.437	0.66	−0.069	0.044	−0.090	0.026	−3.411	<0.001	−0.141	−0.038	0.026	0.025	1.031	0.30	−0.023	0.074
	PSS													0.425	0.020	21.455	<0.001	0.386	0.463
	Age	0.053	0.022	2.448	0.01	0.011	0.096	0.043	0.021	2.064	0.04	0.002	0.084	0.035	0.020	1.745	0.08	−0.004	0.074
	Nationality	0.159	0.068	2.347	0.02	0.030	0.292	0.031	0.062	0.504	0.61	−0.088	0.152	0.145	0.062	2.332	0.02	0.026	0.271
	Tobacco use	0.598	0.216	2.766	0.006	0.173	1.016	0.222	0.145	1.529	0.13	−0.056	0.515	0.504	0.219	2.300	0.02	0.068	0.934
	Alcohol use	0.220	0.049	4.493	<0.001	0.125	0.315	0.239	0.044	5.389	<0.001	0.151	0.327	0.118	0.044	2.660	0.008	0.030	0.204

Note: PSQI, Pittsburgh Sleep Quality Index; PSS, Perceived Stress Scale; PA, physical activity; Boot SE, bootstrap standard error; CI, confidence interval.

Table 3. Total, direct, and indirect effects of the mediation analyses investigating perceived stress as a mediator between physical activity and sleep quality.

		β	Boot SE	*p*	Bootstrap 95%CI		P$_M$ (%)
					Lower	Upper	
Male	Total effect	−0.074	0.014	<0.001	−0.101	−0.047	–
	Indirect effect	−0.031	0.005	<0.001	−0.042	−0.021	42.4%
	Direct effect	−0.043	0.013	<0.001	−0.068	−0.018	57.6%
Female	Total effect	−0.012	0.028	0.66	−0.069	0.044	–
	Indirect effect	−0.038	0.011	<0.001	−0.060	−0.016	306.3%
	Direct effect	0.026	0.025	0.30	−0.023	0.074	−206.3%

Note: CI, confidence interval; SE, standard error; P$_M$, percentage of mediation.

4. Discussion

As people have placed increased emphasis on health problems, the interrelationship among PA and sleep has drawn wide attention. Although a large number of previous studies have shown that regular PA contributes to improving SQ [15–20], the results of cross-sectional studies are still inconsistent. For instance, Mitchell's and Youngstedt's studies investigated the relationship between PA and sleep, and no correlation was observed [21–23], while a significant relationship between low PA and poor SQ was found in the studies conducted by Feng [58] and Ma [59]. It has been considered that this nonconformity may be the result of differences in research design, ethnicity, and confounding factors. It is worth noting that Semplonius and Willoughby et al. once reported that moderate PA could indirectly predict SQ through emotion regulation [60]. Additionally, a bidirectional relationship between stress and sleep has been reported [61]. For example, Garbarino et al. found that workers exposed to chronic occupational stress had an increased incidence of sleep problems, and bad sleepers suffered more from occupational stress factors than good sleepers. Previous studies have suggested that stress may be predictive of negative health conditions, including PA and sleep [62,63]. Therefore, we further added stress as a mediating factor in this study. As we expected, the results of our study demonstrated the effects of stress mediating the association between PA and SQ among college students, and these associations still remained after controlling for age, nationality, and tobacco and alcohol use. This may imply that reducing stress by increasing PA could be used as an intervention strategy to improve SQ.

The association between PA and sleep might be explained by several direct and indirect biological pathways such as body composition, metabolic activity, cardiopulmonary function, immunity, and nervous system [64,65]. Schnohr et al. and Atlantis et al. found that increasing PA or accepting other exercise-based intervention could decrease the level of perceived stress [66,67]. Moreover, Zillman and Bryant reported that people with stress were more prone to insufficient exercise [68]. This link may be attributed to the BDNF, which is a neurotrophin having roles in the maintenance of neurons involved in emotional and cognitive function [69]. Additionally, it was also reported that BDNF is modified detrimentally in the stress model [70]. Notably, sleep plays a vital role in cognitive functioning involving the consolidation of neuroplasticity, which also tightly links with BDNF. Stress exposure after inadequate sleep further damaged the hypothalamic–pituitary–adrenal response to an increase in cortisol [71]. Additionally, cortisol might improve sleep disruption and brain clearance. This subsequently increases cognition [72] and, hence, decreased stress may promote sleep. From the above, our results hint at a potential mechanism which may need to be verified in the future.

Interestingly, the mediating effects of stress were quite different between males and females, with 42.4% (partial mediating effect) and 306.3% (complete mediating effect) as percentages of mediation. It has been assumed that gender specificity affects the relationship between PA and sleep through differences in biological features, e.g., sex steroids [73] and cognitive function [74]. Kemp et al. showed that the response to negative stimuli of the frontal cortical measured by electroencephalography and the electrophysiological signal in females were much faster and higher than those in males [75], implying that negative emotions are more acceptable for females. Filkowski et al. observed that the noradrenergic locus coeruleus (LC), which is the arousal center in the brain, was more active in females when facing emotional swings [76], suggesting that stress may result in a larger LC-mediated arousal response in females. These clues provided support for our findings.

The main strength of the present study is considering stress as a mediator between PA and sleep. Another strength is including a large sample of Chinese college students as participants. However, this study has several limitations. Firstly, the directional relations among PA, stress, and sleep cannot be observed, as the present study was a cross-sectional one; more longitudinal studies need to be conducted in future. Secondly, the measurements of PA, perceived stress, and SQ were all self-reported. This may have caused errors in record, recall, and social desirability bias [77], which could have affected the reliability and validity of the study. Thirdly, although confounding factors, i.e., age, nationality, and tobacco and alcohol use, were considered, more detailed information, e.g., body composition, cardiorespiratory fitness, and appetite, need to be included. Lastly, the research was carried out among healthy and well-educated college students so the results may be restricted when generalizing to all Chinese college students.

5. Conclusions

We highlight the mediating effects of stress on the association between PA and sleep in both female and male college students, even after adjusting for age, nationality, and tobacco and alcohol use. This may provide some suggestions that PA would improve sleep by aiding individuals in coping with stress and indicate that stress management might be a non-pharmaceutical therapy for sleep improvement.

Author Contributions: The authors' contributions are as follows: X.F. is the principal investigator; he designed the study and oversaw the implementation of the project. X.Z. and N.W. drafted the manuscript and completed the data analyses. S.K. and S.S.S. participated in the revision of manuscript and improved the quality of manuscript. C.W., M.S., T.H. and K.W. participated in the data collection. All authors have read and agreed to the published version of the manuscript.

Funding: The study was funded by the National Social Fund of China (16CTY012).

Institutional Review Board Statement: The study was conducted according to the guidelines of the Declaration of Helsinki, and approved by the Ethics Committee of Shanghai Jiao Tong University (No. 20170100).

Informed Consent Statement: Informed consent was obtained from all subjects involved in the study.

Data Availability Statement: The data in the study are not publicly available in order to protect the privacy of participants.

Acknowledgments: The authors are grateful to all students participating in this study, for their contributions to the data collection. Special thanks to Benjamin How for his contribution in English language editing.

Conflicts of Interest: The authors declare no conflict of interest.

References

1. Stranges, S.; Tigbe, W.; Gómez-Olivé, F.X.; Thorogood, M.; Kandala, N.-B. Sleep Problems: An Emerging Global Epidemic? Findings from the INDEPTH WHO-SAGE Study among More than 40,000 Older Adults from 8 Countries Across Africa and Asia. *Sleep* **2012**, *35*, 1173–1181. [CrossRef] [PubMed]
2. Chattu, V.K.; Manzar, D.; Kumary, S.; Burman, D.; Spence, D.W.; Pandi-Perumal, S.R. The Global Problem of Insufficient Sleep and Its Serious Public Health Implications. *Healthcare* **2018**, *7*, 1. [CrossRef] [PubMed]
3. Daley, M.; Morin, C.M.; Leblanc, M.; Grégoire, J.-P.; Savard, J. The Economic Burden of Insomnia: Direct and Indirect Costs for Individuals with Insomnia Syndrome, Insomnia Symptoms, and Good Sleepers. *Sleep* **2009**, *32*, 55–64. [CrossRef]
4. Ohayon, M.M. Epidemiology of insomnia: What we know and what we still need to learn. *Sleep Med. Rev.* **2002**, *6*, 97–111. [CrossRef] [PubMed]

5. Samaranayake, C.B.; Arroll, B.; Fernando, A.T. Sleep disorders, depression, anxiety and satisfaction with life among young adults: A survey of university students in Auckland, New Zealand. *N. Z. Med. J.* **2014**, *127*, 13–22. [PubMed]

6. Jiang, X.-L.; Zheng, X.-Y.; Yang, J.; Ye, C.-P.; Chen, Y.-Y.; Zhang, Z.-G.; Xiao, Z.-J. A systematic review of studies on the prevalence of Insomnia in university students. *Public Health* **2015**, *129*, 1579–1584. [CrossRef] [PubMed]

7. Li, L.; Wang, Y.-Y.; Wang, S.; Zhang, L.; Li, L.; Xu, D.-D.; Ng, C.H.; Ungvari, G.S.; Cui, X.; Liu, Z.-M.; et al. Prevalence of sleep disturbances in Chinese university students: A comprehensive meta-analysis. *J. Sleep Res.* **2018**, *27*, e12648. [CrossRef]

8. Kline, C. Sleep quality. In *Encyclopedia of Behavioral Medicine*; Gellman, M.D., Turner, J.R., Eds.; Springer: New York, NY, USA, 2013; pp. 1811–1813. [CrossRef]

9. Magnavita, N.; Garbarino, S. Sleep, Health and Wellness at Work: A Scoping Review. *Int. J. Environ. Res. Public Health* **2017**, *14*, 1347. [CrossRef]

10. Zhai, X.; Ye, M.; Wang, C.; Gu, Q.; Huang, T.; Wang, K.; Chen, Z.; Fan, X. Associations among physical activity and smartphone use with perceived stress and sleep quality of Chinese college students. *Ment. Health Phys. Act.* **2020**, *18*, 100323. [CrossRef]

11. Bruce, E.S.; Lunt, L.; McDonagh, J. Sleep in adolescents and young adults. *Clin. Med.* **2017**, *17*, 424–428. [CrossRef]

12. Ghrouz, A.K.; Noohu, M.M.; Manzar, D.; Spence, D.W.; Bahammam, A.S.; Pandi-Perumal, S.R. Physical activity and sleep quality in relation to mental health among college students. *Sleep Breath.* **2019**, *23*, 627–634. [CrossRef] [PubMed]

13. Lang, C.; Kalak, N.; Brand, S.; Holsboer-Trachsler, E.; Pühse, U.; Gerber, M. The relationship between physical activity and sleep from mid adolescence to early adulthood. A systematic review of methodological approaches and meta-analysis. *Sleep Med. Rev.* **2016**, *28*, 32–45. [CrossRef] [PubMed]

14. Passos, G.S.; Poyares, D.L.R.; Santana, M.G.; Tufik, S.; De Mello, M.T. Is exercise an alternative treatment for chronic insomnia? *Clinics* **2012**, *67*, 653–659. [CrossRef]

15. McCurry, S.M.; Pike, K.C.; Vitiello, M.V.; Logsdon, R.G.; Larson, E.B.; Teri, L. Increasing Walking and Bright Light Exposure to Improve Sleep in Community-Dwelling Persons with Alzheimer's Disease: Results of a Randomized, Controlled Trial. *J. Am. Geriatr. Soc.* **2011**, *59*, 1393–1402. [CrossRef]

16. Richards, K.C.; Lambert, C.; Beck, C.K.; Bliwise, D.L.; Evans, W.J.; Ms, G.K.K.; Kleban, M.H.; Lorenz, R.; Rose, K.; Gooneratne, N.S.; et al. Strength Training, Walking, and Social Activity Improve Sleep in Nursing Home and Assisted Living Residents: Randomized Controlled Trial. *J. Am. Geriatr. Soc.* **2011**, *59*, 214–223. [CrossRef]

17. Mustian, K.M.; Sprod, L.K.; Janelsins, M.; Peppone, L.J.; Palesh, O.G.; Chandwani, K.; Reddy, P.S.; Melnik, M.K.; Heckler, C.; Morrow, G.R. Multicenter, Randomized Controlled Trial of Yoga for Sleep Quality Among Cancer Survivors. *J. Clin. Oncol.* **2013**, *31*, 3233–3241. [CrossRef]

18. Raman, G.; Zhang, Y.; Minichiello, V.J.; D'Ambrosio, C.M.; Wang, C. Tai Chi Improves Sleep Quality in Healthy Adults and Patients with Chronic Conditions: A Systematic Review and Meta-analysis. *J. Sleep Disord. Ther.* **2013**, *2*, 141. [CrossRef]

19. Reid, K.J.; Baron, K.G.; Lu, B.; Naylor, E.; Wolfe, L.; Zee, P.C. Aerobic exercise improves self-reported sleep and quality of life in older adults with insomnia. *Sleep Med.* **2010**, *11*, 934–940. [CrossRef]

20. Carandente, F.; Montaruli, A.; Angeli, A.; Sciolla, C.; Roveda, E.; Calogiuri, G. Effects of endurance and strength acute exercise on night sleep quality. *Int. SportMed J.* **2011**, *12*, 113–124.

21. Mitchell, J.A.; Godbole, S.; Moran, K.; Murray, K.; James, P.; Laden, F.; Hipp, J.A.; Kerr, J.; Glanz, K. No Evidence of Reciprocal Associations between Daily Sleep and Physical Activity. *Med. Sci. Sports Exerc.* **2016**, *48*, 1950–1956. [CrossRef]

22. Youngstedt, S.D.; Perlis, M.L.; O'Brien, P.M.; Palmer, C.R.; Smith, M.T.; Orff, H.J.; Kripke, D.F. No association of sleep with total daily physical activity in normal sleepers. *Physiol. Behav.* **2003**, *78*, 395–401. [CrossRef]

23. Youngstedt, S.D.; Kline, C. Epidemiology of exercise and sleep. *Sleep Biol. Rhythm.* **2006**, *4*, 215–221. [CrossRef] [PubMed]

24. Åkerstedt, T.; Orsini, N.; Petersen, H.; Axelsson, J.; Lekander, M.; Kecklund, G. Predicting sleep quality from stress and prior sleep—A study of day-to-day covariation across sixweeks. *Sleep Med.* **2012**, *13*, 674–679. [CrossRef] [PubMed]

25. American Psychological Association. Stress Relief Is within Reach. Available online: https://www.apa.org/topics/stress (accessed on 29 December 2019).

26. Saleh, D.; Camart, N.; Romo, L. Predictors of Stress in College Students. *Front. Psychol.* **2017**, *8*, 19. [CrossRef]

27. Borjalilu, S.; Mohammadi, A.; Mojtahedzadeh, R. Sources and Severity of Perceived Stress among Iranian Medical Students. *Iran. Red Crescent Med. J.* **2015**, *17*, e17767. [CrossRef]

28. Doolin, J.; Vilches, J.E.; Cooper, C.; Gipson, C.; Sorensen, W. Perceived stress and worldview influence sleep quality in Bolivian and United States university students. *Sleep Health* **2018**, *4*, 565–571. [CrossRef]

29. McEwen, B.S.; Bowles, N.P.; Gray, J.D.; Hill, M.N.; Hunter, R.G.; Karatsoreos, I.N.; Nasca, C. Mechanisms of stress in the brain. *Nat. Neurosci.* **2015**, *18*, 1353–1363. [CrossRef]

30. Prospéro-García, O.; Amancio-Belmont, O.; Meléndez, A.L.B.; Ruiz-Contreras, A.E.; Méndez-Díaz, M. Endocannabinoids and sleep. *Neurosci. Biobehav. Rev.* **2016**, *71*, 671–679. [CrossRef]

31. Cirelli, C. Cellular consequences of sleep deprivation in the brain. *Sleep Med. Rev.* **2006**, *10*, 307–321. [CrossRef]

32. Schultchen, D.; Reichenberger, J.; Mittl, T.; Weh, T.R.M.; Smyth, J.M.; Blechert, J.; Pollatos, O. Bidirectional relationship of stress and affect with physical activity and healthy eating. *Br. J. Health Psychol.* **2019**, *24*, 315–333. [CrossRef]

33. Bland, H.W.; Melton, B.F.; Bigham, L.E.; Welle, P.D. Quantifying the impact of physical activity on stress tolerance in college students. *Coll. Stud. J.* **2014**, *48*, 559–568.

34. Cairney, J.; Kwan, M.Y.W.; Veldhuizen, S.; Faulkner, G.E.J. Who Uses Exercise as a Coping Strategy for Stress? Results from a National Survey of Canadians. *J. Phys. Act. Health* **2014**, *11*, 908–916. [CrossRef] [PubMed]
35. VanKim, N.A.; Nelson, T.F. Vigorous Physical Activity, Mental Health, Perceived Stress, and Socializing among College Students. *Am. J. Health Promot.* **2013**, *28*, 7–15. [CrossRef] [PubMed]
36. Cruz, S.Y.; Fabián, C.; Pagán, I.; Ríos, J.L.; González, A.M.; Betancourt, J.; González, M.J.; Rivera-Soto, W.T.; Palacios, C. Physical activity and its associations with sociodemographic characteristics, dietary patterns, and perceived academic stress in students attending college in Puerto Rico. *Puerto Rico Health Sci. J.* **2013**, *32*, 44–50.
37. Substance Abuse and Mental Health Services Administration. Results from the 2019 National Survey on Drug Use and Health: Detailed Tables. Available online: https://www.samhsa.gov/data/report/2019-nsduh-detailed-tables (accessed on 29 December 2020).
38. Leonard, N.R.; Gwadz, M.V.; Eritchie, A.; Linick, J.L.; Cleland, C.M.; Elliott, L.; Egrethel, M. A multi-method exploratory study of stress, coping, and substance use among high school youth in private schools. *Front. Psychol.* **2015**, *6*, 1028. [CrossRef]
39. Naquin, M.R.; Gilbert, G.G. College Students' Smoking Behavior, Perceived Stress, and Coping Styles. *J. Drug Educ.* **1996**, *26*, 367–376. [CrossRef]
40. Veronda, A.C.; Irish, L.A.; Delahanty, D.L. Effect of smoke exposure on young adults' sleep quality. *Nurs. Health Sci.* **2019**, *22*, 57–63. [CrossRef]
41. Rice, K.G.; Van Arsdale, A.C. Perfectionism, perceived stress, drinking to cope, and alcohol-related problems among college students. *J. Couns. Psychol.* **2010**, *57*, 439–450. [CrossRef]
42. Chen, H.; Bo, Q.-G.; Jia, C.-X.; Liu, X. Sleep Problems in Relation to Smoking and Alcohol Use in Chinese Adolescents. *J. Nerv. Ment. Dis.* **2017**, *205*, 353–360. [CrossRef]
43. Yang, T.; Barnett, R.; Peng, S.; Yu, L.; Zhang, C.; Zhang, W. Individual and regional factors affecting stress and problem alcohol use: A representative nationwide study of China. *Health Place* **2018**, *51*, 19–27. [CrossRef]
44. Taylor, D.J.; Bramoweth, A.D.; Grieser, E.A.; Tatum, J.I.; Roane, B.M. Epidemiology of Insomnia in College Students: Relationship With Mental Health, Quality of Life, and Substance Use Difficulties. *Behav. Ther.* **2013**, *44*, 339–348. [CrossRef] [PubMed]
45. West, A.B.; Bittel, K.M.; Russell, M.; Evans, M.B.; Mama, S.K.; Conroy, D. A systematic review of physical activity, sedentary behavior, and substance use in adolescents and emerging adults. *Transl. Behav. Med.* **2020**, *10*, 1155–1167. [CrossRef] [PubMed]
46. International Physical Activity Questionnaire. Guidelines for Data Processing and Analysis of the International Physical Activity Questionnaire (IPAQ)-Short and Long Forms. Available online: https://www.researchgate.net/file.PostFileLoader.html?id=5641 f4c36143250eac8b45b7&assetKey=AS%3A294237418606593%401447163075131 (accessed on 29 December 2020).
47. Macfarlane, D.; Lee, C.C.; Ho, E.Y.; Chan, K.; Chan, D.T. Reliability and validity of the Chinese version of IPAQ (short, last 7 days). *J. Sci. Med. Sport* **2007**, *10*, 45–51. [CrossRef] [PubMed]
48. Cohen, S.; Kamarck, T.; Mermelstein, R. A global measure of perceived stress. *J. Health Soc. Behav.* **1983**, *24*, 385–396. [CrossRef]
49. Lu, W.; Bian, Q.; Wang, W.; Wu, X.; Wang, Z.; Zhao, M. Chinese version of the Perceived Stress Scale-10: A psychometric study in Chinese university students. *PLoS ONE* **2017**, *12*, e0189543. [CrossRef]
50. Buysse, D.J.; Reynolds, C.F.; Monk, T.H.; Berman, S.R.; Kupfer, D.J. The Pittsburgh sleep quality index: A new instrument for psychiatric practice and research. *Psychiatry Res.* **1989**, *28*, 193–213. [CrossRef]
51. Liu, X.; Tang, M.; Hu, L. Reliability and validity of the Pittsburgh sleep quality index. *Chin. J. Psychiatry* **1996**, *29*, 103–107.
52. Baron, R.M.; Kenny, D.A. The moderator–mediator variable distinction in social psychological research: Conceptual, strategic, and statistical considerations. *J. Pers. Soc. Psychol.* **1986**, *51*, 1173–1182. [CrossRef]
53. Collins, L.M.; Graham, J.J.; Flaherty, B.P. An Alternative Framework for Defining Mediation. *Multivar. Behav. Res.* **1998**, *33*, 295–312. [CrossRef]
54. MacKinnon, D.P. Contrasts in multiple mediator models. In *Multivariate Applications in Substance Use Research: New Methods for New Questions*; Rose, J.S., Chassin, L., Eds.; Psychology Press: New York, NY, USA, 2000; pp. 141–160.
55. Shrout, P.E.; Bolger, N. Mediation in experimental and nonexperimental studies: New procedures and recommendations. *Psychol. Methods* **2002**, *7*, 422. [CrossRef]
56. Zhao, X.; Lynch, J.G., Jr.; Chen, Q. Reconsidering Baron and Kenny: Myths and truths about mediation analysis. *J. Consum. Res.* **2010**, *37*, 197–206. [CrossRef]
57. Rosseel, Y. Lavaan: An R package for structural equation modeling and more. Version 0.5–12 (BETA). *J. Stat. Softw.* **2012**, *48*, 1–36. [CrossRef]
58. Feng, Q.; Zhang, Q.-L.; Du, Y.; Ye, Y.-L.; He, Q.-Q. Associations of Physical Activity, Screen Time with Depression, Anxiety and Sleep Quality among Chinese College Freshmen. *PLoS ONE* **2014**, *9*, e100914. [CrossRef] [PubMed]
59. Ma, C.; Zhou, L.; Xu, W.; Ma, S.; Wang, Y. Associations of physical activity and screen time with suboptimal health status and sleep quality among Chinese college freshmen: A cross-sectional study. *PLoS ONE* **2020**, *15*, e0239429. [CrossRef]
60. Semplonius, T.; Willoughby, T. Long-Term Links between Physical Activity and Sleep Quality. *Med. Sci. Sports Exerc.* **2018**, *50*, 2418–2424. [CrossRef]
61. Van Reeth, O.; Weibel, L.; Spiegel, K.; Leproult, R.; Dugovic, C.; Maccari, S. PHYSIOLOGY OF SLEEP (REVIEW)–Interactions between stress and sleep: From basic research to clinical situations. *Sleep Med. Rev.* **2000**, *4*, 201–219. [CrossRef]
62. Stults-Kolehmainen, M.A.; Sinha, R. The Effects of Stress on Physical Activity and Exercise. *Sports Med.* **2014**, *44*, 81–121. [CrossRef]
63. Strygin, K.N. Sleep and stress. *Ross. Fiziol. Zh. Im. I. M. Sechenova* **2011**, *97*, 422–432.

64. Pedersen, B.K.; Saltin, B. Evidence for prescribing exercise as therapy in chronic disease. *Scand. J. Med. Sci. Sports* **2006**, *16*, 3–63. [CrossRef]
65. Chennaoui, M.; Arnal, P.J.; Sauvet, F.; Léger, D. Sleep and exercise: A reciprocal issue? *Sleep Med. Rev.* **2015**, *20*, 59–72. [CrossRef]
66. Schnohr, P.; Kristensen, T.S.; Prescott, E.; Scharling, H. Stress and life dissatisfaction are inversely associated with jogging and other types of physical activity in leisure time—The Copenhagen City Heart Study. *Scand. J. Med. Sci. Sports* **2005**, *15*, 107–112. [CrossRef] [PubMed]
67. Atlantis, E.; Chow, C.M.; Kirby, A.; Singh, M.A.F. An effective exercise-based intervention for improving mental health and quality of life measures: A randomized controlled trial. *Prev. Med.* **2004**, *39*, 424–434. [CrossRef] [PubMed]
68. Zillman, D.; Bryant, J. *Selective Exposure to Communication*; Routledge: New York, NY, USA, 1985.
69. Cowansage, K.K.; LeDoux, J.; Monfils, M.-H. Brain-Derived Neurotrophic Factor: A Dynamic Gatekeeper of Neural Plasticity. *Curr. Mol. Pharmacol.* **2010**, *3*, 12–29. [CrossRef] [PubMed]
70. Choy, K.H.C.; De Visser, Y.; Nichols, N.R.; Buuse, M.V.D. Combined neonatal stress and young-adult glucocorticoid stimulation in rats reduce BDNF expression in hippocampus: Effects on learning and memory. *Hippocampus* **2008**, *18*, 655–667. [CrossRef]
71. Schlotz, W. Investigating associations between momentary stress and cortisol in daily life: What have we learned so far? *Psychoneuroendocrinology* **2019**, *105*, 105–116. [CrossRef]
72. Pistollato, F.; Cano, S.S.; Elio, I.; Vergara, M.M.; Giampieri, F.; Battino, M. Associations between Sleep, Cortisol Regulation, and Diet: Possible Implications for the Risk of Alzheimer Disease. *Adv. Nutr.* **2016**, *7*, 679–689. [CrossRef]
73. Sallis, J.F.; Zakarian, J.M.; Hovell, M.F.; Hofstetter, C. Ethnic, socioeconomic, and sex differences in physical activity among adolescents. *J. Clin. Epidemiol.* **1996**, *49*, 125–134. [CrossRef]
74. Kato, K.; Iwamoto, K.; Kawano, N.; Noda, Y.; Ozaki, N.; Noda, A. Differential effects of physical activity and sleep duration on cognitive function in young adults. *J. Sport Health Sci.* **2018**, *7*, 227–236. [CrossRef]
75. Kemp, A.H.; Silberstein, R.; Armstrong, S.; Nathan, P.J. Gender differences in the cortical electrophysiological processing of visual emotional stimuli. *NeuroImage* **2004**, *21*, 632–646. [CrossRef]
76. Filkowski, M.M.; Olsen, R.M.; Duda, B.; Wanger, T.J.; Sabatinelli, D. Sex differences in emotional perception: Meta analysis of divergent activation. *NeuroImage* **2017**, *147*, 925–933. [CrossRef]
77. Helmerhorst, H.J.F.; Brage, S.; Warren, J.; Besson, H.; Ekelund, U. A systematic review of reliability and objective criterion-related validity of physical activity questionnaires. *Int. J. Behav. Nutr. Phys. Act.* **2012**, *9*, 103. [CrossRef] [PubMed]

International Journal of
*Environmental Research
and Public Health*

Article

Psychological Analysis among Goal Orientation, Emotional Intelligence and Academic Burnout in Middle School Students

Pablo Usán Supervía [1,*], Carlos Salavera Bordás [2] and Víctor Murillo Lorente [3]

[1] Departament of Psychology, Faculty of Human Sciences and Education, University of Zaragoza, Valentín Carderera n°4, 22003 Huesca, Spain
[2] Departament of Psychology, Faculty of Education, University of Zaragoza, Pedro Cerbuna n°12, 50009 Zaragoza, Spain; salavera@unizar.es
[3] Departament of Physiatry and Nursing, Faculty of Health and Sports Sciences, University of Zaragoza, Plaza Universidad n°3, 22002 Huesca, Spain; vmurillo@unizar.es
* Correspondence: pusan@unizar.es

Received: 14 September 2020; Accepted: 22 October 2020; Published: 4 November 2020

Abstract: During schooling, students can undergo, for more or less long periods of time, different contextual settings that can negatively affect their personal and academic development, leading them not to meet their academic goals. The main objective of this research responds to examine the relationships between the constructs of goal orientations, emotional intelligence, and burnout in students. **Method:** This research comprised 2896 students from 15 Spanish high schools with ages between 12 and 18 years distributed across male (N = 1614; 55.73%) and female (N = 1282; 44.26%) genders. The measurements were made through Perception of Success Questionnaire (POSQ), the Trait Meta Mood Scale (TMMS-24) and the Maslach Burnout Inventory Student Survey (MBI-SS). **Results:** Results showed links between task orientation, high emotional intelligence levels, and adaptive behaviors and between ego orientation, academic burnout and less adaptive behavior. Similarly, it was shown that emotional intelligence can be used to predict goal-oriented behaviors. **Conclusion:** It is argued that the promotion of task orientation among secondary school students can lead to the adoption of adaptive behaviors and this, in turn, improve the development of students toward academic and personal settings.

Keywords: adolescents; students; goal orientation; emotional intelligence; burnout

---

1. Introduction

During their school stage, students are affected by contextual, situational, and personal circumstances that can have a strong effect on their learning processes, especially at the secondary school level, during adolescence, a vital stage in the life of the individual about to enter adulthood [1].

For more or less prolonged periods of time, some students do not possess, or do not use, the necessary strategies, tools, and skills to meet their academic demands. This triggers feelings and perceptions that undermine their motivation and their wish to continue studying [2], physical and psychic exhaustion [3], negative and unsatisfactory behavior and loss of interest in school [4]. All these circumstances can lead to poor academic performance and even failure and premature school dropouts [5].

1.1. Goal Orientation

The achievement goal theory attempts to describe the motives of the people in their life performing different personal behaviors [6]. In this case, this theory would respond to the reasons why students act in one way or another daily at school in achievement environments showing different capacities about their abilities, capabilities and skills which suggests two motivational orientations: one, more self-determined, is task-oriented, while the other, less self-determined, is more ego-oriented.

Those students who are task-oriented have the assumption that the motivation and development of their capacities follow the effort and the dedication that the school demands, and suppose their success at school is related to the assumption of intrinsic motivation in performing tasks [7], coping strategies [8], academic performance and happiness [9], and psychological and emotional well-being [10].

Ego-oriented students aim to show skills, capabilities, and competencies above those shown by their peers, and this is related to extrinsic motivations in undertaking school duties [11], anxiety problems [12], less academic efficacy [13], and even lack of commitment to school activities and dropout [14].

The Achievement Goals Theory considers task or ego orientations depending on how students interpret, respond to and experience goal achievement and all its implications [6].

1.2. Emotional Intelligence

The capacity to be able to analyse the emotional information around the environment is called emotional intelligence [15]. Traditionally, the conceptualization of emotional intelligence has been conceptualized in two different ways [16]. The first of them, following the studies of Mayer and Salovey [15], considers emotional intelligence as an inherent cognitive capacity of the person measured by performance-based tests. The second one, the construct of emotional intelligence is considered to be a personality trait and can be measured by questionnaires installed in the lower part of the personality hierarchy levels [17].

Emotionally intelligent people can understand emotions in their immediate environment, understanding their possible causes and effects as well as developing the necessary strategies to regulate or manage different emotional states. In this way, emotional intelligence is constituted by a series of abilities in the person's subjective wellbeing such as emotional attention, comprehension, and regulation [18]. Therefore, emotional attention responds to the ability to perceive, act and express different emotions adequately; emotional comprehension is the ability to understand emotional changes and moods; and the capacity to regulate emotions adequately is called emotional regulation.

Scientific literature on emotional intelligence in school environments is ample, focusing on different groups and contexts. Most research in recent years analyzed the influence of emotional intelligence in wellbeing of primary school students [19], secondary school students [20] and/or university students [21]. These studies led to significant correlations between emotional intelligence and subjective wellbeing and academic happiness linked with better academic performance [22].

It can be argued that emotions play a fundamental role in the adaptation of adolescents in their school environment, helping them to manage all situations with the potential to affect their personal wellbeing, academic motivation, and academic performance, among others [18].

1.3. Academic Burnout

By adopting more non-adaptative behaviors, some students can lose interest in, and no longer feel committed to, their studies, while doubts and/or contradictions about their own personal capabilities arise, which prevent them from moving forward. As noted, this can lead some students to a complete lack of motivation, and even to drop out of school [23]. The presence of these personal symptoms is well known as *academic burnout* [24,25] and the three main dimensions are emotional exhaustion, cynicism, and self-efficacy. On the other hand, emotional exhaustion is associated with the physical and emotional weariness that students might feel, more or less persistently, during their academic lives; cynicism refers to relaxed attitude, indifference and a little interest in their studies and finally, self-efficacy relates to the student's attitude toward the academic tasks, demands and duties. Finally, academic burnout is associated with high levels of stress [26], low levels of vigor and involvement with and dedication to academic duties [27], low levels of self-efficacy [28] and low levels of academic performance [11].

1.4. Goal Orientation, Emotional Intelligence and Academic Burnout

Scientific literature has paid little specific attention to the relationship between these two constructs in students, but some studies exist that approach the issue from a different perspective.

Currently, goal orientation and emotional intelligence are used interchangeably. Emotional intelligence is regarded as a way to interact with our environment, and the motivations of the subject play a very important role in this; as such, emotions are an integral part of motivation, insofar as this triggers goal-oriented behaviours [29]. As pointed out by Fernández, Anaya and Suárez [30], motivational systems and emotional intelligence interact and support one another in pursuance of the desired goal, and to the detriment of interdependent positions. Buck, Powers and Hull [31] found positive relationships between task orientations and the three dimensions that constitute emotional intelligence, as well as with greater commitment to the school task and academic enjoyment. Froiland and Worrell [32], in analyzing a sample of primary and secondary school students, found a relationship between these variables and the student's engagement with the school. Gargallo, Campos and Almerich [33] highlighted the fact that more intrinsically and emotionally intelligent students performed better and showed greater levels of self-efficacy.

By combining the two goal-oriented variables and academic burnout, several studies have related achievement goals and levels of burnout by showing that motivational task-oriented achievement and academic self-efficacy converge in more self-determined behavior patterns [34] while ego-oriented achievement goals are related to emotional exhaustion and cynicism in a lower self-determined behavior patterns [27].

Finally, all studies agree that emotional intelligence and academic burnout are negatively correlated [35,36]. As an example, in a recent study with the same kind of population as in our research, it was determined that students with high emotional intelligence were less likely to experience school anxiety and more likely exhibiting resilience which in turn, reduced the risk of school burnout [37]. In any case, the greater the degree of emotional intelligence, the lower the levels of emotional exhaustion and cynicism. Conversely, the greater the level of emotional intelligence, the higher the level of academic self-efficacy and subjective wellbeing [36].

1.5. Study Objective

Based on the previous arguments, and following Cera, Almagro, Conde and Sáenz-López [38] it can be argued that few studies have specifically analyzed our target variables in the primary and secondary school environment. Therefore, to increase the scientific literature is needed more studies to approach the understanding of the relationship between our target variables, with the ultimate aim of improving the students' academic and personal development in school environments. The main objective of this research responds to analyze and examine the relationships between the variables of goal orientations, emotional intelligence, and burnout in students.

Four hypotheses are presented:

Hypotheses 1. *Task-oriented students present greater levels of emotional intelligence and lower of academic burnout, following more adaptive behaviour.*

Hypotheses 2. *Ego-oriented students present greater levels of academic burnout and lower of emotional intelligence, following less adaptive behaviours.*

Hypotheses 3. *Emotional attention, comprehension and regulation can be used to predict task-oriented behaviors.*

Hypotheses 4. *Physical/emotional exhaustion related to academic burnout can predict ego-oriented behaviors.*

2. Materials and Methods

Participants: This study was composed of 2896 students, both male (N = 1614; 55.73%) and female (N = 1282; 44.26%) with ages ranging from 12 to 18 years (M = 14.78; DT = 1.71) belonging to 15 public schools which were selected by simple random sampling. The students belonging to the different classes of the schools responded to the inclusion criterion in filling the questionnaires as soon as they knew how to read and understand the items of the questionnaires. Incomplete questionnaires (84) were discarded to be include in the sample, including those from students who decided to drop out half way and students with cognitive disorders who could not fully understand the questionnaire were excluded from the study. The missing data was around 3%, taking valid and suitable questionnaires around 97%.

Measurements: The following questionnaires were completed by the research participants: Perception of Success Questionnaire (POSQ) [39] translated into Spanish language by Martínez, Alonso and Moreno [40]. Composed of 12 items and two subvariables: task-oriented (e.g., "When I'm in class, I perform to the best of my ability"); and ego-oriented (e.g., "When I'm in class, I feel successful when I show the teacher and my classmates that I am the best") distributed on a five-point Likert scale in which the students showed their degree of adequacy with the items presented where "1. Strongly disagree" was the minimum mark and "5. Strongly agree" the maximum mark. Cronbach's alpha values was 0.86 for task-oriented and 0.83 for ego-oriented, respectively

Trait Meta Mood Scale-24 (TMMS-24) [41] adapted in a shortened version by Fernández-Berrocal, Extremera and Ramos [42]. The scale is composed of three subvariables: emotional attention ($\alpha = 0.79$) (e.g.,: 'I pay much attention to my feelings'); emotional comprehension ($\alpha = 0.83$) (e.g., 'I know how to label my emotions'); and emotional regulation ($\alpha = 0.82$) (e.g., 'Even if I am not feeling well, I try to have positive thoughts') distributed on a five-point Likert scale in which the students showed their degree of adequacy with the items presented where "1. Strongly disagree" was the minimum mark and "5. Strongly agree" the maximum mark.

Maslach Burnout Inventory—Student Survey (MBI-SS) [24]. The inventory in Spanish language consists of 15 items and three subvariables: physical/emotional exhaustion ($\alpha = 0.82$) (e.g., "Studying or going to class all day is exhausting"); cynicism ($\alpha = 0.80$) (e.g., "I have become less enthusiastic about my studies") and self-efficacy ($\alpha = 0.79$) (e.g., "I feel stimulated when I achieve my study goals"). The inventory was distributed on a five-point Likert scale where "1. Strongly disagree" was the minimum mark and "5. Strongly agree" the maximum mark.

Procedure: Prior to conducting this study, approval was obtained from the participating ESO schools and from the students' parents/guardians. In agreement with the different educational schools, the questionnaires were handed out to all groups. Parents/guardians voluntarily participated through Helsinki Declaration [43] ethical guidelines. The research protocol was endorsed by Psychology and Sociology Department, Universidad de Zaragoza. The research code is (4620). The schools were in charge of telling us the best day to pass the questionnaires to their students. On that day, we passed the questionnaires to the same school in the different classes (1–4 grade) with the help of their tutors in a single class. All the questionnaires were totally anonymous and voluntary and they could even abandon it at any time if they wished not to continue with it.

Data analysis: To carry out the statistical analysis in the research, the description of the variables and sociodemographic data of the participants was carried out in matters such as sex, age and academic year. In turn, a correlational analysis of the three main constructs of the study was carried out. On the other hand, a stepwise multiple regression (showing the last step) was used with the intention of predicting goal orientations (task and ego) on the variables of emotional intelligence and burnout. For these procedures, IBM SPSS v26.0 (IBM, Armonk, NY, USA) was used. Finally, an equation model was performed between the three variables described under the maximum likelihood method with AMOS v24 (IBM, Armonk, NY, USA). For all these procedures adopted a $p \leq 0.05$ significance level and a 95% confidence.

3. Results

The statistical results of the research are described below.

3.1. Demographic Variables

The research comprised 2896 students, distributed in male (N = 1614; 55.73%) and female (N = 1282; 44.26%) adolescents, who were ranging from 12 to 18 years (M = 14.78; DT = 1.71), from 15 public secondary schools (Table 1).

Table 1. Socio-demographic data by variables of gender, age, and school year.

Socio-Demographic Variables		N	%
Gender (male–female)	Male	1614	55.73
	Female	1282	44.26
Age (12–18)	12	447	15.43
	13	543	18.75
	14	553	19.09
	15	677	23.37
	16	476	16.43
	17	154	5.31
	18	46	1.58
School year (1–4 grade)	Year 1	698	24.10
	Year 2	745	25.72
	Year 3	867	29.93
	Year 4	586	20.23

3.2. Descriptive Variables

In Table 2 we can see how different results were found by gender. Approaching to gender, concerning goal task-oriented, emotional attention and self-efficacy are more pronounced in females. Concerning goal ego-oriented, emotional comprehension and regulation and cynicism values, these variables are in males.

Table 2. Goal orientation, emotional intelligence, and academic burnout descriptive variables.

Psychological Variables	Total		Male		Female	
	x	sd	x	sd	x	sd
Goal Task-oriented	3.80	0.80	3.71	0.82	3.91	0.76
Goal Ego-oriented	2.89	1.03	2.95	0.98	2.82	1.08
Emotional attention	3.44	0.75	3.35	0.76	3.55	0.72
Emotional comprehension	3.41	0.72	3.50	0.70	3.31	0.74
Emotional regulation	3.56	0.75	3.62	0.70	3.49	0.79
Physical/emotional exhaustion	3.19	0.96	3.17	1.01	3.22	0.91
Cynicism	2.14	1.06	2.29	1.08	1.97	0.99
Self-efficacy	3.52	0.77	3.49	0.80	3.55	0.72

X: Mean/SD: Standard Deviation.

3.3. Correlational Analysis of Goal Orientation, Emotional Intelligence and Burnout

Then, a correlational analysis was carried out with the corresponding significant relationships of the variables (see Table 3).

Table 3. Goal orientation, emotional intelligence, and academic burnout correlation analysis.

Psychological Variables	1	2	3	4	5	6	7	8
(1) Goal Task-oriented	1							
(2) Goal Ego-oriented	0.289 **	1						
(3) Emotional attention	0.383 **	0.100 **	1					
(4) Emotional comprehension	0.378 **	0.029 **	0.155 **	1				
(5) Emotional regulation	0.355 **	0.093 **	0.255 **	0.0451 **	1			
(6) Physical/emotional exhaustion	−0.182 **	0.132 **	0.100 **	−0.136 **	−0.198 **	1		
(7) Cynicism	−0.434 **	0.164 **	0.047	−0.172 **	−0.251 **	0.328 **	1	
(8) Self-efficacy	0.520 **	0.256 **	0.060	0.331 **	0.388 **	−0.158 **	−0.489 **	1
X	3.80	2.89	3.44	3.41	3.56	3.19	2.14	3.52
SD	0.80	1.03	0.75	0.72	0.75	0.96	1.06	0.77
Cronbach's α	0.86	0.83	0.79	0.83	0.82	0.82	0.80	0.79

** Significant correlation at 0.01 (bilateral); X: Mean/SD: Standard Deviation.

Task orientation is positively correlated with all three dimensions of emotional intelligence—attention ($r = 0.383$), comprehension ($r = 0.378$) and regulation ($r = 0.355$)—and in a negative way with burnout variables as exhaustion ($r = -0.182$) and cynicism ($r = -0.434$). Besides, ego orientation goal was positively correlated, albeit less strongly, with the three dimensions that constitute emotional intelligence and academic burnout.

Finally, emotional comprehension and regulation correlated negatively with physical/emotional exhaustion and cynicism. However, academic self-efficacy correlated positively with emotional comprehension ($r = 0.331$) and regulation ($r = 0.388$).

3.4. Regression Analysis of Emotional Intelligence and Academic Burnout on Goal Orientation

In this procedure, the variables of emotional intelligence and academic burnout will be used as possible predictors of goal orientation (task and ego). The results of the final step of the multiple regression analysis are shown in Tables 4 and 5, introducing the variables that have a significant effect on the task-oriented and ego-oriented goals.

Table 4. Emotional intelligence and academic burnout on goal orientation (task).

Psychological Variables	B	s.e.	R^2	t	Sig.
Constant	2.172	0.296	0.342	7.334	0.000
Attention	0.095	0.035		1.672	0.006
Comprehension	0.096	0.035		2.701	0.007
Regulation	0.093	0.030		3.608	0.002
Self-efficacy	0.363	0.035		10.303	0.000
Cynicism	−0.165	0.024		−6.780	0.000

Excluded variables: Physical/emotional exhaustion (academic burnout). s.e.: standard error; sig.: significative or not.

Table 5. Emotional intelligence and academic burnout on goal orientation (ego).

Psychological Variables	B	s.e.	R^2	t	Sig.
Constant	1.082	0.243	0.077	4.448	0.000
Comprehension	0.117	0.049		2.377	0.018
Self-efficacy	0.323	0.047		6.953	0.000
Exhaustion	0.086	0.035		2.448	0.015

Excluded variables: Emotional attention and regulation (emotional intelligence)/Cynicism and self-efficacy (academic burnout).

In this procedure created for goal orientation, the predictor variables used were emotional attention, comprehension and regulation, academic self-efficacy (emotional intelligence) and cynicism (with a negative value) (academic burnout). Nagelkerke's R^2 obtained *a* value 0.342 for task-oriented goal (Table 4).

Concerning ego orientation (Table 5), the predictor variables used were emotional comprehension, academic self-efficacy and physical/emotional, and the adjustment value was $R^2 = 0.077$.

3.5. Goal Orientation, Emotional Intelligence and Academic Burnout Structural Equation Model

Finally, Figure 1 shows the result of the analysis undertaken with structural equations and the Maximum Likelihood Method, which confirms the suitability of the model and the constructs considered herein. The model indicates a negative correlation of goal orientation and burnout ($r = -0.64$). Furthermore, goal orientations are correlated in a positive relation with emotional intelligence ($r = 0.46$) and academic burnout negatively correlated with emotional intelligence ($r = -0.74$). The adequacy and validity of the structural equations model carried out was endorsed by the appropriate indices: x^2 (17) = 52.396, $p < 0.001$; x^2/gl = 3.082; CFI = 0.96; NFI = 0.95; TLI = 0.90; RMSEA = 0.070, IC = 95% (0.052–0.089).

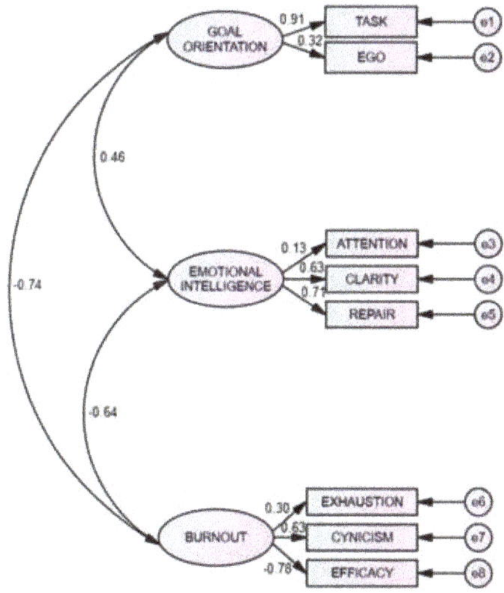

Figure 1. Goal orientation, emotional intelligence and academic burnout structural equation model.

4. Discussion

The purpose of this research was to examine the relation among goal orientation (task and ego), emotional intelligence and academic burnout in adolescent students. Four hypotheses were put forward:

First, that task-oriented students present greater levels of emotional intelligence and lower of academic burnout, following adaptive behaviors.

The supposition was completely confirmed. Our statistical results showed the variables are strongly correlated. In this way, the task orientation of the adolescent students who have the belief that success comes from dedication and effort toward school demand is positively related to emotional attention, comprehension, and regulation, and negatively correlated with the less self-determined physical/emotional exhaustion and cynicism, subvariables pertaining to burnout construct.

The scientific literature shows studies that maintain the line of the results found between the relationship between goal orientation and emotional intelligence. Buck, Powers and Hull [31] advocate positive relationships with the three dimensions that constitute emotional intelligence, along with greater motivation and academic enjoyment. Ferriz, Sicilia and Sáenz [44] highlight the relation between high motivation and emotional intelligence, leading to better academic performance. Froiland and Worrell [32], in analyzing a group of primary and secondary school students, found a relationship between emotional intelligence and task orientation, which are also in relation to the student's commitment to the school. In the same vein, Cera, Almagro, Conde and Sáenz-López [38] add that greater emotional control leads to less self-perceived stress in adolescents. Therefore, it is argued that task-oriented adaptive behaviors and good levels of emotional intelligence lead to greater psychological and emotional wellbeing and less academic burnout in adolescent students [10].

The second hypothesis stated that ego-oriented students present greater levels of academic burnout and lower levels of emotional intelligence, following less adaptive behaviors.

We were only able to partially demonstrate this hypothesis. On the one hand, there was no negative correlation that we predicted in the relationship between emotional intelligence and ego orientations, even showing that the variables were related in a positive way. On the other, the results found maintain the line that the orientation toward the ego was related to academic burnout and non-adaptive behaviors of adolescent students.

Some studies reach similar conclusions concerning the relationship between ego orientation, academic burnout and non-adaptive behaviors. Caballero, Bresó and González [23] argue that less motivated and ego-oriented students suffer greater physical/emotional exhaustion and cynicism. DeFreese and Smith [11] claim that the relationship between goal orientation and extrinsic motivations leads to less academic commitment and engagement. Estrada et al. [35] take into consideration school commitment and task-confronting strategies that lead to non-adaptive behaviors and less psychological wellbeing.

Therefore, the results of the research showed that ego orientation does not have to be related with low levels of emotional intelligence. Following to Saies, Arribas, Cecchini, De Cos and Otaegi [45] extrinsic orientations do not necessarily involve negative attitudes. In this regard, the metacognitive ability of subjects to choose themselves the positive factors of two goal orientations (task and ego) is important when carrying out their school tasks. According to Pekrun [46] managing positive emotions such as hope of success, anticipated enjoyment, subjective normative ability, pride, the feeling of having one's value recognized by significant third parties, etc. can be regarded as positive extrinsic motivations which complement and, to some extent replace, intrinsic motivations. Therefore, students can be motivated complementary towards task orientation and ego orientation according to their motives, aims and reasons for carrying out any activity in any academic and personal situation [45,46].

In this vein, various studies point out that task orientation has a stronger correlation with emotional intelligence than ego, but that the latter correlation exists [47,48]. A relationship between non-adaptive conducts and ego orientation and academic burnout is attested, but the expected negative correlation between ego orientation and low levels of emotional intelligence has not. It is thus shown that in order to pursue their ends and meet their academic targets, students can use their emotional intelligence to find a balance between positive and negative elements in both goal orientations [49].

The third hypothesis stated that emotional attention, comprehension, and regulation can be used to predict task-oriented behaviors. This hypothesis was fully demonstrated, and the exercise also showed that the most determinant factor in this regard was self-efficacy.

Few studies analyzed the predictor variables of goal orientation, but approach the issue from a different perspective. Pérez's [50] analysis of the impact of emotional intelligence on goal orientation argues that emotional comprehension and regulation can be used to predict goal orientation, in line with our own results, which also include emotional attention. Cera et al. [38] found that the dimensions of emotional intelligence can predict more self-determined goal orientations. Inam, Nomaan and Abiodullah [51] suggest that intrinsic motivation, emotional intelligence, and life satisfaction predict self-determined goal orientations.

Finally, the fourth hypothesis stated that physical/emotional exhaustion related to academic burnout can be used to predict ego-oriented behaviors. This hypothesis was not corroborated, although it was shown that physical/emotional exhaustion, emotional comprehension and self-efficacy can be used to predict ego-oriented behavior, in line with previous conclusions, i.e., ego-oriented behaviours carry with them negative implications, although subjects can use them to meet their ends and their academic targets. Schaufeli and Salanova [25] advocate that exhaustion and cynicism can be used to predict extrinsic motivations. In the same vein, Gutiérrez, Ruiz and López [52] emphasize the impact of academic burnout on academic motivation.

5. Conclusions

Based on these results, the effect of goal orientation, emotional intelligence and academic burnout in education seem to be beyond discussion. In addition to other personal and contextual variables, these features constitute a psychological profile that could have an important impact on the academic performance and commitment to school. For example, these findings may lead to the recognition in educators of exhaustion situations and maladaptive responses in their students that affect their school learning as well as poor school performance or lack of motivation in tasks to promote adaptive behaviors such as goal orientation toward the task or emotional intelligence that help prevent burnout or cynicism. At the same time, educational programs carried out by educational professionals (teachers, psychologists, educational communities, etc.) can help improve the variables studied in our study for optimal personal and academic development. Therefore, to advocate for adequate personal and academic development of adolescent students, it is essential to address these issues that to a large extent, will also affect their school development [53].

Finally, as study limitations, the cross-sectional design can be highlighted by taking the data in the schools at a unique moment. This can vary in different trimesters and can affect the scores on the variables studied depending on the academic performance, personal and/or social situation with their classmates or even their happiness or boredom in class. Likewise, the schools responded to a random selection not being able to take all areas of the city with its own features despite the large number of subjects involved in the research. Besides, the variations in scores in the variables of goal orientations, emotional intelligence and academic burnout may change from year to year and it is appropriate to opt for longitudinal studies that measure the evolution of the constructs in subsequent investigations.

Author Contributions: Conceptualization, P.U.S., C.S.B. and V.M.L.; methodology, P.U.S., C.S.B. and V.M.L.; validation, P.U.S., C.S.B. and V.M.L.; formal analysis, P.U.S., C.S.B. and V.M.L.; investigation, P.U.S., C.S.B. and V.M.L.; resources, P.U.S., C.S.B. and V.M.L.; data curation, P.U.S., C.S.B. and V.M.L.; writing—original draft preparation, P.U.S., C.S.B. and V.M.L.; writing—review and editing, P.U.S., C.S.B. and V.M.L.; visualization, P.U.S., C.S.B. and V.M.L.; supervision, P.U.S., C.S.B. and V.M.L.; project administration, P.U.S., C.S.B. and V.M.L.; funding acquisition, P.U.S., C.S.B. and V.M.L. All authors have read and agreed to the published version of the manuscript.

Funding: This study was supported by "Ibercaja" Foundation (Spain) and University of Zaragoza.

Acknowledgments: All participants in this research.

Conflicts of Interest: The authors declare no conflict of interest in this research.

References

1. Reina, M.; Oliva, A. From emotional competence to self-esteem and life-satisfaction in adolescents. *Behav. Psychol.* **2015**, *23*, 345–359.
2. Usán, P.; Salavera, C.; Murillo, V.; Mejías, J.J. Relación entre motivación, compromiso y autoconcepto físico en futbolistas adolescentes de diferentes categorías deportivas. *Cuadernos Psicología Deporte* **2016**, *16*, 183–194.
3. Del Rosal, I.; Dávila, M.A.; Sánchez, S.; Bermejo, M.L. La inteligencia emocional en estudiantes universitarios: Diferencias entre el grado de maestro en educación primaria y los grados en Ciencias. *Int. J. Dev. Educ. Psychol.* **2016**, *1*, 51–62. [CrossRef]
4. Prince, D.; Nurius, P.S. The role of positive academic self-concept in promoting school success. *Child. Youth Serv. Rev.* **2014**, *43*, 145–152. [CrossRef]

5. Baena, A.; Granero, A. Modelo de predicción de la satisfacción con la educación física y la escuela. *Revista Psicodidáctica* **2015**, *20*, 177–192. [CrossRef]

6. Ames, C. Achievement goals, motivational climate and motivational processes. In *Motivation in Sport and Exercise*; Roberts, G.C., Ed.; Human Kinetics: Champaign, IL, USA, 2002; pp. 161–176.

7. Porto, R.C.; Gonçalves, M.P. Motivação e envolvimento acadêmico: Um estudo com estudantes universitários. *Psicol. Esc. Educ.* **2017**, *21*, 515–522. [CrossRef]

8. Salavera, C.; Usán, P. Relación entre los estilos de humor y la satisfacción con la vida en estudiantes de Secundaria. *Eur. J. Investig. Health Psychol. Educ.* **2017**, *7*, 87–97. [CrossRef]

9. Chen, W. The relations between perceived parenting styles and academic achievement in Hong Kong: The mediating role of students' goal orientations. *Learn Individ. Differ.* **2015**, *37*, 48–54. [CrossRef]

10. Gaeta, M.; Cavazos, J.; Sánchez, A.P.; Rosario, P.; Högemann, J. Propiedades psicométricas de la versión mexicana del Cuestionario para la Evaluación de Metas Académicas (CEMA). *Rev. Latinoam. Psicol.* **2015**, *47*, 16–24. [CrossRef]

11. DeFreese, J.D.; Smith, A.L. Teammate social support, burnout, and self determinated motivation in collegiate athletes. *Psychol. Sport Exerc.* **2013**, *14*, 258–265. [CrossRef]

12. Gillet, N.; Vallerand, R.; Paty, B. Situational motivational profiles and performance with elite performers. *J. Appl. Soc. Psychol.* **2013**, *43*, 1200–1210. [CrossRef]

13. Dahling, J.J.; Ruppel, C.L. Learning goal orientation buffers the effects of negative normative feedback on test self-efficacy and reattempt interest. *Learn Individ. Differ.* **2016**, *50*, 296–301. [CrossRef]

14. Baadte, C.; Schnotz, W. Feedback effects on performance, motivation and mood: Are they moderated by the learner's selfconcept? *Scan. J. Educ. Res.* **2014**, *58*, 570–591. [CrossRef]

15. Mayer, J.D.; Salovey, P.; Caruso, D.R. Models of emotional intelligence. In *Handbook of Intelligence*; Sternberg, R.J., Ed.; Cambridge University Press: Cambridge, UK, 2008; pp. 396–420.

16. Romano, L.; Tang, X.; Hietajärvi, L.; Salmela-Aro, K.; Fiorilli, C. Students' Trait Emotional Intelligence and Perceived Teacher Emotional Support in Preventing Burnout: The Moderating Role of Academic Anxiety. *Int. J. Environ. Res. Public Health* **2020**, *17*, 4771. [CrossRef]

17. Petrides, K. Psychometric properties of the trait emotional intelligence questionnaire (TEIQue). In *Handbook of Individual Differences in Cognition*; Springer Science and Business Media LLC: Berlin, Germany, 2009; pp. 85–101. [CrossRef]

18. Fernández-Berrocal, P.; Extremera, N. Ability emotional intelligence, depression, and well-being. *Emot. Rev.* **2016**, *8*, 311–315. [CrossRef]

19. Ferragut, M.; Fierro, A. Inteligencia emocional, bienestar personal y rendimiento académico en preadolescentes. *Rev. Latinoam. Psicol.* **2012**, *44*, 95–104.

20. Pulido, F.; Herrera, F. Influencia de la felicidad en el rendimiento académico en primaria: Importancia de las variables sociodemográficas en un contexto pluricultural. *Revista Española Orientación Psicopedagogía* **2019**, *30*, 41–56. [CrossRef]

21. Páez, M.L.; Castaño, J.J. Inteligencia emocional y rendimiento académico en estudiantes universitarios. *Psicología Caribe* **2015**, *32*, 268–285.

22. Broc, M.A. Academic Performance and Other Psychological, Social and Family Factors in Compulsory Secondary Education Students in a Multicultural Context. *Int. J. Sociol. Educ.* **2018**, *7*, 1–23. [CrossRef]

23. Caballero, C.; Bresó, E.; González, G. Burnout en estudiantes universitarios. *Psicología Caribe* **2015**, *32*, 424–441. [CrossRef]

24. Schaufeli, W.; Martínez, I.; Marques-Pinto, A.; Salanova, M.; Bakker, A. Burnout and engagement in university students: A crossnational study. *J. Cross Cult. Psychol.* **2002**, *33*, 464–468. [CrossRef]

25. Schauffeli, W.; Salanova, M. Efficacy or inefficacy, that's the question: Burnout and work engagement, and their relationships with efficacy believes. *Anxiety Stress Coping* **2007**, *20*, 177–196. [CrossRef] [PubMed]

26. Herrera, L.; Mohamed, L.; Cepero, S. Cansancio emocional en estudiantes universitarios. *Rev. Educ. Hum.* **2016**, *9*, 173–191.

27. Salanova, M.; Martínez, I.M.; Llorens, S. *Una Mirada Más "Positiva" a la Salud Ocupacional Desde la Psicología Organizacional Positiva en Tiempos de Crisis: Aportaciones Desde el Equipo de Investigación WoNT*; Consejo General de Colegios Oficiales de Psicólogos: Madrid, Spain, 2014; Volume 35, pp. 22–30.

28. Rahamati, Z. El estudio de burnout académico en estudiantes con alto y bajo nivel de autosuficiencia. *Rev. Proc.* **2015**, *1*, 49–55.
29. Goleman, D. *Inteligencia Emocional*; Editorial Kairós SA: Madrid, Spain, 2012.
30. Fernández, A.P.; Anaya, D.; Suárez, J.M. Características motivacionales y estrategias de autorregulación motivacional de los estudiantes de secundaria. *Revista Psicodidáctica* **2012**, *17*, 95–112. [CrossRef]
31. Buck, R.; Powers, S.R.; Hull, K.S. Measuring emotional and cognitive empathy using dynamic, naturalistic, and spontaneous emotion displays. *Emotion* **2017**, *17*, 1120–1136. [CrossRef]
32. Froiland, J.M.; Worrell, F.C. Intrinsic motivation, learning goals, engagement, and achievement in a diverse high school. *Psychol. Schools* **2016**, *53*, 321–336. [CrossRef]
33. Gargallo, B.; Campos, C.; Almerich, G. Aprender a aprender en la universidad: Efectos sobre las estrategias de aprendizaje y el rendimiento académico. *Cult. Educ.* **2016**, *9*, 1–40. [CrossRef]
34. Osorio, G.; Perrello, S.; Prado, C. Burnout académico en una muestra de estudiantes universitarios mexicanos. *Enseñanza Investigación Psicología* **2020**, *2*, 27–37.
35. Ardiles, R.; Alfaro, P.; Moya, M. La inteligencia emocional como factor amortiguador del burnout académico y potenciador del engagement académico. *Revista Electrónica Investigación Docencia Universitaria* **2019**, *1*, 109–128.
36. Estrada, H.; De la Cruz, A.; Bahamón, J.; Pérez, J.; Cáceres, M. Burnout académico y su relación con el bienestar psicológico en estudiantes universitarios. *Rev. Espac.* **2018**, *39*, 7–23.
37. Fiorilli, C.; Farina, E.; Buonomo, I.; Costa, S.; Romano, L.; Larcan, R.; Petrides, K.V. Trait Emotional Intelligence and School Burnout: The Mediating Role of Resilience and Academic Anxiety in High School. *Int. J. Environ. Res. Public Health* **2020**, *17*, 3058. [CrossRef]
38. Cera, E.; Almagro, B.; Conde, C.; Sáenz-López, P. Inteligencia emocional y motivación en educación física en secundaria. *Retos* **2015**, *27*, 8–13. [CrossRef]
39. Roberts, G.; Treasure, D.; Balagué, G. Achievement goals in sport: The development and validation of the Perception of Success Questionnaire. *J. Sport Sci.* **1998**, *16*, 337–347. [CrossRef]
40. Martínez, C.; Alonso, N.; Moreno, J.A. Análisis factorial confirmatorio del "Cuestionario de percepción de éxito (POSQ)" en alumnos adolescentes de educación física. In Proceedings of the IV Congreso de la Asociación Española de Ciencias del Deporte, La Coruña, Spain, 24–27 October 2006; pp. 757–761.
41. Salovey, P.; Mayer, J.D.; Goldman, S.L.; Turvey, C.; Palfai, T.P. Emotional attention, clarity and repair: Exploring emotional intelligence using the Trait Meta-Mood Scale. In *Emotion, Disclosure, & Health*; Pennebaker, J.W., Ed.; American Psychological Association: Washington, DC, USA, 1995; pp. 125–154.
42. Fernández-Berrocal, P.; Extremera, N.; Ramos, N. Validity and reliability of the Spanish modified version of the Trait Meta-Mood Scale. *Psychol. Rep.* **2004**, *94*, 751–755. [CrossRef]
43. Asociación Médica Mundial (AMM). *Declaración de Helsinki—Principios Éticos Para las Investigaciones Con Los Seres Humanos*; Asociación Médica Mundial (AMM): Seoul, Korea, 2002.
44. Ferriz, R.; Sicilia, A.; Sáenz, P. Predicting satisfaction in physical education classes: A study based on self determination theory. *Open Educ. J.* **2013**, *6*, 1–7. [CrossRef]
45. Saies, E.; Arribas, S.; Cecchini, J.A.; De-Cos, I.; Otaegi, O. Diferencias en orientación de meta, motivación autodeterminada, inteligencia emocional y satisfacción con los resultados deportivos entre piragüistas expertos y novatos. *Cuadernos Psicología Deporte* **2014**, *14*, 21–30.
46. Pekrun, R. The impact of emotions on learning and achievement: Towards a theory of cognitive/motivational mediators. *Appl. Psychol. Int. Rev.* **2002**, *41*, 359–376. [CrossRef]
47. Pérez, N.; Castejón, J. La Inteligencia emocional como predictor del rendimiento académico en estudiantes universitarios. *Ansiedad Estrés* **2007**, *13*, 119–129.
48. Jiménez, M.; López, E. Impacto de la inteligencia emocional percibida, actitudes sociales y expectativas del profesor en el rendimiento académico. *Electron. J. Res. Educ. Psychol.* **2013**, *11*, 75–98.
49. Couto, S. *Desarrollo de la Relación Entre Inteligencia Emocional y Los Problemas de Convivencia: Estudio Clínico y Experimental*; Visión libros: Madrid, Spain, 2011.
50. Pérez, A. Inteligencia Emocional y Motivación del Estudiante Universitario. Ph.D. Thesis, Universidad de Las Palmas de Gran Canaria, Las Palmas de Gran Canaria, Spain, September 2012.
51. Inam, A.; Nomaan, S.; Abiodullah, M. Parents' Parenting Styles and Academic Achievement of Underachievers and High Achievers at Middle School Level. *Bull. Educ. Res.* **2016**, *38*, 57–74.

and mental health care have not embraced the BPS model or have even argued against it [12,13]. The argumentation is based on the grounds that its boundaries are not sufficiently specific, especially in terms of the psychological approach to be used [14]; that it is not an appropriately developed model at all, and no proper methods are available for investigating its dimensions [15]. Further counterarguments are that its three main domains are not sufficiently integrated [5], and there is not enough empirical evidence to support it [16].

A life course approach represents an attempted synthesis of models of disease causation integrating biological and social risk processes [17]. This approach allows the study of long-term effects on disease risks of physical and social exposures during gestation, childhood, adolescence, young adulthood, and later adult life. It includes studies of the biological, behavioural, and psychosocial pathways that operate across an individual's life course, as well as across generations, influencing the development of diseases [18]. Life course epidemiology builds and tests theoretical models that postulate pathways linking exposures across the life course to health outcomes later in life [17].

The aim of the present study was to test the validity of the BPS model of health from a life course perspective [18] by using data from a nationwide survey of students in Hungary. In our hypothesised model shown in Figure 1, a life course model of health (including its mental and physical aspects together) was constructed by dividing exposures into biological, psychological (teenage activities and lifestyle), and social (sociodemographic variables and social support) domains in accordance with the BPS model. Investigated exposures were arranged in a temporal (based on their occurrence from childhood to present) and inter-related manner as recommended [17]. Mental health, as an outcome, was measured by various psychological constructs such as the dynamic feeling of confidence measured by sense of coherence, psychological distress, and health locus of control. Physical health was approximated by body mass index, and perceived health, including physical and mental aspects, was used as a summary measure of health. The model was tested separately for women and men since some of the outcome measures have been known to differ by gender.

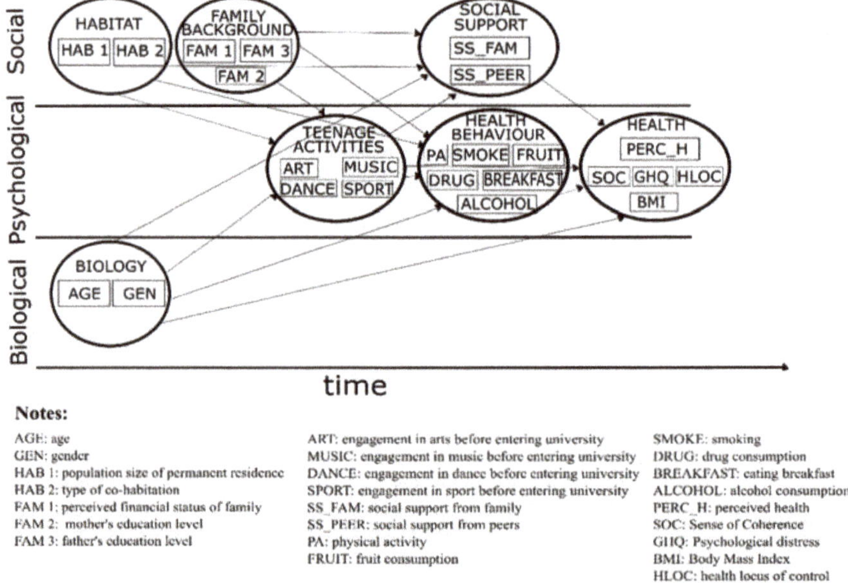

Notes:

AGE: age	ART: engagement in arts before entering university	SMOKE: smoking
GEN: gender	MUSIC: engagement in music before entering university	DRUG: drug consumption
HAB 1: population size of permanent residence	DANCE: engagement in dance before entering university	BREAKFAST: eating breakfast
HAB 2: type of co-habitation	SPORT: engagement in sport before entering university	ALCOHOL: alcohol consumption
FAM 1: perceived financial status of family	SS_FAM: social support from family	PERC_H: perceived health
FAM 2: mother's education level	SS_PEER: social support from peers	SOC: Sense of Coherence
FAM 3: father's education level	PA: physical activity	GHQ: Psychological distress
	FRUIT: fruit consumption	BMI: Body Mass Index
		HLOC: health locus of control

Figure 1. The hypothetical life course model of health for students in higher education.

2. Materials and Methods

2.1. Participants

Sampling was done by stratified systematic sampling as described elsewhere in detail [19]. The sampling frame included students of 27 faculties of the six largest universities and colleges in Hungary (N = 30,901 students) of whom 5% were systematically sampled. 1059 questionnaires were received, yielding a response rate of 68.6%. Of those, 1010 questionnaires (65.4%) were eligible for analysis. The mean age of the students was 23.3 years (age range: 20–49; 97.2% below the age of 30; standard deviation: 2.88); 67% of the respondents were women.

2.2. Tools of Measurement

The questionnaire included domains on demographic (age, gender, population size of permanent residence, type of cohabitation meaning whom the respondent lived with), socioeconomic indicators (parents' education, perceived financial status of family), health, health behaviour (physical activity, diet, smoking, alcohol and drug use), and social support from family and peers. Social support was measured by seven questions, each on a 1–3 scale. Overall sum scores ranged from 7 to 21 (Cronbach's alpha = 0.846). The maximum score of 21 indicated no lack of social support, scores of 18 to 20 indicated a moderate lack of social support, and scores of 17 showed that individuals perceived a severe lack of social support. One item addressed perceived support from peers at the university. Questions not cited separately were adapted from the tool of the Hungarian National Health Interview Survey (HNHIS) 2003 [20] for the purpose of comparisons.

A set of four questions asked respondents about their extracurricular activities before entering higher education, such as dance, art, music, and any kind of sporting activity outside of school-based mandatory physical education. These latter questions were piloted in an earlier small-scale survey (unpublished).

The following variables were used to assess health:

2.2.1. Perceived Health

Perceived health was assessed by a standard question with five response categories from "very bad" to "very good" as recommended by the WHO [21]. Responses were analysed as categorical variables.

2.2.2. Psychological Distress (PD)

PD was measured by the 12-item General Health Questionnaire (GHQ). Items are answered on a 1 to 4-point Likert scale. Cases were detected by scoring in the simplest manner, the so-called usual (0-0-1-1) method which assigns a score of 1 to each existing symptom, while symptoms absent are scored 0. Summarising these yields a score between 0 and 12; notable psychological distress was indicated by scores above 4 [22]. Responses were analysed as binary variables. The Cronbach's alpha of the scale was 0.869.

2.2.3. Sense of Coherence (SoC)

Antonovsky's SoC was quantified by the abbreviated 13-item Orientation to Life questionnaire [23]. Items are answered on a 7-point Likert scale, where the scores of the items 1, 2, 3, 7 and 10 have to be reversed before calculating the total sum score, which varies between 13 and 91. Higher scores indicate stronger levels of sense of coherence. Scores were analysed as continuous variables. The Cronbach's alpha of the scale was 0.829.

Hungarian versions of GHQ and SoC were validated and published earlier [24].

2.2.4. Health Locus of Control (HLoC)

HLoC was approximated by the question of "how much can you do for your health" with four categories of answer from "nothing" to "very much". Responses were analysed as categorical variables.

2.2.5. Body Mass Index (BMI)

The respondents were asked about their body height without shoes and their body weight without clothes. These data were used to calculate BMI with the usual formula (weight (kg)/height (m^2)), which was analysed as a continuous variable.

2.3. Data Collection

Since students at six universities in six cities were included, nationwide data collection was needed for the research. Since neither mailed paper questionnaires alone nor online questionnaires alone were predicted to provide a sufficiently high response rate [25], the optimal method for data collection had to be found. Details of this process were described elsewhere [19]. Briefly, a pilot study was carried out to choose the method which produced the highest response rate in the most cost-effective manner. Based on its results, a combination of postal and Internet-based questionnaires was used for data collection to which a small up-front gift was enclosed, and conditional incentives upon response were promised.

2.4. Construction of the Model

Data were analysed by Stata 10.0 and SPSS 22 software. Since the distribution of the values of psychological distress (test value = 0.122; $p < 0.001$) and sense of coherence (test value = 0.043; $p = 0.001$) were not normal according to the Kolmogorov–Smirnov test, medians were used for point estimation and interquartile range (IQR) for interval estimation of these variables. Scores for SoC and psychological distress were stratified by gender, and medians were compared by the Kruskal–Wallis test. The chi-square and Fisher's exact tests were used to investigate gender differences in terms of categorical variables.

Health as a latent outcome variable was defined by five measured variables of which three related to mental health: SoC as a measure of a dynamic feeling of confidence, GHQ as an approximate measure of psychological distress, and health locus of control approximating the perception of how much control the respondent had over his/her health. Body mass index was used as a measure of physical health, and perceived health was included as a subjective measure of general health. Explanatory variables were grouped into the following latent variables: health behaviour (current physical activity, alcohol, drug consumption, fruit consumption, smoking, eating breakfast); social support (from family and from peers); teenage activities (engagement in sport, arts, dance or music before entering higher education); family background (father's and mother's education, economic status of family); habitat (population size of permanent residence, type of cohabitation in terms of whom the respondent lived with); and biology (age, self-identified gender). Latent variables were arranged on a time axis (time) from childhood to the present on which habitat, family background, and biological factors were defined as (from the present) most distal determinants. Teenage activities were expected to impact upon current health behaviour and social support, whereas all these together thought to impact upon current health status (HEALTH) as shown in Figure 1. We tested this model, including all data using the methods described below, separately for men and women.

At first, canonical correlation analysis was carried out to select variables to be included in the model. Outcome variables constituted the first set, and all other variables were included in the second set in order to determine relationships among these two sets.

The next step was to fit a multivariate analysis of covariance (MANCOVA), keeping in mind its limitation, namely that it is not appropriate for establishing hierarchical relations. The reason to apply MANCOVA was that it fit our purpose better as several groups could be compared with respect to more outcome variables at the same time. We also wished to remove the effect of some concomitant variables called covariates. The effect of a covariate could serve to reduce the error variance of the outcomes. On the other hand, MANCOVA provided the advantage of modelling differences in variances and means over time between groups. The steps of modelling are described in detail below.

2.5. Multivariate Analysis of Covariance (MANCOVA)

The full dataset with 1010 subjects was used for the analysis. Only those outcomes and groups of determinant variables, as well as covariates, were included that remained significant in the previous analysis. The data matrix was first examined for missing data and outliers on the outcomes using the Missing Value Analysis procedure in SPSS 22. Boxplots of the outcomes on the seven groups of determinant variables were also explored. Overall, 116 cases were removed as outliers so that the final sample size was reduced to 894. The proportion of missing values was less than 5% for almost all variables and was replaced by the mean. The assumptions of MANCOVA were tested by a series of univariate analyses of variance (ANOVA). Levene's test was used to check the equality of variance, and Box's M test was used for the homogeneity of covariance assumption. In light of the rather large number of groups, multivariate normality could be assumed because of the central limit theorem. The assumption of low measurement error of the covariates was tested by calculating Cronbach's alpha coefficient. Interaction terms between covariates and groups of determinants were checked by performing univariate covariance analyses and were considered significant at 5%. F-ratio, Wilks's lambda, and eta squared were used to interpret the results of MANCOVA [26].

3. Results

Of those who responded, 65.2% rated their health as very good or good. There was no significant difference in terms of perceived health between men and women ($p = 0.470$).

Regarding psychological distress (GHQ), almost one-quarter (23.6%) of the respondents scored above the cut-off value (4) that indicates unfavourable mental status. The proportion of women with notable psychological distress was significantly higher compared to men (26.5% vs. 17.6%; $p = 0.002$). The median score of men was significantly more favourable, reflecting lower distress than that of women ($p < 0.001$). The median for SoC was 62 points (IQR: 16, min: 21, max: 87), with no significant difference between men and women ($p = 0.862$). There was a marked gender difference in psychological distress, therefore, the model of health was analysed separately for men and women.

3.1. Variable Selection for Modelling "Health"

Based on the results of the canonical correlation analysis, two major dimensions were formulated. The first dimension may be designated as "mental health" that was mainly correlated with drugs (except smoking), social support, and sport in teenage years. The perceived financial status of the family and maternal education were also influential on this dimension (mental health). The second dimension approximated "physical health" that showed a strong relationship with age, drugs (except marijuana) and dancing in teenage years. Based on these results, other variables were omitted from further investigations. Wilk's lambda statistic was 0.601, and the F-statistic also proved to be significant ($p < 0.001$) of all canonical correlations (r1 = 0.518; r2 = 0.282).

As described above, health as a group of outcome variables (HEALTH in Figure 1) was approximated by body mass index (BMI), sense of coherence (SoC), psychological distress (GHQ), health locus of control (HLOC), and perceived health (PERC_H). A reasonable but modest correlation was found among these outcome variables except for body mass index, so this was excluded from the final model. Health locus of control was also excluded from the outcomes in the final model as its correlation with covariates was weak, and it also violated the assumption of equality between-group variances ($p < 0.001$; F = 1.679). Accordingly, sense of coherence, psychological distress, and perceived health were included in the final model.

In addition, six groups of determinant variables (variables of health behaviour such as smoking and fruit consumption, social support from family and support from peers, use of sedatives with or without a prescription and physical activity), and four covariates such as age, teenage experience with dance, sport, and arts were defined. After the inclusion of covariates, multivariate outcomes became much stronger in most cases, and some of the error variances were also reduced. The assumption of homogeneity of the covariances was fulfilled (Box's M: 339.11; F(234,6233) = 1.064; p = 0.243). Reliability analysis for covariates yielded a reasonably high Cronbach's alpha coefficient (0.768).

3.2. Result of MANCOVA for Men

As can be seen from Figure 1, the hypothetical model posited the determinants of health at different time points (see "time" axis). Living conditions, family background, age, and gender were set as the most distal (farthest from the present) determinants, followed by teenage activities (activities before entering university). Health behaviour and social support were experienced while at university. All these impacted the latent variable of actual health as approximated by the three measured variables described above.

All the most distant determinant variables disappeared from the multivariate analysis of covariance for men (R squared = 0.767), leaving only sport and dance as teenage activities the most distant determinants, of which sport impacted indirectly (through interaction with actual physical activity), dance impacted directly on health (Table 1).

Table 1. Effects of health determinants on indicators of health in men.

Variable	Sense of Coherence [1]	Psychological Distress [2]	Perceived Health [3]	Overall Effect on the Multivariate Outcome	Partial Eta Squared [4]
	F-ratios from the Tests of between Subject Effects			p-Value	According to Wilks' Lambda
Intercept	71.254 **	62.273 **	65.218 **	**<0.001**	0.558
Social support from peers at the university	0.702	0.234	0.328	0.855	0.005
Social support from family, friends	3.528 *	0.757	0.258	**0.048**	0.031
Use of sedatives with prescription	0.199	0.367	0.919	0.937	0.005
Use of sedatives without prescription	0.160	0.770	0.725	0.957	0.007
Fruit consumption	2.058	1.666	1.150	0.378	0.021
Physical activity	1.506	1.143	2.262 *	**0.022**	0.052
Smoking	0.693	1.167	2.490 *	**0.022**	0.052
Arts	0.000	1.843	2.258	0.306	0.015
Dance	1.615	2.079	8.679 **	**0.034**	0.035
Sport	0.317	0.442	0.034	0.724	0.005
Age	0.058	0.021	0.699	0.814	0.004
Social support from peers × sport	1.550	1.780	0.252	0.302	0.015
Smoking × sport	0.314	1.178	1.075	0.507	0.019
Physical activity × sport	1.492	2.625 *	1.551	**0.013**	0.039
Social support from family, friends × age	2.024	0.845	0.193	0.341	0.014

Notes: Significant overall effects are marked in bold. [1]: R squared for sense of coherence = 0.331; [2]: R squared for psychological distress = 0.227; [3]: R squared for perceived health = 0.297; [4]: overall R squared = 0.767; * significant at $p < 0.05$; ** significant at $p < 0.01$.

Actual physical activity, smoking, dance during teenage years and social support from their family had a significant positive impact on health measured by sense of coherence, psychological distress, and perceived health in male students (Figure 2).

3.3. Result of MANCOVA for Women

MANCOVA yielded a different result for women (R squared = 0.551). Of the most distant determinant variables, age had a significant direct impact on current health. Dance during teenage years had a significant direct impact on current health, while teenage sport impacted only indirectly through peer support and smoking. Current use of sedatives without prescription, fruit consumption, smoking, social support from family and peers alike had a direct, significant impact on health (Table 2).

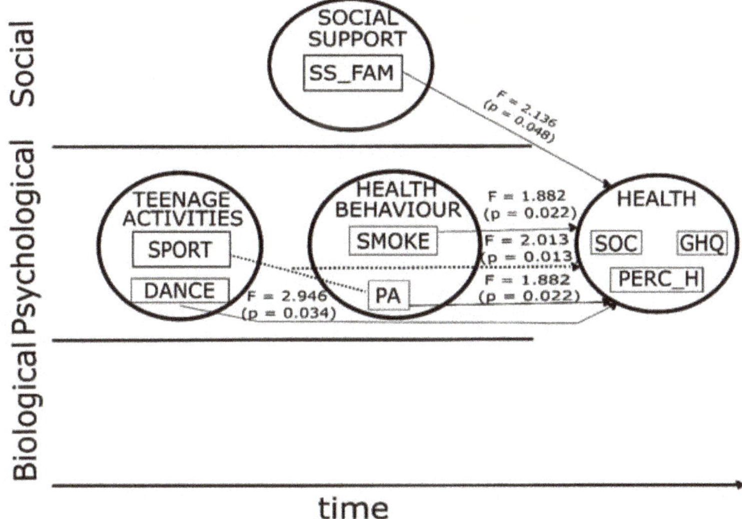

Notes: The values on the arrows represent F-ratio and p-value. Dotted line represents interaction.

DANCE: engagement in dance before entering university
SPORT: engagement in sport before entering university
SS_FAM: social support from family
PA: physical activity
SMOKE: smoking
PERC_H: perceived health
SOC: Sense of Coherence
GHQ: Psychological distress

Figure 2. Best fit model for the relationship between health and its determinants in male students.

All determinants—except current physical activity—found significant in men were also significant in women as well, but in addition, age, social support from peers, fruit consumption, and use of sedatives without prescription (drug use) also remained in the model. In contrast to male students, female students' health seemed to be more obviously multifactorial (compare Figure 3 to Figure 2).

Table 2. Effects of health determinants on indicators of health in women.

Variable	Sense of Coherence [1]	Psychological Distress [2]	Perceived Health [3]	Overall Effect on the Multivariate Outcome	Partial Eta Squared [4]
	F-Ratios from the Tests of between Subject Effects			p-Value	According to Wilks' Lambda
Intercept	82.360	130.549	42.212	**<0.001**	0.401
Social support from peers at the university	4.827 **	0.183	0.764	**0.035**	0.012
Social support from family, friends	1.626	6.084 **	2.011	**0.021**	0.013
Use of sedatives with prescription	1.294	0.999	2.791 *	0.149	0.012
Use of sedatives without prescription	4.761 **	2.545 *	3.085 *	**0.001**	0.020
Fruit consumption	2.094	3.658 **	4.500 **	**0.001**	0.023
Physical activity	2.698 *	1.887	1.831	0.084	0.014
Smoking	1.819	1.320	2.830 *	**0.018**	0.017
Arts	1.295	1.432	0.101	0.565	0.004
Dance	0.074	7.438 **	0.078	**0.024**	0.017
Sport	0.078	0.882	2.119	0.364	0.006
Age	7.353 **	1.377	5.580 *	**0.014**	0.019
Social support from peers × sport	0.529	4.288 *	1.754	**0.021**	0.013
Smoking × sport	1.783	1.007	3.579 **	**0.030**	0.016
Physical activity × sport	2.211	1.289	0.405	0.378	0.010
Social support from family, friends × age	0.263	4.182 *	1.241	0.135	0.009

Notes: Significant overall effects are marked in bold. [1]: R squared for sense of coherence = 0.331; [2]: R squared for psychological distress = 0.227; [3]: R squared for perceived health = 0.297; [4]: overall R squared = 0.767; * significant at $p < 0.05$; ** significant at $p < 0.01$.

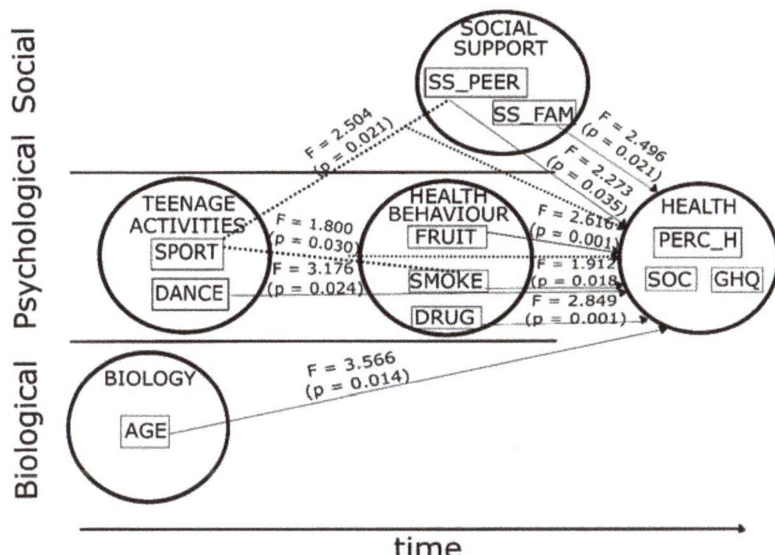

Notes: The values on the arrows represent F-ratio and p-value. Dotted lines represent interactions.

AGE: age
DANCE: engagement in dance before entering university
SPORT: engagement in sport before entering university
SS_FAM: social support from family
SS_PEER: social support from peers
FRUIT: fruit consumption

SMOKE: smoking
DRUG: drug consumption
PERC_H: perceived health
SOC: Sense of Coherence
GHQ: Psychological distress

Figure 3. Best fit model for the relationship between health and its determinants in female students.

4. Discussion

The biopsychosocial model of health with a limited life course perspective was found to be a useful point of departure to investigate the relationship between various health determinants and health outcomes in a large sample of university students by multivariate analysis of covariance. Considerably different sets of health determinants were identified for the two genders. Male students' health was determined by fewer variables that are more proximal in time (closer to the present), more centred around physical activity, and less influenced by social relations. Female students' health, as opposed to that of men, was influenced by age; more determinants were grouped around the ingestion of various substances and social support. Our final models did not contradict our hypothetical model (shown in Figure 1) and were in line with other findings regarding gender differences in the genesis of health [27,28]. Regarding the possible reasons why are there fewer variables involved in men's health and more that influence women's health, only speculation can be offered. Gender is a socioeconomic determinant of health according to the World Health Organization [29]. Women's quality of life and opportunities have still been lagging behind that of men, which is why gender equality and women's empowerment has been one of the 17 sustainable development goals of the United Nations [30]. Disadvantage and inequality due to being a woman results in greater difficulties to satisfy needs that tend to call for greater efforts to mobilise resources. Therefore, a gender approach can be helpful in a number of areas, for example, it can accelerate the prevention of noncommunicable diseases [31–35].

One of the strengths of our model derives from the fact that determinants of health were measured by a rather comprehensive set of variables. Another strength is the inclusion of positive (sense of coherence) and negative (psychological distress) measures of mental health along with perceived health, a reliable measure of general health [36]. The BPS model in a life-course perspective allowed the creation of an easy-to-comprehend visual guide (model). The generalizability of our results is supported by the rather large dataset from a nationwide sample of students on track to become teachers at the largest universities of the country, but limitations following from self-reported assessment tools have to be taken into consideration when interpreting the results.

Optimal modelling should include longitudinally generated temporal [17] and hierarchical [37] data. One shortcoming is the cross-sectional nature of data collection in our study. However, this was countered by specific items that related to earlier periods in the students' lives. Data were not representative for all students in higher education in Hungary, but since the effect size measures of the model were acceptable, especially in light of the complexity of the latent outcome variable (health) and the high number of explanatory variables, general conclusions may be drawn. The generalisability of our results should be confirmed in other populations, including older persons. Men and women were not different in terms of sense of coherence, but—though women had greater social support and used more coping techniques—a greater proportion of women were overstressed compared to men. Having a wide range of coping techniques is of special importance for students preparing to become teachers because teaching—along with working in prisons and being a police officer—is one of the most stressful occupations in modern times [38]. The difference between the genders in terms of social support is in line with previous studies, where perceived social support from family or friends was more important for women compared to men [39]. Health behaviour, reflecting a particular combination of more or less effective coping techniques [40], has been shown to be a strong predictor of morbidity and mortality [41,42]. Interventions to change health behaviour at universities need to be targeted, for which our results offer a practical guide in addition to arguing for the usefulness of the biopsychosocial model of health in research aimed at uncovering the relationship between health determinants and outcomes. We hope that our study will be one of those that help "turn meaningful epidemiologic questions into studies that provide useful epidemiologic answers" [43], especially in terms of whether gender-specificity or gender-equality in health-promoting interventions are needed and how to develop proper interventions based on evidence [44]. Answers for these questions will be needed to achieve Goal 5 specifying gender equality of the Sustainable Development Goals of the United Nations [30].

5. Conclusions

The positive effects of social support on health, along with gender differences in health, are well known, but less is known about the specifics of gender differences on health behaviour among university students. A multivariate model of the determinants of health in a life course perspective was investigated with data obtained from college students at major Hungarian universities. Notable differences between health determinants of men and women were uncovered. Male students' health is determined by a fewer number of variables that are more proximal (closer to the present), more focused on physical activity, and less influenced by social relations. The health of female students is influenced by more determinants, age being one of them; determinants are related to oral consumption of various substances and social support. Our final models are in agreement with our hypothetical model and the findings of others, adding further evidence to the notion that gender differences exist in health determinants that should be taken into account when planning interventions.

Author Contributions: Conceptualisation, K.K. and I.V.-B.; methodology, K.K. and S.K.; formal analysis and review; É.B. and S.K.; resources, data collection, K.K.; preparation of the first draft, editing, review, É.B.; editing, review, K.K.; review, I.V.-B. and R.Á.; visualization, S.K.; funding acquisition, R.Á. All authors have read and agreed to the submitted version of the manuscript.

Funding: Data collection for funding was provided by the Ministry of Education, Hungary [NKFP1-00003/2005]. Writing of the manuscript was carried out in the framework of GINOP-2.3.2-15-2016-00005 project. This project is co-financed by the European Union under the European Regional Development Fund. The funders had no role in the design of the study, in the collection, analyses, or interpretation of data; in the writing of the manuscript, or in the decision to publish the results.

Institutional Review Board Statement: The study was conducted according to the guidelines of the Declaration of Helsinki and approved by the Commission on Research Ethics of the Medical and Health Science Centre of the University of Debrecen, Hungary (DEOEC RKEB/IKEB: 2506-2006).

Informed Consent Statement: The students were informed in writing that participation was anonymous and voluntary at the beginning of the questionnaire. No personal data were collected, so a consent form was not requested to be signed by the ethics committee. If participants posted back the questionnaire or filled out the online form, they were treated as having given consent.

Data Availability Statement: The data supporting the findings of this study are available from the corresponding author upon reasonable request.

Conflicts of Interest: The authors declare no conflict of interest.

References

1. Engel, G. The need for a new medical model: A challenge for biomedicine. *Science* **1977**, *196*, 129–136. [CrossRef]
2. Borrell-Carrió, F.; Suchman, A.L.; Epstein, R.M. The biopsychosocial model 25 years later: Principles, practice, and scientific inquiry. *Ann. Fam. Med.* **2004**, *2*, 576–582. [CrossRef]
3. McCollum, L.; Pincus, T. A biopsychosocial model to complement a biomedical model: Patient questionnaire data and socioeconomic status usually are more significant than laboratory tests and imaging studies in prognosis of rheumatoid arthritis. *Rheum. Dis. Clin. N. Am.* **2009**, *35*, 699–712. [CrossRef] [PubMed]
4. Adler, R.H. Engel's biopsychosocial model is still relevant today. *J. Psychosom. Res.* **2009**, *67*, 607–611. [CrossRef]
5. Hatala, A.R. The status of the "biopsychosocial" model in health psychology: Towards an integrated approach and a critique of cultural conceptions. *Open J. Med. Psychol.* **2012**, *1*, 51–62. [CrossRef]
6. Rodgers, R.F.; Paxton, S.J.; McLean, S.A. A biopsychosocial model of body image concerns and disordered eating in early adolescent girls. *J. Youth Adolesc.* **2014**, *43*, 814–823. [CrossRef] [PubMed]
7. World Health Organization. Towards a common language for functioning, disability and health: ICF. In *The International Classification of Functioning, Disability and Health*; World Health Organization: Geneva, Switzerland, 2002.
8. Becker, S.P.; Langberg, J.M.; Byars, K.C. Advancing a biopsychosocial and contextual model of sleep in adolescence: A review and introduction to the special issue. *J. Youth Adolesc.* **2015**, *44*, 239–270. [CrossRef]
9. Mitra, S.; Shakespeare, T. Remodeling the ICF. *Disabil. Health J.* **2019**, *12*, 337–339. [CrossRef]
10. McDermott, S.; Turk, M.A. Support for reevaluation of the International Classification of functioning, disability, and health (ICF): A view from the US. *Disabil. Health J.* **2019**, *12*, 137–138. [CrossRef] [PubMed]
11. Henningsen, P. Still modern? Developing the biopsychosocial model for the 21st century. *J. Psychosom. Res.* **2015**, *79*, 362–363. [CrossRef] [PubMed]
12. Ghaemi, S.N. The rise and fall of the biopsychosocial model. *Br. J. Psychiatry* **2009**, *195*, 3–4. [CrossRef]
13. Alvarez, A.S.; Pagani, M.; Meucci, P. The clinical application of the biopsychosocial model in mental health: A research critique. *Am. J. Phys. Med. Rehabil.* **2012**, *91*, S173–S180. [CrossRef]
14. Ghaemi, S.N. The biopsychosocial model in psychiatry: A critique. *Existenz* **2011**, *6*, 1–8.
15. McLaren, N. A critical review of the biopsychosocial model. *Aust. N. Z. J. Psychiatry* **1998**, *32*, 86–92. [CrossRef] [PubMed]
16. Pilgrim, D. The biopsychosocial model in Anglo-American psychiatry: Past, present and future? *J. Ment. Health* **2002**, *11*, 585–594. [CrossRef]
17. Kuh, D.; Ben-Shlomo, Y.; Lynch, J.; Hallqvist, J.; Power, C. Life course epidemiology. *J. Epidemiol. Community Health* **2003**, *57*, 778–783. [CrossRef] [PubMed]
18. Ben-Shlomo, Y.; Kuh, D. A life course approach to chronic disease epidemiology: Conceptual models, empirical challenges and interdisciplinary perspectives. *Int. J. Epidemiol.* **2002**, *31*, 285–293. [CrossRef]
19. Balajti, I.; Daragó, L.; Ádány, R.; Kósa, K. College students' response rate to an incentivized combination of postal and web-based health survey. *Eval. Health Prof.* **2010**, *33*, 164–176. [CrossRef]
20. Országos Epidemiológiai Központ. *Önkitöltős kérdőív Országos Lakossági Egészségfelmérés 2003*; Self-Administered Questionnaire Hungarian National Health Interview Survey 2003; Országos Epidemiológiai Központ: Budapest, Hungary, 2003.
21. De Bruin, A.; Picavet, H.S.J.; Nossikov, A. *Health Interview Surveys: Towards International Harmonization of Methods and Instruments. WHO Regional Publications European*; Series No. 58; World Health Organization Regional Office for Europe: Copenhagen, Denmark, 1996.
22. Goldberg, D.P.; Gater, R.; Sartorius, N.; Ustun, T.B.; Piccinelli, M.; Gureje, O.; Rutter, C. The validity of two versions of the GHQ in the WHO Study of Mental Illness in General Health Care. *Psychol. Med.* **1997**, *27*, 191–197. [CrossRef]

23. Antonovsky, A. *Unraveling the Mystery of Health*; Jossey-Bass Publishers: San Francisco, CA, USA, 1987.
24. Balajti, I.; Vokó, Z.; Ádány, R.; Kósa, K. A rövidített koherencia-érzés és az általános egészség (GHQ) kérdőívek magyar nyelvű változatának validálása (Validation of the Hungarian versions of the abbreviated sense of coherence (SoC) and the general health questionnaire (GHQ-12)). *Mentálhig. Pszichoszomat* **2007**, *8*, 147–161. [CrossRef]
25. Bälter, K.A.; Bälter, O.; Fondell, E.; Lagerros, Y.T. Web-based and mailed questionnaires: A comparison of response rates and compliance. *Epidemiology* **2005**, *16*, 577–579. [CrossRef] [PubMed]
26. Field, A. *Discovering Statistics Using SPSS*; SAGE Publications Ltd.: London, UK, 2009.
27. Vlassoff, C. Gender differences in determinants and consequences of health and illness. *J. Health Popul. Nutr.* **2007**, *25*, 47–61.
28. Denton, M.; Prus, S.; Walters, V. Gender differences in health: A Canadian study of the psychosocial, structural and behavioural determinants of health. *Soc. Sci. Med.* **2004**, *58*, 2585–2600. [CrossRef]
29. Commission on Social Determinants of Health. *Closing the Gap in a Generation: Health Equity through Action on the Social Determinants of Health*; Final Report of the Commission on Social Determinants of Health; World Health Organization: Geneva, Switzerland, 2008; Available online: http://whqlibdoc.who.int/publications/2008/9789241563703_eng.pdf (accessed on 25 June 2021).
30. Sustainable Development Goals. United Nations 2015. Available online: https://sdgs.un.org/goals/goal5 (accessed on 25 May 2021).
31. World Health Organization. *Why Using a Gender Approach Can Accelerate Noncommunicable Disease Prevention and Control in the WHO European Region*; WHO Regional Office for Europe: Copenhagen, Denmark, 2019; Available online: https://www.euro.who.int/__data/assets/pdf_file/0003/399063/GenderApproachAndNCDsPreventionAndControl-eng.PDF (accessed on 25 June 2021).
32. A Look into Masculine Identity in Mexican Young Men. Available online: http://editorial.upnvirtual.edu.mx/index.php/publicaciones/descargas/category/1-pdf?download=423:a-look-into-masculine-identity (accessed on 25 June 2021).
33. García-Vega, E. Acerca del género y la salud. *Pap. Psicól.* **2011**, *32*, 282–288.
34. Aznar, M.; Pilar, M. Género y salud. *Suma Psicol* **2008**, *15*, 75–94.
35. Valls Llobet, C.; Banqué, M.; Fuentes, M.; Ojuel, J. Morbilidad diferencial entre mujeres y hombres. *Anu. Psicol.* **2008**, *39*, 9–22. [CrossRef]
36. Mackenbach, J.P.; Simon, J.G.; Looman, C.W.; Joung, I.M. Self-assessed health and mortality: Could psychosocial factors explain the association? *Int. J. Epidemiol.* **2002**, *31*, 1162–1168. [CrossRef]
37. Greenland, S. Principles of multilevel modelling. *Int. J. Epidemiol.* **2000**, *29*, 158–167. [CrossRef]
38. Johnson, S.; Cooper, C.; Cartwright, S.; Donald, I.; Taylor, P.; Millet, C. The experience of work-related stress among occupations. *J. Manag. Psychol.* **2005**, *20*, 178–187. [CrossRef]
39. Lee, C.Y.; Goldstein, S.E. Loneliness, Stress, and Social Support in Young Adulthood: Does the Source of Support Matter? *J. Youth Adolesc.* **2016**, *45*, 568–580. [CrossRef]
40. Lehrer, P.M.; Woolfolk, R.L. *Principles and Practice of Stress Management*, 3rd ed.; Sime, W.E., Ed.; The Guilford Press: New York, NY, USA, 2007.
41. World Health Organization. *2008–2013 Action Plan for the Global Strategy for the Prevention and Control of Noncommunicable Diseases*; World Health Organization: Geneva, Switzerland, 2008.
42. Stringhini, S.; Dugravot, A.; Shipley, M.; Goldberg, M.; Zins, M.; Kivimäki, M.; Marmot, M.; Sabia, S.; Singh-Manoux, A. Health behaviours, socioeconomic status, and mortality: Further analyses of the British Whitehall II and the French GAZEL prospective cohorts. *PLoS Med.* **2011**, *8*, e1000419. [CrossRef] [PubMed]
43. Kaufman, J.S.; Hernán, M.A. Epidemiologic methods are useless: They can only give you answers. *Epidemiology* **2012**, *23*, 785–786. [CrossRef] [PubMed]
44. Tang, K.C.; Ehsani, J.P.; McQueen, D.V. Evidence based health promotion: Recollections, reflections, and reconsiderations. *J. Epidemiol. Community Health* **2003**, *57*, 841–843. [CrossRef] [PubMed]

International Journal of
Environmental Research and Public Health

MDPI

Article

The Effect of Changes in Physical Self-Concept through Participation in Exercise on Changes in Self-Esteem and Mental Well-Being

Inwoo Kim [1] and Jihoon Ahn [2],*

[1] Department of Sports Culture, Dongguk University, Seoul 04620, Korea; iwkim@dongguk.edu
[2] Department of Physical Education, Seoul National University, Seoul 08826, Korea
* Correspondence: dkswlgns88@snu.ac.kr; Tel.: +82-2-880-7798

Abstract: The aim of the present study was to examine the impact of the changes in physical self-concept induced by exercise participation on the changes in global self-esteem and mental well-being using a structural model analysis. A total of 189 university students in Seoul, Korea, participated in the present study for two waves. The participants responded through a survey measuring physical self-concept, self-esteem, and mental well-being before and after a six-week exercise course. Regression analysis was used to calculate the amount of change in each variable, and the calculated residual scores were used for correlation analysis and structural model analysis. The amounts of changes in the variables are significantly correlated with each other and there was a complementary mediating effect of the changes in self-esteem on the pathway from the changes in physical self-concept to the changes in mental well-being. Physical self-concept changed by exercise participation might directly and positively influence mental well-being, and it can indirectly influence the changes in mental well-being via the improvement of self-esteem.

Keywords: exercise participation; physical self-concept; self-esteem; mental well-being

Citation: Kim, I.; Ahn, J. The Effect of Changes in Physical Self-Concept through Participation in Exercise on Changes in Self-Esteem and Mental Well-Being. *Int. J. Environ. Res. Public Health* **2021**, *18*, 5224. https://doi.org/10.3390/ijerph18105224

Academic Editors: Luciana Zaccagni and Emanuela Gualdi-Russo

Received: 9 April 2021
Accepted: 13 May 2021
Published: 14 May 2021

Publisher's Note: MDPI stays neutral with regard to jurisdictional claims in published maps and institutional affiliations.

1. Introduction

Physical activity and exercise are known to be beneficial to health through preventing diseases, enhancing physical abilities, diminishing depression, and promoting happiness [1]. Positive psychologists who have focused on psychological and subjective well-being also emphasized that exercise is one of the important factors in a happy life [2], which has led to a growing interest in the relationship between physical activity and happiness.

Even though physical activity is widely known to be an important factor in reducing negative emotions and leading to a happy life, many people still do not exercise. In particular, the lack of exercise among young adults has been remarkable. According to a study, the level of physical activity among university students was lower than that of adolescents [3], and there is also research showing that the sedentary lifestyle of college students limits the amount of physical activity and causes many health problems [4]. This period is the transition from adolescence to adulthood, which is a very important period for the formation of health behaviors that can be maintained throughout life [5]. Therefore, it is important to encourage university students to participate in sports by emphasizing that exercise is a crucial element of a happy life.

According to existing research related to the correlation between well-being and exercise participation, physical activity can positively influence mental health by reducing negative depression and anxiety and enhancing self-esteem and self-concept [6]. Especially physical self-concept as an important element of the global self-concept has been widely researched in the fields of sports and exercise psychology related to exercise and well-being.

Physical self-concept is defined as one's perception or evaluation of their physical ability and physical appearance, and it is one of the sub-factors of the global self-esteem

119

with social self-concept and emotional self-concept [7]. Fox and Corbin [7] developed the Physical Self-Perception Profile (PSPP), and identified four sub-factors of the PSPP, including sports competence, attractive body, physical strength, and physical condition. Marsh, Richards, Johnson, Roche and Tremayne [8] redefined the PSPP in terms of eleven subdomains by dividing the four factors in the PSPP and adding self-esteem and general physical self-concept. They developed the Physical Self-Description Questionnaire (PSDQ). Thus, physical self-concept is explained by several sub-factors and it explains one's general self-perception at the same time in three levels of hierarchical structure.

Many studies have reported that the physical self-concept is positively related with happiness and well-being. Fox [9,10] identified that self-perception of body is essential for mental health and well-being, and Morales-Rodríguez and colleagues [11] reported that physical self-concept is one of the psychosocial factors of the psychological well-being in university students. Roh [12] assured that the perception of one's physical state is correlated with health perception and psychological well-being, and Kim and Oh [13] also reported by a meta-analysis that there was a positive correlation between physical self-perception and happiness. In other words, existing research has shown that there is a general consensus that developing a positive self-concept is helpful for a happy life.

Although the relationship between physical self-concept and happiness is being discussed from a positive psychological perspective, existing research has been conducted with a limited selection of some elements of happiness, such as psychological well-being, or life satisfaction. However, positive psychologists have defined happiness as a multidimensional concept that includes subjective well-being defined as affirmation and satisfaction of life [14] and psychological well-being [15] pursuing self-actualizing happiness. Moreover, Keyes [16] redefined subjective well-being as emotional well-being and suggested an integrated concept of mental health by adding social well-being to the existing psychological well-being. A recent study insists that, to fully understand mental health, it is necessary to consider the multidimensional nature of well-being, and many existing studies that do not take this into account have missed the key facets of well-being [17]. In other words, the concept of mental health, which includes all aspects of emotional, psychological, and social aspects, is widely applied in studies related to happiness, while existing studies on physical self-concept and well-being did not consider this, so they might have missed important factors in mental health.

To examine the relationship between mental health and physical self-concept, it might be necessary to examine one's self-esteem. According to early research, developing the physical self-concept could induce the enhancement of one's self-esteem because it is one of the sub-factors of global self-esteem [7,18]. A study from Garn and colleagues [19] reported that students' physical self-concept could influence their general self-esteem, and Harther [20] has also supported that physical self-esteem is a major factor influencing global self-esteem. Such implications mean that the change of one's perception of physical worth could induce positive changes of overall self-perception or self-esteem.

Physical activity or exercise may play a role in developing physical self-worth. Babic, Morgan, Plotnikoff, Lonsdale, White, and Lubans [21] revealed by a meta-analysis that various perceptions about the body were strongly related to physical activity, and that strategic participation in exercise could develop certain factors of the physical self-concept. In addition, a recent Spanish study identified that physical activity can help adolescents develop a positive self-concept and improve psychological well-being through satisfaction with their body and improvement of their physical self-concept [22]. On the contrary, it was also said that the improvement of self-perception about sports competence and physical fitness can act in the direction of promoting participation in exercise. Thus, it can be said that physical activity can be an antecedent variable of the physical self-concept or an outcome variable [23], which might mean that the causal relationship between the two variables remains ambiguous. This suggests that a longitudinal design, not a cross-sectional, is needed to examine the causal relationship between exercise participation and physical self-concept.

To sum up, physical activity and exercise are known to contribute to the improvement of happiness and quality of life by positively influencing the physical self-concept and self-esteem. However, despite these advantages, many people still do not exercise, especially university students, who are at a crucial time in the formation of lifelong health behaviors, do not participate enough in exercise. Existing studies have a limitation in that they did not apply the multidimensional concept of happiness defined from positive psychology. Additionally, since the causality of participation in exercise and physical self-concept is unclear, a longitudinal study is needed to examine that the physical self-concept changes through exercise and affects self-esteem and happiness.

The purpose of the present study was to examine the effect of physical self-concept changed by participation in exercise on the changes in self-esteem and mental health in university students. To consider the multidimensional aspect of happiness, we applied the concept of mental health including emotional well-being, psychological well-being, and social well-being suggested by Keyes [16]. Moreover, a longitudinal design was used to examine the changes of other variables due to participation in the exercise. This study extends the knowledges about how the changes in self-awareness through exercise can affect various aspects of mental health, and through this, it provides implications for the importance and direction of participation in exercise for young students' happy life.

2. Methods

2.1. Participants

A total of 189 students from a university located in Seoul, Korea, who are enrolled in physical fitness classes participated in this study. The participants were sampled through the convenient sampling method, and only the students who did not exercise at all outside of class participated in the survey. There were three classes taught by three lecturers but the teaching methods and the contents of all classes consisted equally with various kinds of exercises, such as weightlifting, aerobic exercise, core training, and stretching. The students' majors varied, but none of them majored in physical education or sports science, and they had no experience taking physical fitness classes in the past. The survey was conducted twice before and after the classes, and 161 students participated in both surveys (102 males and 59 females; M age = 21.8, SD age = 2.1). There were no significant differences between classes and gender in the means of each variable.

2.2. Study Procedure

Among the students who voluntarily enrolled in the physical fitness classes, the students who did not exercise outside of the class were encouraged to participate in the study. Prior to conducting the survey, permissions were obtained from the instructor and the internal review board of the researchers' university, and the questionnaire was distributed only to students who voluntarily agreed to participate in the survey. The first survey was conducted before starting the classes, and the second survey was conducted after the physical fitness classes, which were conducted twice a week for 6 weeks and 100 min per session. The questionnaire was composed of questions that measure physical self-concept, self-esteem, and mental health, and it takes about 15 min to complete.

2.3. Measures

Physical self-concept. The Korean version of the Physical Self-Description Questionnaire (K-PSDQ) [24] translated from the original version developed by Marsh, Richards, Johnson, and Roche [8] was used to measure the participants' physical self-concept. The K-PSDQ consists of 4 questions per factor, a total of 40 questions for 10 factors, but only four factors that can change through physical fitness classes were selected and used in the study, including strength, endurance, flexibility, and sports competence. Examples of the questions for each factor include: "I am stronger than most people my age" for strength, "I can run a long way without stopping" for endurance, "I am quite good at bending, twisting, and turning my body" for flexibility, and "I am good at sports" for sports competence. A confirmatory factor analysis on the four-factor model with 16 items was conducted, and the result revealed that the model fit was acceptable with the fit indices (χ^2 = 186.546, df = 98, $p < 0.001$, CFI = 0.962, TLI = 0.954, RMSEA = 0.076). Each item was coded on a 6-point Likert scale (1 = absolutely not, 6 = very much so). The internal consistency of the scale was also acceptable, with the Cronbach's alpha coefficients ranging from 0.881 to 0.946.

Self-esteem. To measure self-esteem of the participants, the Korean version of the Rosenberg Self-Esteem Scale (K-RSE) was used [25]. The K-RSE consists of 9 items with 2 factors, excluding item 8 from the original scale, which has been verified as inappropriate due to cultural differences in several studies [25]. Each question was rated on a 4-point Likert scale ranging from 1 (not at all) to 4 (very much), and negative items were reverse-coded so that a high score means high self-esteem. As a result of the confirmatory factor analysis, the construct validity of the two-factor model of the K-RSE was verified with the fit indices (χ^2 = 54.758, df = 26, $p = 0.001$, TLI = 0.952, CFI = 0.965, RMSEA = 0.083), and Cronbach's alpha coefficients were 0.87 and 0.875.

Mental health. The Korean version of the Mental Health Continuum-Short Form (K-MHC–SF) was used to measure the participants' mental health [26]. The K-MHC–SF is a 14-item scale consisting of three factors: emotional well-being (3 items), social well-being (5 items), and psychological well-being (6 items). Examples of the questions for each factor include: "I felt a sense of happiness" (emotional well-being), "I felt that our society was becoming a better place to live" (social well-being), "I felt that my life had a sense of direction or meaning" (psychological well-being), and each question was answered with the frequency experienced over the past month coded on a 6-point Likert scale (1 = not at all, 2 = once or twice, 3 = about once a week, 4 = about two or three times a week, 5 = almost every day, 6 = every day). As a result of conducting a confirmatory factor analysis on 14 items of 3 factors to verify the construct validity of the scale, two items for social well-being with a standardized regression coefficient of less than 0.5 were deleted. The results of the second confirmatory factor analysis for 12 items revealed that the construct validity was acceptable (χ^2 = 105.036, df = 51, $p < 0.001$, TLI = 0.944, CFI = 0.956, RMSEA = 0.081), and the internal consistency of the scale was also acceptable, with the Cronbach's alpha coefficients ranging from 0.839 to 0.935.

2.4. Analysis

In this study, the AMOS 18.0 and SPSS 22.0 programs were used to analyze the collected data. First, a descriptive statistical analysis was conducted to understand the overall trends and characteristics (mean and standard deviation, etc.) of the participants. Surveys were conducted twice at intervals of six weeks, and to analyze the individual level of changes in the three variables (physical self-concept, self-esteem, and mental health), residual scores were generated and used for the main analysis. Based on the suggestion from Smith and Beaton [27], after inputting the measurement values in the first wave to the independent variable of the regression equation and the measurement values in the second wave to the dependent variable, the standardized residuals were calculated by regression analysis. Pearson's correlation analysis was performed to examine the correlations between the standardized residuals, which mean the amount of change in each variable.

Structural equation modeling was used to examine the structural relationship between the amount of change in the variables. Specifically, we tried to test a model in which changes in self-esteem mediate the relationship between changes in physical self-concept and changes in the three well-beings of mental health. To verify the mediating effect, we used the χ^2 difference test and bootstrapping, which is a method to verify the confidence level of the indirect effect. At this time, the confidence interval was set to 95% and the number of repetitions was set to 2000.

3. Results

3.1. Descriptive Statistics

Table 1 shows the results of descriptive statistics for the general tendency and normality of the variables measured in this study. Although some SD values were less than 1.0, the skewness and kurtosis were below the reference value (≤ 2, ≤ 7, respectively) confirming the normality of the data [28]. The means of all variables were higher in the second wave than in the first wave, and as a result of the paired sample t-test, all the differences were statistically significant.

Table 1. Descriptive statistics and paired sample t-test.

	Wave 1				**Wave 2**				t
	M	**SD**	**Skewness**	**Kurtosis**	**M**	**SD**	**Skewness**	**Kurtosis**	
sport competence	3.01	1.13	0.09	−0.72	3.44	1.15	−0.18	−0.52	−5.72 **
flexibility	3.04	1.13	0.12	−0.62	3.36	1.19	0.25	−0.60	−4.12 **
endurance	2.76	1.24	0.33	−0.74	3.17	1.22	0.12	−0.62	−5.96 **
strength	2.78	0.91	−0.02	−0.54	3.25	0.97	−0.10	−0.30	−7.00 **
positive self-esteem	3.14	0.50	−0.01	−0.04	3.24	0.49	−0.25	0.06	−2.84 **
negative self-esteem	3.34	0.62	−0.85	0.21	3.44	0.60	−1.04	1.00	−2.37 *
emotional well-being	4.20	0.87	−0.39	0.59	4.54	0.90	−0.50	1.00	−4.95 **
social well-being	3.38	1.01	0.18	0.05	3.71	1.06	0.03	−0.26	−4.30 **
psychological well-being	4.02	0.89	−0.21	−0.01	4.33	0.93	−0.33	−0.03	−4.73 **

$* p < 0.05$, $** p < 0.01$.

3.2. Correlations between the Amount of Change in the Measured Variables

The correlations between the amount of change in variables were analyzed and represented in Table 2. As mentioned above, the standardized residuals of the measurement variables were generated and used as the amount of change in each variable. Table 2 shows that flexibility had a significant correlation with all variables except negative self-esteem and emotional well-being, and muscle strength was significantly correlated with all variables except negative self-esteem. All other variables were found to have a significant correlation between each other.

Table 2. Correlation matrix.

	ΔPhysical Self-Concept				**ΔSelf-Esteem**		**ΔMental Health**		
	1.	**2.**	**3.**	**4.**	**5.**	**6.**	**7.**	**8.**	**9.**
1. sport competence	1								
2. flexibility	0.32 **	1							
3. endurance	0.58 **	0.34 **	1						
4. strength	0.63 **	0.47 **	0.45 **	1					
5. positive self-esteem	0.32 **	0.25 **	0.32 **	0.33 **	1				
6. negative self-esteem	0.23 *	0.04	0.19 *	0.11	0.45 **	1			
7. emotional well-being	0.38 **	0.11	0.28 **	0.26 **	0.39 **	0.36 **	1		
8. social well-being	0.23 **	0.18 **	0.27 **	0.26 **	0.21 **	0.24 **	0.39 **	1	
9. psychological well-being	0.40 **	0.26 **	0.28 **	0.38 **	0.45 **	0.40 **	0.66 **	0.42 **	1

$* p < 0.05$, $** p < 0.01$.

3.3. Testing the Adequacy of the Research Model

The structural equation model analysis was conducted in which the relationship between changes in physical self-concept and changes in mental health is mediated by the amount of change in self-esteem (Figure 1). The research model showed an acceptable fit to the data with the model fit indices ($\chi^2 = 39.805$, df = 24, $p < 0.05$, TLI = 0.946, CFI = 0.964, RMSEA = 0.064, SRMR = 0.046). Each path coefficient is shown in Figure 1.

It was identified that the amount of change in physical self-concept significantly predicted the amount of change in self-esteem ($\beta = 0.506$, $p < 0.01$), which is the mediating variable, and the amount of change in mental well-being ($\beta = 0.287$, $p < 0.05$), which is the dependent variable. In addition, the amount of change in self-esteem was found to significantly predict the amount of change in mental well-being ($\beta = 0.565$, $p < 0.01$).

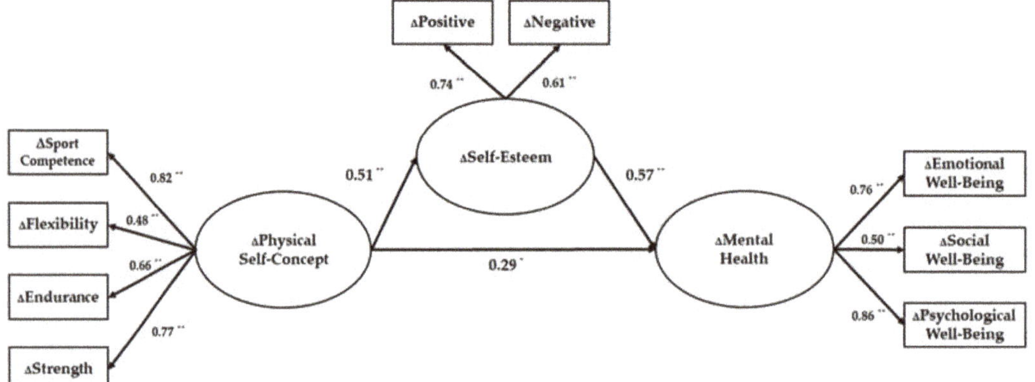

Figure 1. Relationship between change in physical self-concept, change in self-esteem, and change in mental health, * $p < 0.05$, ** $p < 0.01$.

3.4. Mediating Effect of the Amount of Change in Self-Esteem

To verify the mediating effect of the amount of change in self-esteem in the relationship between the amount of change in physical self-concept and the amount of change in mental well-being, the complete mediation model was set as a competition model, and a χ^2 difference test with the research model was conducted. Table 3 shows the results of comparing the fit of the research model and the fit of the fully mediated model constraining the effect of the change in physical self-concept on the change in mental well-being to "0".

Table 3. Comparison of fit indices between the research model and the competition model.

	χ^2	df	p	TLI	CFI	RMSEA	SRMR
Research model	39.805	24	<0.05	0.946	0.964	0.064	0.046
Competition model	45.159	25	<0.05	0.934	0.954	0.071	0.054

As shown in Table 3, there was a statistically significant difference in the value of χ^2 between the research model and the competition model ($\Delta\chi^2 = 5.353$, $\Delta df = 1$, $p = 0.021$). This means that the amount of change in self-esteem partially mediated the relationship between the amount of change in physical self-concept and the amount of change in mental well-being.

To verify the specific mediating effect, an indirect effect analysis through the bootstrapping method was performed in the 95% confidence interval. As shown in Table 4, the indirect effect ($\beta = 0.299$, $p = 0.012$) of the change of physical self-concept on the change of mental well-being was verified to be significant.

Table 4. Direct, indirect, and total effect.

Path	Direct Effect	Indirect Effect	Total Effect
ΔPhysical self-concept → ΔSelf-esteem	0.506 *	-	0.506 *
ΔSelf-esteem → Δ Mental well-being	0.565 **	-	0.565 **
ΔPhysical self-concept → Δ Mental well-being	0.287 *	0.286 *	0.573 **

* $p < 0.05$, ** $p < 0.01$.

4. Discussion

Physical self-concept is one of the important factors contributing to the overall self-esteem which is known as an essential element of a happy life. Considering that participation in exercise helps form a physical self-concept, it can be concluded that exercise can lead to happiness. Although several studies that partially support the process of participation in exercise leading to well-being have been implied, there has been no case of empirically examining this process in a single study. In addition, existing studies have shown that the concept of happiness is not an integrated perspective. Therefore, this study attempted to infer the effect of the physical self-concept changed through participation in exercise for a certain period on the changed self-esteem and mental well-being in university students by collecting short-term longitudinal data. In addition, this study attempted to discuss happiness from a more integrated perspective by using the concept of mental well-being, including psychological well-being, emotional well-being, and social well-being.

This study calculated the amount of change in physical self-concept, self-esteem, and mental well-being using data collected by repeated measurements before and after a six-week exercise participation. Based on significant correlations between the amount of change of the variables, a structural equation model analysis was conducted to examine the relationships between variables, and the following implications were derived based on the analysis.

As a result of the analysis, the amount of change in the sub-factors of the physical self-concept was found to have a positive correlation with the amount of change in self-esteem, and the amount of change in the physical self-concept was found to positively predict the amount of change in self-esteem. This is consistent with the results of a study by Garn et al. [19] that showed a positive correlation between physical self-concept and total self-esteem of students participating in a physical education class. In addition, it also supports the study from Ryu and Lee [29] who reported that self-esteem can be predicted through the physical self-concept of the elderly participating in physical activities. According to the hierarchical model of Marsh and colleagues [18], global self-esteem is formed through multidimensional self-concepts including physical self-concept. This study is meaningful in that it confirmed the relationship between the amount of change of each variable through longitudinal data, unlike previous studies using cross-sectional data. Thus, the results of this study imply that the more positively the physical self-concept changes through exercise, the more the self-esteem is improved.

The results from this study also revealed that the amount of change in self-esteem and mental well-being were positively correlated, and the path coefficient leading from the amount of change in self-esteem to the amount of change in mental well-being was also significant. These results indicate that the improved self-esteem through exercise participation positively affects changes in mental well-being, which is in line with some related existing studies [29–31]. A study reporting that adult women's participation in sports contributes to the improvement of self-esteem and subjective happiness [32], and another study [33] which revealed that self-esteem of participants in live sports positively predicts psychological well-being are also consistent with the results of this study. However, these studies did not consider the effect of self-esteem changes through exercise on various aspects of happiness because they selected only one of the factors of psychological well-being [29,31,33] or subjective well-being [32]. The results of this study that the positive changes of self-esteem through exercise participation can enhance complete happiness may be helpful in forming a positive attitude toward participation in exercise among college students.

The amount of change in various sub-factors of the physical self-concept, such as sports competence, appearance, endurance, and muscle strength were found to be positively correlated with the change in mental well-being, and the change in physical self-concept positively predicted the change in mental well-being. This result is partially consistent with the results of theoretical studies [10,19], and empirical studies [12,34] identified that the physical self-concept is one of the positive factors predicting life satisfaction. The notable implications of this study are that the psychological, subjective, and social well-being level can be improved as the positive perception of one's body increases.

As a result of χ^2 difference verification and indirect effect analysis, the amount of change in self-esteem was found to complementarily mediate the relationship between the amount of change in physical self-concept and the amount of change in mental well-being. It can be said that this is the result of empirically verifying the hierarchical relationship between physical self-concept and self-esteem shown in Fox and Corbin's physical self-perception profile [7], Marsh and colleagues' [18] physical self-concept model, and a related study that self-esteem is one of strong predictors for hedonic and eudemonic well-being [35]. That is, based on the existing hierarchical model, this study empirically found that step-by-step changes in physical self-concept through participation in exercise can lead to changes in mental well-being through changes in self-esteem. In addition, the results of this study, which imply that changes in physical self-concept can have a direct effect on changes in mental well-being, support the conclusions presented in a recent meta-analysis by Kim and Oh [13]. Moreover, it can be said that it complements the limitations of existing studies which were mainly cross-sectional studies.

5. Conclusions

This study was conducted to verify the effects of positive changes in physical self-concept through exercise participation on changes in self-esteem and mental health in university students through a short-term longitudinal data analysis and to present the effect and direction of exercise for the pursuit of happiness. Participation in exercise for six weeks led to positive changes in the level of physical self-concept, self-esteem, and mental well-being of college students, and it was verified that there was a complementary mediating effect through the structural relationship analysis between the amount of change of each variable. These results are meaningful in that the causality of the hierarchical relationship leading to participation in exercise, physical self-concept, self-esteem, and mental well-being was verified. Thus, the present study suggests the direction that it is effective to aim for a change in physical self-concept to pursue a happy life through exercise participation for young adults who are in an important period for the formation of healthy behaviors that can be maintained throughout life.

The limitations of this study and suggestions for further research are as follows. First, the participants of this study consist only of college students enrolled in a university physical fitness class. To generalize the research results, it is necessary to recruit participants of a wider range of ages and conduct repeated research. In addition, in this study, since the liberal arts lecture called physical fitness was used as an exercise method, it was not possible to suggest an effective method for improving the physical self-concept in terms of exercise methods, intensity, and frequency. If follow-up studies using various exercise methods are conducted, a more specific direction can be presented to promote well-being through improvement of the physical self-concept.

Author Contributions: Conceptualization, I.K. and J.A.; methodology and analysis, I.K.; writing—original draft preparation, I.K.; writing—review and editing, I.K. and J.A. All authors have read and agreed to the published version of the manuscript.

Int. J. Environ. Res. Public Health **2021**
, *18*, 5224

Funding: This research received no external funding.

Institutional Review Board Statement: The present study was conducted according to the guidelines of the Declaration of Helsinki and approved by the Institutional Review Board of Seoul National University (IRB No. 2103/002-003).

Informed Consent Statement: Informed consent was obtained from all subjects involved in the study.

Data Availability Statement: The data presented in this study are available on request from the corresponding author.

Conflicts of Interest: The authors declare no conflict of interest.

References

1. WHO. *Global Recommendations on Physical Activity for Health*; Geneva World Heal Organ: Geneva, Switzerland, 2010; p. 60.
2. Seligman, M.E. *Flourish: A Visionary New Understanding of Happiness and Well-Being*; Simon and Schuster: New York, NY, USA, 2012.
3. Nelson, M.C.; Neumark-Stzainer, D.; Hannan, P.J.; Sirard, J.R.; Story, M. Longitudinal and secular trends in physical activity and sedentary behavior during adolescence. *Pediatrics* **2006**, *118*, e1627–e1634. [CrossRef] [PubMed]
4. Plotnikoff, R.C.; Costigan, S.A.; Williams, R.L.; Hutchesson, M.J.; Kennedy, S.G.; Robards, S.L.; Allen, J.; Collins, C.E.; Callister, R.; Germov, J. Effectiveness of interventions targeting physical activity, nutrition and healthy weight for university and college students: A systematic review and meta-analysis. *Int. J. Behav. Nutr. Phys. Act.* **2015**, *12*, 1–10. [CrossRef] [PubMed]
5. Castro, O.; Bennie, J.; Vergeer, I.; Bosselut, G.; Biddle, S.J.H. How Sedentary Are University Students? A Systematic Review and Meta-Analysis. *Prev. Sci.* **2020**, *21*, 332–343. [CrossRef]
6. Dale, L.P.; Vanderloo, L.; Moore, S.; Faulkner, G. Physical activity and depression, anxiety, and self-esteem in children and youth: An umbrella systematic review. *Ment. Health Phys. Act.* **2019**, *16*, 66–79. [CrossRef]
7. Fox, K.R.; Corbin, C.B. The physical self-perception profile: Devlopment and preliminary validation. *J. Sport Exerc. Psychol.* **1989**, *11*, 408–430. [CrossRef]
8. Marsh, H.W.; Richards, G.E.; Johnson, S.; Roche, L.; Tremayne, P. Physical Self-Description Questionnaire: Psychometric properties and a miiltitrait-meltimethod analysis of relations to existing instruments. *J. Sport Exerc. Psychol.* **1994**, *16*, 270–305. [CrossRef]
9. Fox, K.R. *The Physical Self: From Motivation to Well-Being*; Human Kinetics: Champaign, IL, USA, 1997.
10. Fox, K.R. Self-esteem, self-perceptions and exercise. *Int. J. Sport Psychol.* **2000**, *13*, 228–240.
11. Morales-Rodríguez, F.M.; Espigares-López, I.; Brown, T.; Pérez-Mármol, J.M. The Relationship between psychological well-being and psychosocial factors in university students. *Int. J. Environ. Res. Public Health* **2020**, *17*, 4778. [CrossRef]
12. Roh, S.Y. The influence of physical self-perception of female college students participating in Pilates classes on perceived health state and psychological wellbeing. *J. Exerc. Rehabil.* **2018**, *14*, 192–198. [CrossRef]
13. Kim, Y.W.; Oh, S.H. Meta-analysis of the Relationship between Exercise Participants' Physical Self-concept and Happiness. *Korean J. Phys. Edu.* **2017**, *56*, 179–191. [CrossRef]
14. Diener, E. Subjective well-being. *Psychol. Bull.* **1984**, *95*, 542–575. [CrossRef]
15. Ryff, C.D. Happiness is everything, or is it? Explorations on the meaning of psychological well-being. *J. Pers. Soc. Psychol.* **1989**, *57*, 1069–1081. [CrossRef]
16. Keyes, C.L.M. Social well-being. *Soc. Psychol. Q.* **1998**, *61*, 121–140. [CrossRef]
17. Swami, V.; Weis, L.; Barron, D.; Furnham, A. Positive body image is positively associated with hedonic (emotional) and eudaimonic (psychological and social) well-being in British adults. *J. Soc. Psychol.* **2018**, *158*, 541–552. [CrossRef] [PubMed]
18. Marsh, H.W.; Martin, A.J.; Jackson, S. Introducing a short version of the physical self-description questionnaire: New strategies, short-form evaluative criteria, and applications of factor analyses. *J. Sport Exerc. Psychol.* **2010**, *32*, 438–482. [CrossRef] [PubMed]
19. Garn, A.C.; McCaughtry, N.; Martin, J.; Shen, B.; Fahlman, M. A Basic Needs Theory investigation of adolescents' physical self-concept and global self-esteem. *Int. J. Sport Exer. Psychol.* **2012**, *10*, 314–328. [CrossRef]
20. Harter, S. *Self-Perception Profile for Adolescents: Manual and Questionnaires*; Univeristy of Denver, Department of Psychology: Denver, CO, USA, 2012.
21. Babic, M.J.; Morgan, P.J.; Plotnikoff, R.C.; Lonsdale, C.; White, R.L.; Lubans, D.R. Physical activity and physical self-concept in youth: Systematic review and meta-analysis. *Sports Med.* **2014**, *44*, 1589–1601. [CrossRef] [PubMed]
22. Fernández-Bustos, J.G.; Infantes-Paniagua, Á.; Cuevas, R.; Contreras, O.R. Effect of Physical Activity on Self-Concept: Theoretical Model on the Mediation of Body Image and Physical Self-Concept in Adolescents. *Front. Psychol.* **2019**, *10*, 1537. [CrossRef] [PubMed]
23. Garn, A.C.; Morin, A.J.S.; White, R.L.; Owen, K.B.; Donley, W.; Lonsdale, C. Moderate-to-vigorous physical activity as a predictor of changes in physical self-concept in adolescents. *Health Psychol.* **2020**, *39*, 190–198. [CrossRef]
24. Kim, B.J. Development and Validation of the Korean Version of the Physical Self-Description Questionnaire. *Korean J. Sport Psychol.* **2001**, *12*, 69–90.
25. Choi, S.M.; Cho, Y.I. Measurement Invariance of Self-Esteem between Male and Female Adolescents and Method Effects Associated with Negatively Worded Items. *Korean J. Psychol.* **2013**, *32*, 571–589.

26. Lim, Y.J.; Ko, Y.G.; Shin, H.C. Psychometric Evaluation of the Mental Health Continuum-Short Form (MHC-SF) in South Koreans. *Korean J. Psychol.* **2012**, *31*, 369–386.
27. Smith, P.; Beaton, D. Measuring change in psychosocial working conditions: Methodological issues to consider when data are collected at baseline and one follow-up time point. *Occup. Environ. Med.* **2008**, *65*, 288–296. [CrossRef]
28. Curran, P.J.; West, S.G.; Finch, J.F. The robustness of test statistics to nonnormality and specification error in confirmatory factor analysis. *Psychol. Methods* **1996**, *1*, 16–29. [CrossRef]
29. Ryu, M.K.; Lee, K.P. The Relation Analysis among Physical Self-Concept, Self-esteem and Quality of Life in Participation Leisure Sport Old Adults. *Korean J. Sports Sci.* **2018**, *27*, 467–477. [CrossRef]
30. Wong, M.Y.C.; Chung, P.K.; Leung, K.M. Examining the Exercise and Self-Esteem Model Revised with Self-Compassion among Hong Kong Secondary School Students Using Structural Equation Modeling. *Int. J. Environ. Res. Public Health* **2021**, *18*, 3661. [CrossRef]
31. Michèle, J.; Guillaume, M.; Alain, T.; Nathalie, B.; Claude, F.; Kamel, G. Social and leisure activity profiles and well-being among the older adults: A longitudinal study. *Aging Ment. Health* **2019**, *23*, 77–83. [CrossRef] [PubMed]
32. Bake, W.C.; Kim, S.K. The Causal Analysis of Relationship among Women's Sport Participation Variables, Self-Esteem and Subjective Well-being. *Korean J. Phys. Edu.* **2012**, *43*, 249–260.
33. Kim, D.K.; Kim, J.S.; Shin, J.H. Verification of Relationship Model among Self-Esteem, Immersion, Psychological Well-Bing and Happiness of Sports for all Participants. *Korean J. Sports Sci.* **2012**, *21*, 335–346.
34. Kang, H.W. An Investigation of Mediated Effect of Health Education Participation of University Students Physical Self-concept in Psychological Happiness Orientation and Relations Among Life-Satisfaction-Effect of PRECEDE model. *Korean J. Sport. Leis. Stud.* **2012**, *48*, 25–38. [CrossRef]
35. Pandey, R.; Tiwari, G.K.; Parihar, P.; Rai, P.K. Positive, not negative, self-compassion mediates the relationship between self-esteem and well-being. *Psychol. Psychother. Theory Res. Pract.* **2021**, *94*, 1–15. [CrossRef] [PubMed]

International Journal of
*Environmental Research
and Public Health*

Review

Physical Activity Promotes Health and Reduces Cardiovascular Mortality in Depressed Populations: A Literature Overview

Martino Belvederi Murri *,†, **Federica Folesani** †, **Luigi Zerbinati, Maria Giulia Nanni, Heifa Ounalli, Rosangela Caruso and Luigi Grassi**

Department of Biomedical and Specialty Surgical Sciences, Institute of Psychiatry, University of Ferrara, Via Fossato di Mortara 64a, 44121 Ferrara, Italy; flsfrc@unife.it (F.F.); zrblgu@unife.it (L.Z.); nnnmgl@unife.it (M.G.N.); nllhfe@unife.it (H.O.); crsrng@unife.it (R.C.); luigi.grassi@unife.it (L.G.)

* Correspondence: martino.belvederimurri@unife.it; Tel.: +39-0532-455-813; Fax: +39-0532-212-240

† These authors contributed equally to the manuscript.

Received: 18 June 2020; Accepted: 27 July 2020; Published: 31 July 2020

Abstract: Major depression is associated with premature mortality, largely explained by heightened cardiovascular burden. This narrative review summarizes secondary literature (i.e., reviews and meta-analyses) on this topic, considering physical exercise as a potential tool to counteract this alarming phenomenon. Compared to healthy controls, individuals with depression consistently present heightened cardiovascular risk, including "classical" risk factors and dysregulation of pertinent homeostatic systems (immune system, hypothalamic–pituitary–adrenal axis and autonomic nervous system). Ultimately, both genetic background and behavioral abnormalities contribute to explain the link between depression and cardiovascular mortality. Physical inactivity is particularly common in depressed populations and may represent an elective therapeutic target to address premature mortality. Exercise-based interventions, in fact, have proven effective reducing cardiovascular risk and mortality through different mechanisms, although evidence still needs to be replicated in depressed populations. Notably, exercise also directly improves depressive symptoms. Despite its potential, however, exercise remains under-prescribed to depressed individuals. Public health may be the ideal setting to develop and disseminate initiatives that promote the prescription and delivery of exercise-based interventions, with a particular focus on their cost-effectiveness.

Keywords: depression; physical activity; exercise; mortality; cardiovascular diseases; cardiovascular risk factors

---

1. Introduction

Depression is associated with a shortened life expectancy, but this phenomenon might be counteracted by exploiting the numerous benefits of physical exercise. In recent years, research made us aware of a disconcerting finding: individuals who receive a diagnosis of a major depression live on average ten years less than non-depressed subjects [1]. This phenomenon has been largely attributed to the impact of physical diseases, rather than suicide or accidental deaths [2]. In particular, cardiovascular diseases seem to be responsible for the largest quota of premature mortality [3]. This is relatively unsurprising, given that several lines of evidence associate depression with increased cardiovascular risk at the population level [4]. Above all, depression itself is considered a risk factor for myocardial infarction and coronary death [5]; in addition, the American Heart Association included it as the only "psychosocial" risk factor among various indicators of adverse outcomes for individuals with acute coronary syndrome [6].

Several lines of evidence actually show that multiple, biologically plausible mechanisms may pave the way from depression to cardiovascular diseases. To mention the most robustly replicated findings, depression is associated with altered activity of the autonomic nervous system (ANS) of the heart [7], as well as with altered levels of pro-inflammatory markers [8,9].

The premature mortality of depressed individuals, overall, should not just pertain the field of psychiatry but also constitute a public health concern, considering the high prevalence of this disorder. According to a recent report, in fact, 4.4% of the world's population suffers from a clinically relevant depressive disorder [10]. This alarming figure should make the depressed population a priority target for public health strategies to prevent cardiovascular mortality [2], and, based on several lines of evidence, interventions based on physical activity or physical exercise may be particularly fit to tackle this issue [11,12], while in the meantime they also improve the mental health of patients suffering from depression [13,14].

This narrative review has the aim of summarizing recent secondary literature (i.e., reviews and meta-analyses) that examines: (1) the cardiovascular risk and mortality of individuals suffering from depression; (2) the effects of physical activity or physical exercise on cardiovascular risk and cardiovascular mortality. We will conclude by briefly discussing commonly held misconceptions that may prevent clinicians from prescribing physical activity to depressed patients and potential implications for public health.

2. Depression Increases Mortality

Several epidemiological studies have shown that depression is associated with premature mortality. The effect of depression on mortality was specifically examined by a recent meta-analysis pooling the results of 293 studies, with an observation period ranging from less than 1 year to more than 10 years. Compared with subjects from the general population, depressed individuals had an unadjusted relative risk (RR) for mortality of 1.64 (95% CI 1.56–1.72) over the study period [15]. Several studies accounted for the role of different confounders, such as demographic variables, smoking, exercise, weight, and severity of the comorbid disorder. When only adjusted estimates were considered, the relative risk was lower, but still considerable (RR 1.52; 95% CI = 1.45–1.59). Results of single-cohort studies may also be informative of the impact of depression on mortality: among 5,103,699 Danish residents who were followed up from 1995 to 2013, those with depression displayed a double mortality rate compared to non-depressed subjects, corresponding to a shorter life expectancy by 14.0 years for men and 10.1 years for women [1]. Interestingly, while depression is more frequent among women, the phenomenon of premature mortality has been mostly observed among males [16].

The relationship between depression and mortality, however, is complex and likely to be influenced by several factors. Socioeconomic status, for instance, heavily impacts on health and could be partly responsible of the observed association [17]. Nonetheless, depression has been found associated with increased mortality both in low- and high-income countries [18].

The presence of physical diseases is another fundamental factor to consider when examining the association between depression and mortality. Indeed, physically ill individuals are expected to be frequently depressed, as well as display shortened life expectancy. In other words, observing a significant association between depression and mortality may not necessarily entail a causal relationship, since depression may constitute a mere consequence of the concomitant physical diseases (confounding by indication) [19]. Overall, cardiovascular diseases are of particular interest, since they are the leading cause of mortality in the general population, accounting for approximately one-third of all deaths [20], as well as among depressed subjects [19]. Available data suggest that the additional presence of depression in populations with cardiovascular diseases does indeed explain an increase in mortality, even when the severity of the primary disease is accounted for [15,21,22]. However, caution is required before interpreting such relationships as strictly causal [19]. Indeed, the severity of depression might be important to determine the prognosis of comorbid physical diseases. Even though mortality rates appeared to be higher among subjects with major, rather than subthreshold depression, this difference may not be meaningful [23].

In conclusion, the association between depression and premature mortality is a fairly robust epidemiological finding, albeit it might not necessarily entail a direct causal relationship [19].

3. Which Mechanisms Are Involved in the Higher Cardiovascular Risk of Depression?

Cardiovascular diseases are the main cause of premature death in depressive disorders, with major depression being associated with a hazard ratio of cardiovascular mortality of 1.63 (95% CI: 1.25–2.13) according to a recent meta-analysis [24]. More specifically, depression is as an independent risk factor for myocardial infarction and coronary heart disease [25] as well as coronary death [5], sudden cardiac death and the recurrence of arrhythmias such as atrial fibrillation [26]. Several lines of research have attempted to disentangle the multiple mechanisms that may mediate this phenomenon. Not surprisingly, the higher cardiovascular mortality rates observed among depressed subjects appear to be multi-factorial, involving biological as well as psychosocial factors. However, several of them appear to be amenable to modification.

3.1. Biological Factors

Subjects with major depression are prone to develop nearly all "traditional" cardiovascular risk factors, as illustrated in Table 1. Major depression is associated with a higher probability of becoming overweight and developing obesity since adolescence, especially among females [27,28], as well as a higher prevalence of type II diabetes [29], and hypertension [30]. Eventually, a higher prevalence of metabolic syndrome, hyperglycemia and hypertriglyceridemia was found in depressed individuals than control subjects in a meta-analysis of 18 studies [31], even controlling for confounding variables such as age, gender, and smoking habit. Surprisingly, however, discordant findings were observed in terms of serum lipid profile in depressed individuals: depression appeared to be associated with lower serum LDL when considered as a continuous measure and higher serum LDL when considered as a categorical measure [32]. The authors suggested the possibility of an association with lower LDL levels at the depression onset, with an increase in LDL over the course of the disease, in association with weight gain and metabolic syndrome.

Depressed individuals display a dysregulation of homeostatic systems [4]. In particular, dysregulation of the HPA axis is one of the most consistent biological alterations associated with depression [33], particularly in old age [34], with possible differences according to the melancholic or atypical subtype [35]. Depressed individuals display dysfunctions of the Autonomic Nervous System (ANS) in terms of a lower heart rate variability (HRV) that has been considered as a reliable predictor of negative cardiovascular outcomes [7,36]. The activity of the immune system is also abnormal in major depression, usually featuring an exaggerated pro-inflammatory state, both in the periphery and central nervous system [37–41]. Cross-sectional studies show that compared to healthy controls, depressed individuals consistently display higher levels of specific cytokines and chemokines in the peripheral blood, even after excluding physical comorbidities known to affect inflammatory markers [9]. A dysregulated immune system activity may derive from genetic predisposition [42] as well as altered neuro-immunological mechanisms [43–45] and behavioral factors such as tobacco use, thus the direction of the causal relationship between inflammatory markers and depressive symptoms appears to be extremely complex and likely bidirectional. Recent studies, for instance, suggest that abnormal TNF-alpha levels may directly cause the onset of depressive symptoms in adolescents [46]. Thus, depression and cardiovascular health have a complex, intertwined relationship. Both dimensions are regulated by the products of various pleiotropic genes involved in metabolism, HPA axis activity, inflammation, neurotransmission and circadian rhythms [47]. Animal models have been particularly useful to study the mechanisms underlying the comorbidity between depression and cardiovascular diseases and may provide further indications on their interplay. Recent evidence points to a role of chronic social stress as a trigger for depression, mediated by ANS dysfunction, endothelial damage and increases in pro-inflammatory cytokines [48,49].

3.2. Psychosocial Factors

People affected by depression are more prone to engage in unhealthy lifestyles, which in turn play a role in the pathogenesis of cardiovascular diseases. Depression appears to be associated with a tendency for unhealthy diets [50], and a higher risk of alcohol use disorder, in a bidirectional way [51]. Depression in adolescents (13–19 years old) predicts the taking up of cigarette smoking [52]; in turn, a history of depression is associated with lower probabilities of short- and long-term smoking cessation [53]. Depressed individuals also show a lower adherence to medications, especially for chronic disease such as cardiovascular ones [54]. In the elderly, the role of depression and physical diseases, especially cardiovascular disease, in impacting level of functioning and disability [55] is also mediated by low physical activity and time spent watching TV [56].

Last but not least, one of the most important and modifiable cardiovascular risk factors is physical inactivity. Compared with healthy individuals, those with depression are less likely to engage in physical activity and display a greater tendency of sedentary behaviors [12]; they also show reduced physical fitness [57], which is independently associated with cardiovascular events [58,59]. Recently, a large cross-sectional study, conducted on 1,237,194 adult people from the US confirmed the finding of a robust, U-shaped association between engagement in physical exercise and mental health in general, which held with adjustment for a number of sociodemographic and physical variables [60]. The authors observed that better mental health was associated with practicing specific sports, especially team sports, and it was proportional to the frequency and intensity of physical exercise, being worse at the extremes of minimal and excessive physical activity. A recent large study from Scotland on patients aged ≥ 16 years with cardiovascular diseases observed that meeting physical activity levels as recommended by the UK Chief Medical Officer was associated with better mental well-being [61].

Table 1. Association between depression and cardiovascular risk factors.

Cardiovascular Risk Condition	Studies	Association between Depression and Risk Factor
Obesity and overweight	[27]	13 prospective studies on adolescents (of which, 7 evaluating depression leading to obesity and 6 obesity leading to depression). Bi-directional relationship, stronger for depression leading to obesity. Depression or depressive symptoms in adolescents is associated with an increased risk of 70% (RR 1.70, 95% CI: 1.40; 2.07) of becoming obese, while obesity in adolescents is associated with an increased risk of 40% (RR 1.40, 95% CI: 1.26; 1.70) of becoming depressed.
Type II diabetes	[29]	16 studies comparing major depressive disorder (clearly defined) to the general population in terms of the prevalence of type II diabetes. Major depression was associated with a higher risk for type II diabetes (RR 1.49; 95% CI = 1.29–1.72; $p < 0.001$) (when comparing age- and gender-matched populations: RR 1.36; 95% CI = 1.28–1.44; $p < 0.001$).
Metabolic profile	[31]	18 cross-sectional studies. Higher prevalence of metabolic syndrome in depressed (30.5%) than control individuals (OR 1.54, 95% CI = 1.21–1.97, $p = 0.001$); higher risk for hyperglycemia (OR 1.33; 95% CI = 1.03–1.73, $p = 0.03$) and hypertriglyceridemia (OR 1.17, 95% CI = 1.04–1.30, $p = 0.008$). Controlling for confounding factors.
	[32]	18 cohort studies. Lower LDL (mean difference = −4.29; 95% CI = −8.19, −0.40, $p = 0.03$) in depression when serum LDL considered as a continuous measure. Lower depression when low LDL (OR 0.90; 95% CI = 0.80–1.01, $p = 0.08$) when serum LDL considered as a categorical measure.
Hypertension	[30]	9 prospective studies, 22.367 participants, mean follow-up period 9.6 years. Increased risk of hypertension incidence with adjusted RR 1.42 (95% CI = 1.09–1.86, $p = 0.009$).
Inflammation	[9]	82 case-control studies. Elevated plasma levels of some cytokines and chemokines in depressed subjects (IL-6, TNF-α, IL-10, sIL-2R, CCL2, IL-13, IL-18, IL-12, sTINFR-2) (g = −0.477, $p = 0.043$).
Autonomic dysfunction	[7]	29 case-control studies. Lower HRV in depressed individuals (g = −0.349; CI 95% = −0.505, −0.193, $p < 0.001$).
	[36]	18 studies. Depression is associated with a lower HRV (g = −0.301, $p < 0.001$); negative correlation between depression severity and HRV (r = −0.354, $p < 0.001$).
Behavioral Factors		
Unbalanced diet	[50]	3 cross-sectional studies. 2 out of 3 studies support an association between depression and unhealthy diets.
Alcohol consumption	[51]	7 studies (2 out of 7 prospective studies). Increased risk of alcohol use disorder in depressed individuals (adjusted OR 2.09; 95% CI = 1.29–3.38).
Tobacco smoking	[52]	12 prospective studies. Depression predicted onset of smoking in adolescents (RR 1.41; 95% CI = 1.21–1.63, $p < 0.001$).
	[53]	42 clinical trials on smoking cessation. History of depression is associated with lower odds of short-term (OR 0.83; 95% CI = 0.72–0.95, $p = 0.009$) and long-term abstinence (OR 0.81; 95% CI = 0.67–0.97, $p = 0.023$).
Compliance to therapy	[54]	31 studies cross-sectional studies on chronic diseases. Depressed individuals are more often non-adherent to prescribed medications (OR 1.76; 95% CI = 1.22–2.57).
Sedentary behaviors	[60]	Cross-sectional study on more than 1 million individuals in US on mental health burden and its association with physical exercise.
	[12]	24 cross-sectional studies. Depressed individuals tend to engage less in physical activity (standardized mean difference = −0.251; 95% CI = −0.03, 0.15, $p < 0.001$) and more in sedentary behavior (standardized mean difference = 0.09; 95% CI = 0.01–0.18, $p = 0.02$).

4. Physical Activity May Narrow the Mortality Gap of Depression

Targeting physical inactivity may be one preferential strategy to narrow the mortality gap associated with depression. Several lines of evidence directly or indirectly support this view. This modifiable risk factor may be even the key for public health interventions [62,63]. Engaging in physical activity in depressed individuals, in fact, may not only have a direct favorable impact on the premature mortality rate associated with the disease, by reducing cardiovascular risk and cardiovascular mortality [64,65], but may also directly treat symptoms of depression [66]. Here, we make an approximation by conflating studies examining physical activity with those evaluating physical exercise, but we encourage potential readers in gaining a more detailed insight into their different implications. The former is a structured form of physical activity with a specific purpose directed toward physical health, while the latter is a generic form of activity [67]. Figure 1 depicts the effects of exercise in the relationship between depression, cardiovascular risk and mortality.

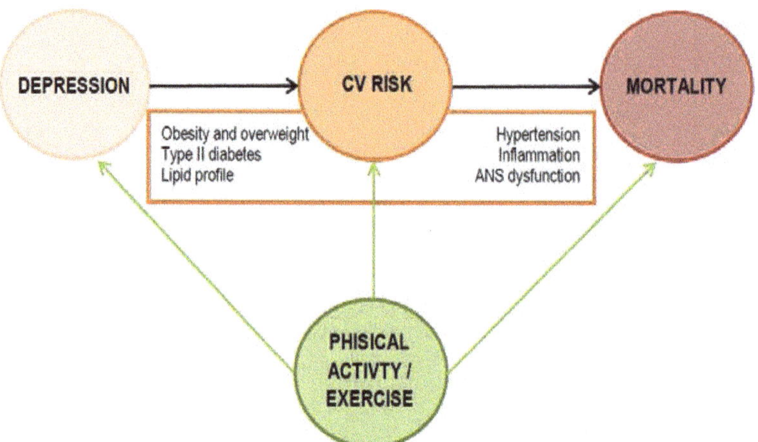

Figure 1. Effects of exercise in the relationship between depression, cardiovascular risk and mortality. ANS: Autonomic Nervous System.

Several meta-analyses have evaluated the role of exercise on cardiovascular risk factors (Table 2). First, exercise improves cardiorespiratory fitness at different ages [68,69], even in the elderly [70] and when performed in a reduced-time session of moderate activity [71].

Second, exercise helps in weight loss together with a balanced diet, and counterbalances the negative effects associated with overweight and obesity [72–74]. These effects appear to be related to the influence of exercise on the metabolic profile associated with obesity and overweight: physical activity reduces fasting insulin levels [69,75] and visceral adiposity [76]. The latter, especially, is a strong predictor of morbidity and mortality, and exercise alone has proven to be more effective than a hypocaloric diet alone in weight reduction. An exercise session, moreover, acutely influences the levels of appetite-regulative hormones [77], even though its long-term effects are not yet been examined in detail.

Third, exercise contributes to achieve better glycemic metabolism in type II diabetes, as it reduces insulin resistance and HbA1c levels [69,78–80]. Physical activity also ameliorates the lipid profile: it reduces triglyceride levels, while increasing HDL levels [69].

Fourth, blood pressure control is better achieved with exercise [81–83], as it reduces both systolic and diastolic blood pressure in pre- and hypertensive subjects.

Fifth, physical activity normalized body homeostatic system activity. It modulates immune responses: exercise decreases C-reactive protein (CRP) and IL-6 levels in type II diabetes [84], and reduces CRP and fibrinogen levels in patients with coronary artery disease [85]. Furthermore, regular physical activity appears to benefit the immune system, slowing down its aging over time [86].

The ANS is also influenced by exercise, which improves HRV in type II diabetes [87], and cardiovascular diseases such as coronary artery disease and heart failure [88–90], suggesting an improvement effect on parasympathetic activity and a reduction effect on sympathetic activity. Further, the HPA axis, the main hormonal stress-response system, is influenced by physical activity: regular exercise, in fact, appears to ameliorate the stress response accelerating the return to a resting state [91–93].

The impact of structured physical activity on behaviors at risk for cardiovascular diseases was also examined in different studies; however, no effects on alcohol use nor smoking habit were observed [94,95] and the effects on sedentary behaviors appeared to be uncertain [96–98].

Table 2. Effects of exercise on cardiovascular risk factors.

Cardiovascular Risk Factor	Studies	Impact of Physical Exercise
Obesity and overweight	[76]	117 studies. Exercise has better effects than a hypocaloric diet alone in reducing visceral adiposity ($p = 0.08$). However, it has less effects on total weight loss than diet alone.
	[77]	20 trials. Appetite-regulative hormone levels are acutely influenced by exercise.
Type II diabetes	[78]	27 prospective randomized or controlled trials of aerobic exercise training in adult subjects with type II diabetes, with a minimum duration of 2 weeks. Reduction in HbA1c% (mean difference = -0.71%; 95% CI = -1.11, -0.31, $p = 0.0005$) and insulin resistance (mean difference = -1.02, 95% CI = -1.77, -0.28, $p = 0.007$).
Lipid profile	[69]	160 RCTs. Exercise reduces triglycerides ($p = 0.02$), and increases HDL ($p < 0.001$).
Hypertension	[81]	93 RCTs. Reduction in systolic blood pressure and diastolic blood pressure. Different effects for different types of exercise and different blood pressure levels (greater for hypertensive patients).
Inflammation	[84]	14 RCTs. Exercise reduces CRP (-14% from baseline, 95% CI = -1.09, -0.23) and IL-6 levels (-18% from baseline, 95% CI = -1.44, -0.32) in type II diabetes.
	[85]	23 trials. Exercise reduces CRP (SMD = -0.500; 95% CI = -0.844, -0.157, $p = 0.004$) and fibrinogen levels (SMD = -0.544; 95% CI = -1.058, -0.030, $p = 0.038$) in coronary artery disease.
	[86]	Exercise enhances immune competency and slows down the aging of the immune system.
Autonomic dysfunction	[87]	15 trials. Improvements in HRV in type II diabetes after at least 3 month of an exercise program.
	[89]	16 RCTs. Exercise training leads to an improvement in HRV in coronary artery disease.
	[88]	19 studies (RCTs, quasi-RCTs and controlled trials of exercise training in adult patients with heart failure). Exercise improves HRV.

Notes. RCTs: Randomized Controlled Trials.

Cardiac rehabilitation (CR) represents an evident example of how useful exercise is for the secondary prevention of cardiovascular diseases: CR consists of integrated programs comprising physical, psychological and social interventions, with exercise playing a pivotal role. In particular, CR among coronary heart disease patients was effective reducing cardiovascular mortality (RR 0.74, 95% CI 0.64–0.86), reducing the long-term risk of myocardial infarction (RR 0.67, 95% CI 0.50–0.90), and the need for rehospitalizations (RR 0.82, 95% CI 0.70–0.96) [99]. Several guidelines on CR have been provided over the years, with notable differences across countries: nonetheless, the central element of CR seems to be aerobic endurance training [100].

The role of physical activity on mortality was estimated by a recent systematic umbrella review: physical activity reduced the risk of all-cause mortality (hazard ratios between 0.6 and 0.7), according to the intensity and frequency [101]. Moreover, significant effects were observed specifically on cardiovascular mortality, and even exercising at a lower intensity than recommendations by guidelines [101]. Resistance training was also associated with significant reductions in all-cause mortality (HR: 0.79; 95%CI: 0.69–0.91) but not cardiovascular mortality [102]. Overall, the impact on mortality in specific conditions such as coronary heart disease, stroke, heart failure, and diabetes was assessed as comparable to that of first-line medications [11].

An important additional effect of physical exercise on mortality might depend on its favorable effects on symptoms of depression. This has been observed in patients with cardiovascular diseases associated with depression. For instance, exercise was effective at reducing depression in patients after acute coronary events [103] and older adults with cardiovascular diseases [104]. In cardiovascular populations, evidence suggests the effectiveness of both high-intensity training [105] and low-intensity practices based on stress reduction, meditation or muscle relaxation [106]. However, part of the effects of exercise may depend on improving feelings of energy and fatigue [107]. Moreover, evidence supporting the antidepressant effect of exercise is rapidly growing in populations with primary major depressive disorder [108–113]. Studies are providing methodologically robust evidence on the efficacy of exercise alone for mild and moderate depression, and together with other treatments for severe depression [13,114,115]. The effects of depression on cardiovascular health, in fact, seem also dependent on the severity of depressive symptoms. By influencing the course and severity of depression, exercise may thus indirectly dampen such negative impact. In addition, it is still unknown whether exercise might actually impact on alcohol and substance use and cigarette smoking in mental health populations [116–119].

Finally, research is starting to investigate whether depressed patients undergo similar mechanisms of cardiovascular risk reduction that are observed in the general population and in cardiac patients. For instance, scheduling exercise sessions during Cognitive Behavioral Therapy was shown to positively modulate the immune system in depressed patients [120]. Similarly, cortisol levels appear to return to counteract HPA axis dysregulation in depressed patients [121]. Other improvements in neuroendocrine activity, autonomic nervous system activity, inflammatory markers and oxidative stress have been replicated in depressed populations, although further research is still needed before drawing firm conclusions [122–124].

5. How to Prescribe Physical Activity to Depressed Individuals

Physical activity has been included as a treatment for depression in the context of some clinical guidelines for depression, although its importance still remains downplayed for obscure reasons [13,125], whereas it is considered as an important treatment option to reduce mortality, especially mortality associated with cardiovascular conditions [126,127].

Despite the available evidence on its efficacy for depression, exercise remains under-prescribed. The reasons may be many, but among them, clinicians are often unaware of available indications on how to deliver physical activity to depressed individuals [128–133]. Actually, one of the first obstacles to overcome in order to improve the prescription of physical activity has been identified in physician prejudices and resistance based on the belief that patients will not adhere [13,134,135], whereas depressed patients usually display good adherence to exercise programs [136]; dropout rates from RCTs that include an exercise protocol are usually low and not different from those of control groups [115]. However, the presence of an instructor or other types of supervision may be crucial to motivate patients with severe mental illnesses to adhere to exercise programs [137]. Among depressed individuals, supervision is suggested at least in the initial phases [110]. Another reason why physicians may be reluctant to prescribe exercise to depressed individuals might depend on the perception that insufficient evidence is available on its efficacy, or on difficulties identifying the right "dose" to indicate to patients. In fact, historically, it has been difficult to identify a consistent threshold in terms of frequency and duration that achieves a meaningful reduction in cardiovascular disease incidence and mortality [13,101,138]. Moreover, the existing recommendations are mainly based on guidelines derived from the general or cardiovascular populations [114,139]. Some indications, however, may be translated to depressed populations in the absence of more specific data: cardiovascular benefits are immediately evident even adding small amounts of physical activity to the daily routine. Sedentary individuals, such as depressed patients, may display a steep risk decline even adding very short bouts (e.g., 10 min or less) of moderate–vigorous physical activity. Finally, reducing sedentary time or engaging in light physical activity also reduces cardiovascular risk, although it may require more time per day [138]. In sum, little physical activity is always "better than nothing" when it comes to cardiovascular risk reduction. Similar indications may become available regarding specific effects on mood and other depressive symptoms, with preliminary evidence suggesting that resistance and mixed training may yield higher efficacy than aerobic-only training [65].

By all means, however, the pleasure associated with exercise performances should be taken into account when prescribing physical activity [140]. In this so-called affect-based exercise, the goal of physical activity programs is mainly focused on the performance of activities associated with pleasant feelings, which may in turn also favor adherence to the exercise treatment. Ladwig and colleagues [140] suggest encouraging the patient to evaluate the pleasure associated with practice on a Likert rating scale regarding personal feelings, and then autonomously regulate the intensity and duration of exercise in order to maintain a satisfactory rating score on the aforementioned scale. Despite anhedonic experiences which are commonly observed in depression, these patients may still perceive exercise as pleasant [141–143]. The positive affective response obtained with exercise is also associated with treatment response, predicting both the improvement of depressive symptoms as well as the adherence to the exercise program [144,145].

Moreover, another barrier to the prescription and delivery of physical activity may depend on the need to involve different professionals and not necessarily physicians and other health professionals. However, for depressive and other mental disorders, it is highly recommended that the professionals involved have experience in the mental health field [146]. A collaborative and integrated approach with various disciplines is also highly recommended.

Taking the public health perspective, several interventions have been promoted to increase the physical activity level of the general population [62,63]. Some have proven effective, such as those involving telephone-assisted interventions, as well as changes in the workplace environment [138]. Furthermore, it has been observed that public health interventions for the promotion of physical activity have a high probability of being cost-effective in the general population [147] and among patients with mental disorders [148]. In this context, primary care might be a preferential setting to improve the physical activity habits of patients, benefiting especially patients with cardiovascular risk factors [149,150]. However, barriers limiting the prescription of exercise by clinicians need to be addressed. Some strategies deriving from behavioral economics have been provided to help overcome decision biases concerning physical activity [149].

Limitations

This narrative review entails some limitations, the most evident being the lack of a systematic approach to the literature review. However, given the complexity of the topic and the heterogeneous methodological approaches (including epidemiology, mechanisms, as well as trial results on multiple outcomes), we deemed it useful to present the public health audience, as well as clinicians, with an overview of extant secondary literature, rather than focusing on more specific aspects.

6. Conclusions

The premature mortality of individuals with depression is a major unsolved issue not only for the mental health field, but also for public health. This phenomenon largely depends on a detrimental effect of depression on cardiovascular health, because this disorder leads to developing or exacerbating unhealthy lifestyles as well as causing imbalances across different body homeostatic systems. Among modifiable cardiovascular risk factors, physical inactivity may be the preferential target for clinical and public health interventions which may ultimately reduce the mortality gap. In fact, delivering physical exercise or physical activity may not only improve depression severity, but also directly tackle the constitutive elements of cardiovascular risk. Nonetheless, several challenges remain to be addressed by further research: (1) to provide more robust, direct evidence on the reduction in cardiovascular risk and mortality in depressed subjects; (2) to further tailor exercise- and physical activity-based interventions for depressed populations; (3) to extend the knowledge on, and tackle barriers to, exercise prescription by clinicians, and to provide them with streamlined indications to increase the prescription of exercise; (4) to assess the cost-effectiveness of exercise-based interventions; (5) to elaborate and assess public health strategies based on this effective, inexpensive and safe behavior.

Author Contributions: Project administration and supervision, L.G.; resources and writing, M.B.M., F.F. and L.Z.; review and editing, M.G.N., H.O. and R.C. All authors have read and agreed to the published version of the manuscript.

Funding: This research received no external funding.

Conflicts of Interest: All authors declare they have no conflict of interest.

References

1. Laursen, T.M.; Musliner, K.L.; Benros, M.E.; Vestergaard, M.; Munk-Olsen, T. Mortality and life expectancy in persons with severe unipolar depression. *J. Affect. Disord.* **2016**, *193*, 203–207. [CrossRef]
2. Walker, E.R.; McGee, R.E.; Druss, B.G.; Reisinger, E.; McGee, R.E.; Druss, B.G. Mortality in Mental Disorders and Global Disease Burden Implications: A Systematic Review and Meta-analysis. *JAMA Psychiatr.* **2015**, *72*, 334–341. [CrossRef]
3. Goldstein, B.I.; Carnethon, M.R.; Matthews, K.A.; McIntyre, R.S.; Miller, G.E.; Raghuveer, G.; Stoney, C.M.; Wasiak, H.; McCrindle, B.W. Major Depressive Disorder and Bipolar Disorder Predispose Youth to Accelerated Atherosclerosis and Early Cardiovascular Disease: A Scientific Statement from the American Heart Association. *Circulation* **2015**, *132*, 965–986. [CrossRef]
4. Dhar, A.K.; Barton, D.A. Depression and the link with cardiovascular disease. *Front. Psychiatr.* **2016**, *7*, 33. [CrossRef]
5. Wu, Q.; Kling, J.M. Depression and the Risk of Myocardial Infarction and Coronary Death. *Medicine* **2016**, *95*, e2815. [CrossRef]
6. Lichtman, J.H.; Froelicher, E.S.; Blumenthal, J.A.; Carney, R.M.; Doering, L.V.; Frasure-Smith, N.; Freedland, K.E.; Jaffe, A.S.; Leifheit-Limson, E.C.; Sheps, D.S.; et al. Depression as a risk factor for poor prognosis among patients with acute coronary syndrome: Systematic review and recommendations: A scientific statement from the american heart association. *Circulation* **2014**, *129*, 1350–1369. [CrossRef] [PubMed]
7. Alvares, G.A.; Quintana, D.S.; Hickie, I.B.; Guastella, A.J. Autonomic nervous system dysfunction in psychiatric disorders and the impact of psychotropic medications: A systematic review and meta-analysis. *J. Psychiatr. Neurosci.* **2016**, *41*, 89–104. [CrossRef] [PubMed]
8. Leighton, S.P.; Nerurkar, L.; Krishnadas, R.; Johnman, C.; Graham, G.J.; Cavanagh, J. Chemokines in depression in health and in inflammatory illness: A systematic review and meta-Analysis. *Mol. Psychiatr.* **2018**, *23*, 48–58. [CrossRef] [PubMed]
9. Köhler, C.A.; Freitas, T.H.; Maes, M.; de Andrade, N.Q.; Liu, C.S.; Fernandes, B.S.; Stubbs, B.; Solmi, M.; Veronese, N.; Herrmann, N.; et al. Peripheral cytokine and chemokine alterations in depression: A meta-analysis of 82 studies. *Acta Psychiatr. Scand.* **2017**, *135*, 373–387. [CrossRef]
10. World Health Organization. *Depression and Other Common Mental Disorders: Global Health Estimates*; World Health Organization: Geneva, Switzerland, 2017.
11. Naci, H.; Ioannidis, J.P.A. Comparative effectiveness of exercise and drug interventions on mortality outcomes: Metaepidemiological study. *Br. J. Sports Med.* **2015**, *49*, 1414–1422. [CrossRef]
12. Schuch, F.; Vancampfort, D.; Firth, J.; Rosenbaum, S.; Ward, P.; Reichert, T.; Bagatini, N.C.; Bgeginski, R.; Stubbs, B. Physical activity and sedentary behavior in people with major depressive disorder: A systematic review and meta-analysis. *J. Affect. Disord.* **2017**, *210*, 139–150. [CrossRef] [PubMed]
13. Ekkekakis, P.; Belvederi Murri, M. Exercise as antidepressant treatment: Time for the transition from trials to clinic? *Gen. Hosp. Psychiatr.* **2017**, *49*, A1–A5. [CrossRef] [PubMed]
14. Kvam, S.; Kleppe, C.L.; Nordhus, I.H.; Hovland, A. Exercise as a treatment for depression: A meta-analysis. *J. Affect. Disord.* **2016**, *202*, 67–86. [CrossRef] [PubMed]
15. Cuijpers, P.; Vogelzangs, N.; Twisk, J.; Kleiboer, A.; Li, J.; Penninx, B.W. Comprehensive meta-analysis of excess mortality in depression in the general community versus patients with specific illnesses. *Am. J. Psychiatr.* **2014**, *171*, 453–462. [CrossRef]
16. Cuijpers, P.; Vogelzangs, N.; Twisk, J.; Kleiboer, A.; Li, J.; Penninx, B.W. Is excess mortality higher in depressed men than in depressed women? A meta-analytic comparison. *J. Affect. Disord.* **2014**, *161*, 47–54. [CrossRef]
17. Lynch, J.; Smith, G.D.; Harper, S.; Hillemeier, M.; Ross, N.; Kaplan, G.A.; Wolfson, M. Is income inequality a determinant of population health? Part 1: A systematic review. *Milbank Q.* **2004**, *82*, 5–99. [CrossRef]
18. Brandão, D.J.; Fontenelle, L.F.; da Silva, S.A.; Menezes, P.R.; Pastor-Valero, M. Depression and excess mortality in the elderly living in low- and middle-income countries: Systematic review and meta-analysis. *Int. J. Geriatr. Psychiatr.* **2019**, *34*, 22–30. [CrossRef]
19. Machado, M.O.; Veronese, N.; Sanches, M.; Stubbs, B.; Koyanagi, A.; Thompson, T.; Tzoulaki, I.; Solmi, M.; Vancampfort, D.; Schuch, F.B.; et al. The association of depression and all-cause and cause-specific mortality: An umbrella review of systematic reviews and meta-analyses. *BMC Med.* **2018**, *16*, 1–13. [CrossRef]

20. Roth, G.A.; Johnson, C.; Abajobir, A.; Abd-Allah, F.; Abera, S.F.; Abyu, G.; Ahmed, M.; Aksut, B.; Alam, T.; Alam, K.; et al. Global, Regional, and National Burden of Cardiovascular Diseases for 10 Causes, 1990 to 2015. *J. Am. Coll. Cardiol.* **2017**, *70*, 1–25. [CrossRef]

21. Leung, Y.W.; Flora, D.B.; Gravely, S.; Irvine, J.; Carney, R.M.; Grace, S.L. The impact of premorbid and postmorbid depression onset on mortality and cardiac morbidity among patients with coronary heart disease: Meta-analysis. *Psychosom. Med.* **2012**, *74*, 786. [CrossRef]

22. Gathright, E.C.; Goldstein, C.M.; Josephson, R.A.; Hughes, J.W. Depression increases the risk of mortality in patients with heart failure: A meta-analysis. *J. Psychosom. Res.* **2017**, *94*, 82–89. [CrossRef] [PubMed]

23. Cuijpers, P.; Vogelzangs, N.; Twisk, J.; Kleiboer, A.; Li, J.; Penninx, B.W. Differential mortality rates in major and subthreshold depression: Meta-analysis of studies that measured both. *Br. J. Psychiatr.* **2013**, *202*, 22–27. [CrossRef] [PubMed]

24. Correll, C.U.; Solmi, M.; Veronese, N.; Bortolato, B.; Rosson, S.; Santonastaso, P.; Thapa-Chhetri, N.; Fornaro, M.; Gallicchio, D.; Collantoni, E.; et al. Prevalence, incidence and mortality from cardiovascular disease in patients with pooled and specific severe mental illness: A large-scale meta-analysis of 3,211,768 patients and 113,383,368 controls. *World Psychiatr.* **2017**, *16*, 163–180. [CrossRef]

25. Gan, Y.; Gong, Y.; Tong, X.; Sun, H.; Cong, Y.; Dong, X.; Wang, Y.; Xu, X.; Yin, X.; Deng, J.; et al. Depression and the risk of coronary heart disease: A meta-analysis of prospective cohort studies. *BMC Psychiatr.* **2014**, *14*, 1–11. [CrossRef] [PubMed]

26. Shi, S.; Liu, T.; Liang, J.; Hu, D.; Yang, B. Depression and Risk of Sudden Cardiac Death and Arrhythmias: A Meta-Analysis. *Psychosom. Med.* **2017**, *79*, 153–161. [CrossRef]

27. Mannan, M.; Mamun, A.; Doi, S.; Clavarino, A. Is there a bi-directional relationship between depression and obesity among adult men and women? Systematic review and bias-adjusted meta analysis. *Asian J. Psychiatr.* **2016**, *21*, 51–66. [CrossRef]

28. Solmi, M.; Köhler, C.A.; Stubbs, B.; Koyanagi, A.; Bortolato, B.; Monaco, F.; Vancampfort, D.; Machado, M.O.; Maes, M.; Tzoulaki, I.; et al. Environmental risk factors and nonpharmacological and nonsurgical interventions for obesity: An umbrella review of meta-analyses of cohort studies and randomized controlled trials. *Eur. J. Clin. Investig.* **2018**, *48*. [CrossRef]

29. Vancampfort, D.; Mitchell, A.J.; De Hert, M.; Sienaert, P.; Probst, M.; Buys, R.; Stubbs, B. Type 2 diabetes in patients with major depressive disorder: A meta-analysis of prevalence estimates and predictors. *Depress. Anxiety* **2015**, *32*, 763–773. [CrossRef]

30. Meng, L.; Chen, D.; Yang, Y.; Zheng, Y.; Hui, R. Depression increases the risk of hypertension incidence: A meta-analysis of prospective cohort studies. *J. Hypertens.* **2012**, *30*, 842–851. [CrossRef]

31. Vancampfort, D.; Correll, C.U.; Wampers, M.; Sienaert, P.; Mitchell, A.J.; De Herdt, A.; Probst, M.; Scheewe, T.W.; De Hert, M. Metabolic syndrome and metabolic abnormalities in patients with major depressive disorder: A meta-analysis of prevalences and moderating variables. *Psychol. Med.* **2014**, *44*, 2017–2028. [CrossRef]

32. Persons, J.E.; Fiedorowicz, J.G. Depression and serum low-density lipoprotein: A systematic review and meta-analysis. *J. Affect. Disord.* **2016**, *206*, 55–67. [CrossRef] [PubMed]

33. Stetler, C.; Miller, G.E. Depression and hypothalamic-pituitary-adrenal activation: A quantitative summary of four decades of research. *Psychosom. Med.* **2011**, *73*, 114–126. [CrossRef] [PubMed]

34. Belvederi Murri, M.; Pariante, C.; Mondelli, V.; Masotti, M.; Atti, A.R.; Mellacqua, Z.; Antonioli, M.; Ghio, L.; Menchetti, M.; Zanetidou, S.; et al. HPA axis and aging in depression: Systematic review and meta-analysis. *Psychoneuroendocrinology* **2014**, *41*, 46–62. [CrossRef] [PubMed]

35. Gold, P.W. The organization of the stress system and its dysregulation in depressive illness. *Mol. Psychiatr.* **2015**, *20*, 32–47. [CrossRef]

36. Kemp, A.H.; Quintana, D.S.; Gray, M.A.; Felmingham, K.L.; Brown, K.; Gatt, J.M. Impact of Depression and Antidepressant Treatment on Heart Rate Variability: A Review and Meta-Analysis. *Biol. Psychiatr.* **2010**, *67*, 1067–1074. [CrossRef]

37. Poole, L.; Dickens, C.; Steptoe, A. The puzzle of depression and acute coronary syndrome: Reviewing the role of acute inflammation. *J. Psychosom. Res.* **2011**, *71*, 61–68. [CrossRef]

38. Alexopoulos, G.S. Mechanisms and treatment of late-life depression. *Transl. Psychiatr.* **2019**, *9*, 1–16. [CrossRef]

39. McNutt, M.D.; Liu, S.; Manatunga, A.; Royster, E.B.; Raison, C.L.; Woolwine, B.J.; Demetrashvili, M.F.; Miller, A.H.; Musselman, D.L. Neurobehavioral effects of interferon-α in patients with hepatitis-C: Symptom dimensions and responsiveness to paroxetine. *Neuropsychopharmacology* **2012**, *37*, 1444–1454. [CrossRef]

40. Zunszain, P.A.; Hepgul, N.; Pariante, C.M. Inflammation and Depression. *Curr. Top. Behav. Neurosci.* **2013**, *14*, 135–151.

41. Baumeister, D.; Akhtar, R.; Ciufolini, S.; Pariante, C.M.; Mondelli, V. Childhood trauma and adulthood inflammation: A meta-analysis of peripheral C-reactive protein, interleukin-6 and tumour necrosis factor-α. *Mol. Psychiatr.* **2016**, *21*, 642–649. [CrossRef]

42. Barnes, J.; Mondelli, V.; Pariante, C.M. Genetic Contributions of Inflammation to Depression. *Neuropsychopharmacology* **2017**, *42*, 81–98. [CrossRef] [PubMed]

43. Pariante, C.M. Why are depressed patients inflamed? A reflection on 20 years of research on depression, glucocorticoid resistance and inflammation. *Eur. Neuropsychopharmacol.* **2017**, *27*, 554–559. [CrossRef] [PubMed]

44. Enache, D.; Pariante, C.M.; Mondelli, V. Markers of central inflammation in major depressive disorder: A systematic review and meta-analysis of studies examining cerebrospinal fluid, positron emission tomography and post-mortem brain tissue. *Brain. Behav. Immun.* **2019**, *81*, 24–40. [CrossRef] [PubMed]

45. Escelsior, A.; Sterlini, B.; Belvederi Murri, M.; Valente, P.; Amerio, A.; di Brozolo, M.R.; da Silva, B.P.; Amore, M. Transient receptor potential vanilloid 1 antagonism in neuroinflammation, neuroprotection and epigenetic regulation: Potential therapeutic implications for severe psychiatric disorders treatment. *Psychiatr. Genet.* **2020**, *30*, 39–48. [CrossRef]

46. Moriarity, D.P.; Kautz, M.M.; Mac Giollabhui, N.; Klugman, J.; Coe, C.L.; Ellman, L.M.; Abramson, L.Y.; Alloy, L.B. Bidirectional associations between inflammatory biomarkers and depressive symptoms in adolescents: Potential causal relationships. *Clin. Psychol. Sci.* **2020**. [CrossRef]

47. Kahl, K.G.; Stapel, B.; Frieling, H. Link between depression and cardiovascular diseases due to epigenomics and proteomics: Focus on energy metabolism. *Prog. Neuropsychopharmacol. Biol. Psychiatr.* **2019**, *89*, 146–157. [CrossRef]

48. Carnevali, L.; Montano, N.; Statello, R.; Sgoifo, A. Rodent models of depression-cardiovascular comorbidity: Bridging the known to the new. *Neurosci. Biobehav. Rev.* **2017**, *76*, 144–153. [CrossRef]

49. Carnevali, L.; Montano, N.; Tobaldini, E.; Thayer, J.F.; Sgoifo, A. The contagion of social defeat stress: Insights from rodent studies. *Neurosci. Biobehav. Rev.* **2020**, *111*, 12–18. [CrossRef]

50. Quirk, S.E.; Williams, L.J.; O'Neil, A.; Pasco, J.A.; Jacka, F.N.; Housden, S.; Berk, M.; Brennan, S.L. The association between diet quality, dietary patterns and depression in adults: A systematic review. *BMC Psychiatr.* **2013**, *13*, 1. [CrossRef]

51. Boden, J.M.; Fergusson, D.M. Alcohol and depression. *Addiction* **2011**, *106*, 906–914. [CrossRef]

52. Chaiton, M.O.; Cohen, J.E.; O'Loughlin, J.; Rehm, J. A systematic review of longitudinal studies on the association between depression and smoking in adolescents. *BMC Public Health* **2009**, *9*, 356–366. [CrossRef] [PubMed]

53. Hitsman, B.; Papandonatos, G.D.; McChargue, D.E.; Demott, A.; Herrera, M.J.; Spring, B.; Borrelli, B.; Niaura, R. Past major depression and smoking cessation outcome: A systematic review and meta-analysis update. *Addiction* **2013**, *108*, 294–306. [CrossRef] [PubMed]

54. Grenard, J.L.; Munjas, B.A.; Adams, J.L.; Suttorp, M.; Maglione, M.; McGlynn, E.A.; Gellad, W.F. Depression and medication adherence in the treatment of chronic diseases in the United States: A meta-analysis. *J. Gen. Intern. Med.* **2011**, *26*, 1175–1182. [CrossRef] [PubMed]

55. Grassi, L.; Caruso, R.; Da Ronch, C.; Härter, M.; Schulz, H.; Volkert, J.; Dehoust, M.; Sehner, S.; Suling, A.; Wegscheider, K.; et al. Quality of life, level of functioning, and its relationship with mental and physical disorders in the elderly: Results from the MentDis_ICF65+ study. *Health Qual. Life Outcomes* **2020**, *18*, 1–12. [CrossRef]

56. Da Ronch, C.; Canuto, A.; Volkert, J.; Massarenti, S.; Weber, K.; Dehoust, M.C.; Nanni, M.G.; Andreas, S.; Sehner, S.; Schulz, H.; et al. Association of television viewing with mental health and mild cognitive impairment in the elderly in three European countries, data from the MentDis-ICF65+ project. *Ment. Health Phys. Act.* **2015**, *8*, 8–14. [CrossRef]

57. Papasavvas, T.; Bonow, R.O.; Alhashemi, M.; Micklewright, D. Depression Symptom Severity and Cardiorespiratory Fitness in Healthy and Depressed Adults: A Systematic Review and Meta-Analysis. *Sports Med.* **2016**, *46*, 219–230. [CrossRef]

58. García-Hermoso, A.; Cavero-Redondo, I.; Ramírez-Vélez, R.; Ruiz, J.R.; Ortega, F.B.; Lee, D.C.; Martínez-Vizcaíno, V. Muscular Strength as a Predictor of All-Cause Mortality in an Apparently Healthy Population: A Systematic Review and Meta-Analysis of Data From Approximately 2 Million Men and Women. *Arch. Phys. Med. Rehabil.* **2018**, *99*, 2100–2113. [CrossRef]

59. Kodama, S.; Saito, K.; Tanaka, S.; Maki, M.; Yachi, Y.; Asumi, M.; Sugawara, A.; Totsuka, K.; Shimano, H.; Ohashi, Y.; et al. Cardiorespiratory fitness as a quantitative predictor of all-cause mortality and cardiovascular events in healthy men and women: A meta-analysis. *J. Am. Med. Assoc.* **2009**, *301*, 2024–2035. [CrossRef]

60. Chekroud, S.R.; Gueorguieva, R.; Zheutlin, A.B.; Paulus, M.; Krumholz, H.M.; Krystal, J.H.; Chekroud, A.M. Association between physical exercise and mental health in 1.2 million individuals in the USA between 2011 and 2015: A cross-sectional study. *Lancet Psychiatr.* **2018**, *5*, 739–746. [CrossRef]

61. Salman, A.; Sellami, M.; Al-Mohannadi, A.S.; Chun, S. The associations between mental well-being and adherence to physical activity guidelines in patients with cardiovascular disease: Results from the scottish health survey. *Int. J. Environ. Res. Public Health* **2019**, *16*, 3596. [CrossRef]

62. Heath, G.W.; Parra, D.C.; Sarmiento, O.L.; Andersen, L.B.; Owen, N.; Goenka, S.; Montes, F.; Brownson, R.C.; Alkandari, J.R.; Bauman, A.E.; et al. Evidence-based intervention in physical activity: Lessons from around the world. *Lancet* **2012**, *380*, 272–281. [CrossRef]

63. Sallis, J.F.; Bull, F.; Guthold, R.; Heath, G.W.; Inoue, S.; Kelly, P.; Oyeyemi, A.L.; Perez, L.G.; Richards, J.; Hallal, P.C. Progress in physical activity over the Olympic quadrennium. *Lancet* **2016**, *388*, 1325–1336. [CrossRef]

64. Ekkekakis, P. *Routledge Handbook of Physical Activity and Mental Health*; Routledge: Abingdon-on-Thames, UK, 2013.

65. Pedersen, B.K.; Saltin, B. Exercise as medicine—Evidence for prescribing exercise as therapy in 26 different chronic diseases. *Scand. J. Med. Sci. Sports* **2015**, *25*, 1–72. [CrossRef] [PubMed]

66. Gerber, M.; Holsboer-Trachsler, E.; Pühse, U.; Brand, S. Exercise is medicine for patients with major depressive disorders: But only if the "pill" is taken! *Neuropsychiatr. Dis. Treat.* **2016**, *12*, 1977. [CrossRef] [PubMed]

67. Caspersen, C.J.; Powell, K.E.; Christenson, G.M. Physical Activity, Exercise and Physical Fitness Definitions for Health-Related Research. *Public Health Rep.* **1985**, *100*, 126–131. [PubMed]

68. De Carvalho Souza Vieira, M.; Boing, L.; Leitão, A.E.; Vieira, G.; Coutinho de Azevedo Guimarães, A. Effect of physical exercise on the cardiorespiratory fitness of men—A systematic review and meta-analysis. *Maturitas* **2018**, *115*, 23–30. [CrossRef]

69. Lin, X.; Zhang, X.; Guo, J.; Roberts, C.K.; McKenzie, S.; Wu, W.C.; Liu, S.; Song, Y. Effects of exercise training on cardiorespiratory fitness and biomarkers of cardiometabolic health: A systematic review and meta-analysis of randomized controlled trials. *J. Am. Heart Assoc.* **2015**, *4*, e002014. [CrossRef]

70. Hurst, C.; Weston, K.L.; McLaren, S.J.; Weston, M. The effects of same-session combined exercise training on cardiorespiratory and functional fitness in older adults: A systematic review and meta-analysis. *Aging Clin. Exp. Res.* **2019**, *31*, 1701–1717. [CrossRef]

71. Magutah, K.; Thairu, K.; Patel, N. Effect of short moderate intensity exercise bouts on cardiovascular function and maximal oxygen consumption in sedentary older adults. *BMJ Open Sport Exerc. Med.* **2020**, *6*, e000672. [CrossRef]

72. Shaw, K.; Gennat, H.; O'Rourke, P.; Del Mar, C. Exercise for overweight or obesity. *Cochrane Database Syst. Rev.* **2006**. [CrossRef]

73. Ekkekakis, P.; Vazou, S.; Bixby, W.R.; Georgiadis, E. The mysterious case of the public health guideline that is (almost) entirely ignored: Call for a research agenda on the causes of the extreme avoidance of physical activity in obesity. *Obes. Rev.* **2016**, *17*, 313–329. [CrossRef] [PubMed]

74. Fock, K.M.; Khoo, J. Diet and exercise in management of obesity and overweight. *J. Gastroenterol. Hepatol.* **2013**, *28*, 59–63. [CrossRef] [PubMed]

75. Marson, E.C.; Delevatti, R.S.; Prado, A.K.G.; Netto, N.; Kruel, L.F.M. Effects of aerobic, resistance, and combined exercise training on insulin resistance markers in overweight or obese children and adolescents: A systematic review and meta-analysis. *Prev. Med.* **2016**, *93*, 211–218. [CrossRef]

76. Verheggen, R.J.H.M.; Maessen, M.F.H.; Green, D.J.; Hermus, A.R.M.M.; Hopman, M.T.E.; Thijssen, D.H.T. A systematic review and meta-analysis on the effects of exercise training versus hypocaloric diet: Distinct effects on body weight and visceral adipose tissue. *Obes. Rev.* **2016**, *17*, 664–690. [CrossRef] [PubMed]

77. Schubert, M.M.; Sabapathy, S.; Leveritt, M.; Desbrow, B. Acute exercise and hormones related to appetite regulation: A meta-analysis. *Sports Med.* **2014**, *44*, 387–403. [CrossRef] [PubMed]

78. Grace, A.; Chan, E.; Giallauria, F.; Graham, P.L.; Smart, N.A. Clinical outcomes and glycaemic responses to different aerobic exercise training intensities in type II diabetes: A systematic review and meta-analysis. *Cardiovasc. Diabetol.* **2017**, *16*, 37. [CrossRef]

79. Umpierre, D.; Ribeiro, P.A.B.; Kramer, C.K.; Leitão, C.B.; Zucatti, A.T.N.; Azevedo, M.J.; Gross, J.L.; Ribeiro, J.P.; Schaan, B.D. Physical activity advice only or structured exercise training and association with HbA1c levels in type 2 diabetes: A systematic review and meta-analysis. *JAMA* **2011**, *305*, 1790–1799. [CrossRef]

80. Salman, A.; Ukwaja, K.N.; Alkhatib, A. Factors associated with meeting current recommendation for physical activity in scottish adults with diabetes. *Int. J. Environ. Res. Public Health* **2019**, *16*, 3857. [CrossRef]

81. Cornelissen, V.A.; Smart, N.A. Exercise training for blood pressure: A systematic review and meta-analysis. *J. Am. Heart Assoc.* **2013**, *2*, e004473. [CrossRef]

82. Ashton, R.E.; Tew, G.A.; Aning, J.J.; Gilbert, S.E.; Lewis, L.; Saxton, J.M. Effects of short-term, medium-term and long-term resistance exercise training on cardiometabolic health outcomes in adults: Systematic review with meta-analysis. *Br. J. Sports Med.* **2018**, *54*, 341–348.

83. Naci, H.; Salcher-Konrad, M.; Dias, S.; Blum, M.R.; Sahoo, S.A.; Nunan, D.; Ioannidis, J.P.A. How does exercise treatment compare with antihypertensive medications? A network meta-analysis of 391 randomised controlled trials assessing exercise and medication effects on systolic blood pressure. *Br. J. Sports Med.* **2019**, *53*, 859–869. [CrossRef] [PubMed]

84. Hayashino, Y.; Jackson, J.L.; Hirata, T.; Fukumori, N.; Nakamura, F.; Fukuhara, S.; Tsujii, S.; Ishii, H. Effects of exercise on C-reactive protein, inflammatory cytokine and adipokine in patients with type 2 diabetes: A meta-analysis of randomized controlled trials. *Metabolism* **2014**, *63*, 431–440. [CrossRef]

85. Swardfager, W.; Herrmann, N.; Cornish, S.; Mazereeuw, G.; Marzolini, S.; Sham, L.; Lanctôt, K.L. Exercise intervention and inflammatory markers in coronary artery disease: A meta-analysis. *Am. Heart J.* **2012**, *163*, 666–676. [CrossRef] [PubMed]

86. Campbell, J.P.; Turner, J.E. Debunking the Myth of Exercise-Induced Immune Suppression: Redefining the Impact of Exercise on Immunological Health Across the Lifespan. *Front. Immunol.* **2018**, *9*, 648. [CrossRef] [PubMed]

87. Villafaina, S.; Collado-Mateo, D.; Fuentes, J.P.; Merellano-Navarro, E.; Gusi, N. Physical Exercise Improves Heart Rate Variability in Patients with Type 2 Diabetes: A Systematic Review. *Curr. Diab. Rep.* **2017**, *17*, 110. [CrossRef] [PubMed]

88. Pearson, M.J.; Smart, N.A. Exercise therapy and autonomic function in heart failure patients: A systematic review and meta-analysis. *Heart Fail. Rev.* **2018**, *23*, 91–108. [CrossRef]

89. Nolan, R.P.; Jong, P.; Barry-Bianchi, S.M.; Tanaka, T.H.; Floras, J.S. Effects of drug, biobehavioral and exercise therapies on heart rate variability in coronary artery disease: A systematic review. *Eur. J. Prev. Cardiol.* **2008**, *15*, 386–396. [CrossRef]

90. Besnier, F.; Labrunée, M.; Pathak, A.; Pavy-Le Traon, A.; Galès, C.; Sénard, J.M.; Guiraud, T. Exercise training-induced modification in autonomic nervous system: An update for cardiac patients. *Ann. Phys. Rehabil. Med.* **2017**, *60*, 27–35. [CrossRef]

91. Gerber, M.; Ludyga, S.; Mücke, M.; Colledge, F.; Brand, S.; Pühse, U. Low vigorous physical activity is associated with increased adrenocortical reactivity to psychosocial stress in students with high stress perceptions. *Psychoneuroendocrinology* **2017**, *80*, 104–113. [CrossRef]

92. Klaperski, S.; von Dawans, B.; Heinrichs, M.; Fuchs, R. Effects of a 12-week endurance training program on the physiological response to psychosocial stress in men: A randomized controlled trial. *J. Behav. Med.* **2014**, *37*, 1118–1133. [CrossRef]

93. Chen, C.; Nakagawa, S.; An, Y.; Ito, K.; Kitaichi, Y.; Kusumi, I. The exercise-glucocorticoid paradox: How exercise is beneficial to cognition, mood, and the brain while increasing glucocorticoid levels. *Front. Neuroendocrinol.* **2017**, *44*, 83–102. [CrossRef] [PubMed]

94. Hallgren, M.; Vancampfort, D.; Giesen, E.S.; Lundin, A.; Stubbs, B. Exercise as treatment for alcohol use disorders: Systematic review and meta-analysis. *Br. J. Sports Med.* **2017**, *51*, 1058–1064. [CrossRef] [PubMed]

95. Klinsophon, T.; Thaveeratitham, P.; Sitthipornvorakul, E.; Janwantanakul, P. Effect of exercise type on smoking cessation: A meta-analysis of randomized controlled trials. *BMC Res. Notes* **2017**, *10*, 442. [CrossRef] [PubMed]

96. De Vries, N.M.; van Ravensberg, C.D.; Hobbelen, J.S.M.; Olde Rikkert, M.G.M.; Staal, J.B.; Nijhuis-van der Sanden, M.W.G. Effects of physical exercise therapy on mobility, physical functioning, physical activity and quality of life in community-dwelling older adults with impaired mobility, physical disability and/or multi-morbidity: A meta-analysis. *Ageing Res. Rev.* **2012**, *11*, 136–149. [CrossRef] [PubMed]

97. Melanson, E.L. The effect of exercise on non-exercise physical activity and sedentary behavior in adults. *Obes. Rev.* **2017**, *18*, 40–49. [CrossRef]

98. Ruotsalainen, H.; Kyngäs, H.; Tammelin, T.; Kääriäinen, M. Systematic review of physical activity and exercise interventions on body mass indices, subsequent physical activity and psychological symptoms in overweight and obese adolescents. *J. Adv. Nurs.* **2015**, *71*, 2461–2477. [CrossRef]

99. Anderson, L.; Thompson, D.R.; Oldridge, N.; Zwisler, A.D.; Rees, K.; Martin, N.; Taylor, R.S. Exercise-based cardiac rehabilitation for coronary heart disease. *Cochrane Database Syst. Rev.* **2016**. [CrossRef]

100. Price, K.J.; Gordon, B.A.; Bird, S.R.; Benson, A.C. A review of guidelines for cardiac rehabilitation exercise programmes: Is there an international consensus? *Eur. J. Prev. Cardiol.* **2016**, *23*, 1715–1733. [CrossRef]

101. Kraus, W.E.; Powell, K.E.; Haskell, W.L.; Janz, K.F.; Campbell, W.W.; Jakicic, J.M.; Troiano, R.P.; Sprow, K.; Torres, A.; Piercy, K.L. Physical Activity, All-Cause and Cardiovascular Mortality, and Cardiovascular Disease. *Med. Sci. Sports Exerc.* **2019**, *51*, 1270. [CrossRef]

102. Saeidifard, F.; Medina-Inojosa, J.R.; West, C.P.; Olson, T.P.; Somers, V.K.; Bonikowske, A.R.; Prokop, L.J.; Vinciguerra, M.; Lopez-Jimenez, F. The association of resistance training with mortality: A systematic review and meta-analysis. *Eur. J. Prev. Cardiol.* **2019**, *26*, 1647–1665. [CrossRef]

103. Janzon, E.; Abidi, T.; Bahtsevani, C. Can physical activity be used as a tool to reduce depression in patients after a cardiac event? What is the evidence? A systematic literature study. *Scand. J. Psychol.* **2015**, *56*, 175–181. [CrossRef] [PubMed]

104. Gellis, Z.D.; Kang-Yi, C. Meta-analysis of the effect of cardiac rehabilitation interventions on depression outcomes in adults 64 years of age and older. *Am. J. Cardiol.* **2012**, *110*, 1219–1224. [CrossRef] [PubMed]

105. Martland, R.; Mondelli, V.; Gaughran, F.; Stubbs, B. Can high-intensity interval training improve physical and mental health outcomes? A meta-review of 33 systematic reviews across the lifespan. *J. Sports Sci.* **2020**, *38*, 430–469. [CrossRef] [PubMed]

106. Younge, J.O.; Gotink, R.A.; Baena, C.P.; Roos-Hesselink, J.W.; Hunink, M.G.M. Mind-body practices for patients with cardiac disease: A systematic review and meta-analysis. *Eur. J. Prev. Cardiol.* **2015**, *11*, 1385–1398. [CrossRef]

107. Puetz, T.W.; Beasman, K.M.; O'connor, P.J. The effect of cardiac rehabilitation exercise programs on feelings of energy and fatigue: A meta-analysis of research from 1945 to 2005. *Eur. J. Prev. Cardiol.* **2006**, *13*, 886–893. [CrossRef]

108. Toni, G.; Belvederi Murri, M.; Piepoli, M.; Zanetidou, S.; Cabassi, A.; Squatrito, S.; Bagnoli, L.; Piras, A.; Mussi, C.; Senaldi, R.; et al. Physical Exercise for Late-Life Depression: Effects on Heart Rate Variability. *Am. J. Geriatr. Psychiatr.* **2016**, *24*, 989–997. [CrossRef]

109. Kahl, K.G.; Kerling, A.; Tegtbur, U.; Gützlaff, E.; Herrmann, J.; Borchert, L.; Ates, Z.; Westhoff-Bleck, M.; Hueper, K.; Hartung, D. Effects of additional exercise training on epicardial, intra-abdominal and subcutaneous adipose tissue in major depressive disorder: A randomized pilot study. *J. Affect. Disord.* **2016**, *192*, 91–97. [CrossRef]

110. Knapen, J.; Vancampfort, D.; Moriën, Y.; Marchal, Y. Exercise therapy improves both mental and physical health in patients with major depression. *Disabil. Rehabil.* **2015**, *37*, 1490–1495. [CrossRef]

111. Siqueira, C.C.; Valiengo, L.L.; Carvalho, A.F.; Santos-Silva, P.R.; Missio, G.; De Sousa, R.T.; Di Natale, G.; Gattaz, W.F.; Moreno, R.A.; Machado-Vieira, R. Antidepressant efficacy of adjunctive aerobic activity and associated biomarkers in major depression: A 4-week, randomized, single-blind, controlled clinical trial. *PLoS ONE* **2016**, *11*, e0154195. [CrossRef]

112. Stubbs, B.; Rosenbaum, S.; Vancampfort, D.; Ward, P.B.; Schuch, F.B. Exercise improves cardiorespiratory fitness in people with depression: A meta-analysis of randomized control trials. *J. Affect. Disord.* **2016**, *190*, 249–253. [CrossRef]

113. Schuch, F.B.; Vancampfort, D.; Rosenbaum, S.; Richards, J.; Ward, P.B.; Veronese, N.; Solmi, M.; Cadore, E.L.; Stubbs, B. Exercise for depression in older adults: A meta-analysis of randomized controlled trials adjusting for publication bias. *Rev. Bras. Psiquiatr.* **2016**, *38*, 247–254. [CrossRef] [PubMed]

114. Ekkekakis, P. Honey, i shrunk the pooled SMD! Guide to critical appraisal of systematic reviews and meta-analyses using the Cochrane review on exercise for depression as example. *Ment. Health Phys. Act.* **2015**, *8*, 21–36. [CrossRef]

115. Stubbs, B.; Vancampfort, D.; Rosenbaum, S.; Ward, P.B.; Richards, J.; Soundy, A.; Veronese, N.; Solmi, M.; Schuch, F.B. Dropout from exercise randomized controlled trials among people with depression: A meta-analysis and meta regression. *J. Affect. Disord.* **2016**, *190*, 457–466. [CrossRef] [PubMed]

116. Smits, J.A.J.; Zvolensky, M.J.; Davis, M.L.; Rosenfield, D.; Marcus, B.H.; Church, T.S.; Powers, M.B.; Frierson, G.M.; Otto, M.W.; Hopkins, L.B.; et al. The efficacy of vigorous-intensity exercise as an aid to smoking cessation in adults with high anxiety sensitivity: A randomized controlled trial. *Psychosom. Med.* **2016**, *78*, 354. [CrossRef] [PubMed]

117. Patten, C.A.; Bronars, C.A.; Douglas, K.S.V.; Ussher, M.H.; Levine, J.A.; Tye, S.J.; Hughes, C.A.; Brockman, T.A.; Decker, P.A.; DeJesus, R.S.; et al. Supervised, vigorous intensity exercise intervention for depressed female smokers: A pilot study. *Nicot. Tob. Res.* **2017**, *19*, 77–86. [CrossRef] [PubMed]

118. Abrantes, A.M.; Blevins, C.E.; Battle, C.L.; Read, J.P.; Gordon, A.L.; Stein, M.D. Developing a Fitbit-supported lifestyle physical activity intervention for depressed alcohol dependent women. *J. Subst. Abus. Treat.* **2017**, *80*, 88–97. [CrossRef]

119. Wang, D.; Wang, Y.; Wang, Y.; Li, R.; Zhou, C. Impact of physical exercise on substance use disorders: A meta-analysis. *PLoS ONE* **2014**, *9*, e110728. [CrossRef]

120. Euteneuer, F.; Dannehl, K.; Del Rey, A.; Engler, H.; Schedlowski, M.; Rief, W. Immunological effects of behavioral activation with exercise in major depression: An exploratory randomized controlled trial. *Transl. Psychiatr.* **2017**, *7*, e1132. [CrossRef]

121. Beserra, A.H.N.; Kameda, P.; Deslandes, A.C.; Schuch, F.B.; Laks, J.; de Moraes, H.S. Can physical exercise modulate cortisol level in subjects with depression? A systematic review and meta-analysis. *Trends Psychiatr. Psychother.* **2018**, *40*, 360–368. [CrossRef]

122. Kandola, A.; Ashdown-Franks, G.; Hendrikse, J.; Sabiston, C.M.; Stubbs, B. Physical activity and depression: Towards understanding the antidepressant mechanisms of physical activity. *Neurosci. Biobehav. Rev.* **2019**, *107*, 525–539. [CrossRef]

123. Schuch, F.B.; Deslandes, A.C.; Stubbs, B.; Gosmann, N.P.; Silva, C.T.B.; da Fleck, M.P.d.A. Neurobiological effects of exercise on major depressive disorder: A systematic review. *Neurosci. Biobehav. Rev.* **2016**, *61*, 1–11. [CrossRef] [PubMed]

124. Pope, B.S.; Wood, S.K. Advances in understanding mechanisms and therapeutic targets to treat comorbid depression and cardiovascular disease. *Neurosci. Biobehav. Rev.* **2020**. [CrossRef]

125. National Collaborating Centre for Mental Health. *Depression: The Nice Guideline on the Treatment and Management of Depression in Adults*; National Collaborating Centre for Mental Health: London, UK, 2010.

126. Sarris, J.; O'Neil, A.; Coulson, C.E.; Schweitzer, I.; Berk, M. Lifestyle medicine for depression. *BMC Psychiatr.* **2014**, *14*, 1–13. [CrossRef] [PubMed]

127. Cairns, K.E.; Yap, M.B.H.; Pilkington, P.D.; Jorm, A.F. Risk and protective factors for depression that adolescents can modify: A systematic review and meta-analysis of longitudinal studies. *J. Affect. Disord.* **2014**, *169*, 61–75. [CrossRef] [PubMed]

128. Baron, D.A.; Lasarow, S.; Baron, S.H. Exercise for the treatment of depression. In *Physical Exercise Interventions for Mental Health*; Lam, L.C.W., Riba, M., Eds.; Cambridge University Press: Cambridge, UK, 2016; pp. 26–40.

129. Nyström, M.B.T.; Neely, G.; Hassmén, P.; Carlbring, P. Treating Major Depression with Physical Activity: A Systematic Overview with Recommendations. *Cogn. Behav. Ther.* **2015**, *44*, 341–352. [CrossRef] [PubMed]

130. Rethorst, C.D.; Trivedi, M.H. Evidence-based recommendations for the prescription of exercise for major depressive disorder. *J. Psychiatr. Pract.* **2013**, *19*, 204–212. [CrossRef]

131. Stanton, R.; Happell, B.M. An exercise prescription primer for people with depression. *Issues Ment. Health Nurs.* **2013**, *34*, 626–630. [CrossRef]

132. Stanton, R.; Reaburn, P. Exercise and the treatment of depression: A review of the exercise program variables. *J. Sci. Med. Sport* **2014**, *17*, 177–182. [CrossRef] [PubMed]

133. Murri, M.B.; Ekkekakis, P.; Magagnoli, M.; Zampogna, D.; Cattedra, S.; Capobianco, L.; Serafini, G.; Calcagno, P.; Zanetidou, S.; Amore, M. Physical exercise in major depression: Reducing the mortality gap while improving clinical outcomes. *Front. Psychiatr.* **2019**, *9*, 762. [CrossRef] [PubMed]

134. Searle, A.; Calnan, M.; Turner, K.M.; Lawlor, D.A.; Campbell, J.; Chalder, M.; Lewis, G. General Practitioners' beliefs about physical activity for managing depression in primary care. *Ment. Health Phys. Act.* **2012**, *5*, 13–19. [CrossRef]

135. Stanton, R.; Franck, C.; Reaburn, P.; Happell, B. A Pilot Study of the Views of General Practitioners Regarding Exercise for the Treatment of Depression. *Perspect. Psychiatr. Care* **2014**, *51*, 253–259. [CrossRef] [PubMed]

136. Searle, A.; Calnan, M.; Lewis, G.; Campbell, J.; Taylor, A.; Turner, K. Patients' views of physical activity as treatment for depression: A qualitative study. *Br. J. Gen. Pract.* **2011**, *61*, e149–e156. [CrossRef] [PubMed]

137. Firth, J.; Rosenbaum, S.; Stubbs, B.; Gorczynski, P.; Yung, A.R.; Vancampfort, D. Motivating factors and barriers towards exercise in severe mental illness: A systematic review and meta-analysis. *Psychol. Med.* **2016**, *46*, 2869–2881. [CrossRef] [PubMed]

138. Powell, K.E.; King, A.C.; Buchner, D.M.; Campbell, W.W.; DiPietro, L.; Erickson, K.I.; Hillman, C.H.; Jakicic, J.M.; Janz, K.F.; Katzmarzyk, P.T.; et al. The scientific foundation for the physical activity guidelines for Americans. *J. Phys. Act. Health* **2019**, *16*, 1–11. [CrossRef] [PubMed]

139. Garber, C.E.; Blissmer, B.; Deschenes, M.R.; Franklin, B.A.; Lamonte, M.J.; Lee, I.M.; Nieman, D.C.; Swain, D.P. Quantity and quality of exercise for developing and maintaining cardiorespiratory, musculoskeletal, and neuromotor fitness in apparently healthy adults: Guidance for prescribing exercise. *Med. Sci. Sports Exerc.* **2011**, *43*, 1334–1359. [CrossRef] [PubMed]

140. Ladwig, M.A.; Hartman, M.E.; Ekkekakis, P. Affect-based Exercise Prescription: An Idea Whose Time Has Come? *ACSMs. Health Fit. J.* **2017**, *21*, 10–15. [CrossRef]

141. Mata, J.; Hogan, C.L.; Joormann, J.; Waugh, C.E.; Gotlib, I.H. Acute exercise attenuates negative affect following repeated sad mood inductions in persons who have recovered from depression. *J. Abnorm. Psychol.* **2013**, *122*, 45–50. [CrossRef]

142. Mata, J.; Thompson, R.J.; Jaeggi, S.M.; Buschkuehl, M.; Jonides, J.; Gotlib, I.H. Walk on the bright side: Physical activity and affect in major depressive disorder. *J. Abnorm. Psychol.* **2012**, *121*, 297–308. [CrossRef]

143. Weinstein, A.A.; Deuster, P.A.; Francis, J.L.; Beadling, C.; Kop, W.J. The role of depression in short-term mood and fatigue responses to acute exercise. *Int. J. Behav. Med.* **2010**, *17*, 51–57. [CrossRef]

144. Rethorst, C.D.; South, C.C.; Rush, A.J.; Greer, T.L.; Trivedi, M.H. Prediction of treatment outcomes to exercise in patients with nonremitted major depressive disorder. *Depress. Anxiety* **2017**, *34*, 1116–1122. [CrossRef]

145. Suterwala, A.M.; Rethorst, C.D.; Carmody, T.J.; Greer, T.L.; Grannemann, B.D.; Jha, M.; Trivedi, M.H. Affect following first exercise session as a predictor of treatment response in depression. *J. Clin. Psychiatr.* **2016**, *77*, 1036–1042. [CrossRef] [PubMed]

146. Rosenbaum, S.; Hobson-Powell, A.; Davison, K.; Stanton, R.; Craft, L.L.; Duncan, M.; Elliot, C.; Ward, P.B. The Role of Sport, Exercise, and Physical Activity in Closing the Life Expectancy Gap for People with Mental Illness: An International Consensus Statement by Exercise and Sports Science Australia, American College of Sports Medicine, British Association. *Transl. J. Am. Coll. Sports Med.* **2018**, *3*, 72–73.

147. Gebreslassie, M.; Sampaio, F.; Nystrand, C.; Ssegonja, R.; Feldman, I. Economic evaluations of public health interventions for physical activity and healthy diet: A systematic review. *Prev. Med.* **2020**, *136*, 106100. [CrossRef] [PubMed]

148. Park, A.L.; McDaid, D.; Weiser, P.; Von Gottberg, C.; Becker, T.; Kilian, R. Examining the cost effectiveness of interventions to promote the physical health of people with mental health problems: A systematic review. *BMC Public Health* **2013**, *13*, 787. [CrossRef] [PubMed]

149. Shuval, K.; Leonard, T.; Drope, J.; Katz, D.L.; Patel, A.V.; Maitin-Shepard, M.; Amir, O.; Grinstein, A. Physical activity counseling in primary care: Insights from public health and behavioral economics. *CA Cancer J. Clin.* **2017**, *67*, 233–244. [CrossRef] [PubMed]
150. Zanetidou, S.; Belvederi Murri, M.; Menchetti, M.; Toni, G.; Asioli, F.; Bagnoli, L.; Zocchi, D.; Siena, M.; Assirelli, B.; Luciano, C.; et al. Physical Exercise for Late-Life Depression: Customizing an Intervention for Primary Care. *J. Am. Geriatr. Soc.* **2017**, *65*, 348–355. [CrossRef] [PubMed]

International Journal of
Environmental Research and Public Health

Article

Levels and Correlates of Physical Activity in Rural Ingwavuma Community, uMkhanyakude District, KwaZulu-Natal, South Africa

Herbert Chikafu * and Moses J. Chimbari

School of Nursing and Public Health, College of Health Sciences, University of KwaZulu-Natal, Durban 4001, South Africa; chimbari@ukzn.ac.za
* Correspondence: chikafuh@gmail.com

Received: 27 May 2020; Accepted: 5 July 2020; Published: 16 September 2020

Abstract: Physical activity, among others, confers cardiovascular, mental, and skeletal health benefits to people of all age-groups and health states. It reduces the risks associated with cardiovascular disease and therefore, could be useful in rural South Africa where cardiovascular disease (CVD) burden is increasing. The objective of this study was to examine levels and correlates of physical activity among adults in the Ingwavuma community in KwaZulu-Natal (KZN). Self-reported data on physical activity from 392 consenting adults (female, $n = 265$; male, $n = 127$) was used. We used the one-sample t-test to assess the level of physical activity and a two-level multiple linear regression to investigate the relationship between total physical activity (TPA) and independent predictors. The weekly number of minutes spent on all physical activities by members of the Ingwavuma community was 912.2; standard deviation (SD) (870.5), with males having 37% higher physical activity (1210.6 min, SD = 994.2) than females (769.2, SD = 766.3). Livelihood activities constituted 65% of TPA, and sport and recreation contributed 10%. Participants without formal education (20%), those underweight (27%), and the obese (16%) had low physical activity. Notwithstanding this, in general, the Ingwavuma community significantly exceeded the recommended weekly time on physical activity with a mean difference of 762.1 (675.8–848.6) minutes, t (391) = 17.335, $p < 0.001$. Gender and age were significant predictors of TPA in level 1 of the multiple regression. Males were significantly more active than females by 455.4 min ($\beta = -0.25$, $p < 0.001$) and participants of at least 60 years were significantly less active than 18–29-year-olds by 276.2 min ($\beta = -0.12$, $p < 0.05$). Gender, marital status, and health awareness were significant predictors in the full model that included education level, employment status, body mass index (BMI), and physical activity related to health awareness as predictors. The high prevalence of insufficient physical activity in some vulnerable groups, notably the elderly and obese, and the general poor participation in sport and recreation activities are worrisome. Hence we recommend health education interventions to increase awareness of and reshape sociocultural constructs that hinder participation in leisure activities. It is important to promote physical activity as a preventive health intervention and complement the pharmacological treatment of CVDs in rural South Africa. Physical activity interventions for all sociodemographic groups have potential economic gains through a reduction in costs related to the treatment of chronic CVD.

Keywords: cardiovascular disease; healthy ageing; Ingwavuma; KwaZulu-Natal; modifiable behaviour; physical activity; rural; South Africa

---

1. Introduction

The burden of noncommunicable diseases, particularly cardiovascular disease (CVD), is increasing in low-to-medium income countries [1,2], South Africa included [3,4]. The rising burden of CVD in low-

and middle-income countries (LMICs) is ascribed to multiple factors, including rising life expectancy, nutritional transition, poverty, improvement in income levels, and globalisation [5]. In addition to an age-related risk, unhealthy lifestyle practices exacerbate the risk of hypertension, type 2 diabetes mellitus (T2DM), and obesity that are part of CVD aetiology. Alcohol abuse, smoking, unhealthy diet, and physical inactivity also increase the risk of CVD [6,7].

However, health behaviour modification moderates the risk and onset of CVD. Physical activity confers health gains such as enhanced lipoprotein profile that reduces the likelihood of developing coronary heart disease (CHD) [8,9]. While any amount of physical activity, including walking, confers health benefits [10], sufficient moderate-to-vigorous-intensity physical activity augments the management of existing CVD conditions, enhances endothelial function, and lowers the of risk of atherosclerosis, hypertension, obesity, and T2DM [8]. In South Africa, a significant proportion of ischemic heart disease (IHD), ischemic stroke, T2DM, and breast cancer has been associated with physical inactivity [7].

Other than CVD, inactivity is related to numerous adverse outcomes. It is considered a risk factor for mental health disorders, chronic obstructive pulmonary diseases, some cancers, osteoporosis, and all-cause mortality [11–14]. The World Health Organization (WHO) recommends adults to perform at least 150 min of moderate-intensity physical activity or 75 min of vigorous-intensity physical activity or their combined equivalent across five days per week to moderate some of the risks [15]. The guidelines also recommend increasing physical activity to an equivalent of 300 min of moderate-intensity physical activity for additional health benefits. The health benefits of physical activity accrue across sociodemographic categories and health states, albeit relative to the volume of physical activity up to an optimum point [8,16,17].

It is essential to understand the levels and correlates of physical activity in rural South Africa for evidence-based public health interventions against the backdrop of the rising burden of noncommunicable diseases (NCDs) and ageing. Studies conducted in rural South Africa have shown significant inter and intraregional variation in physical activity levels across sociodemographic variables. Notably, physical activity mainly comprises of livelihood activities and varies with gender, age, marital status, and body mass index [18–21]. We, therefore, examined the levels and correlates of physical activity among adults aged 18 and above in Ingwavuma, a rural community in KwaZulu-Natal (KZN) addressing the following specific questions: (a) What is the level of physical activity in the community and does it meet WHO guidelines? (b) What are the components of physical activity in the community? And (c) What are the correlates of physical activity?

2. Methods

2.1. Study Design and Setting

This cross-sectional, observational and analytical study was conducted in Ingwavuma rural community (wards 13, 16, and 17) in Jozini Local Municipality under the uMkhanyakude district municipality in KZN, South Africa. Most households in the uMkhanyakude District depend on small scale subsistence agriculture [22] and social security disbursements for income [23,24]. The uMkhanyakude region experiences erratic rainfall ranging from 600 to 1000 mm per annum [25,26]. Fluctuant rainfall has affected forest cover and agricultural production. Consequently, poverty and food insecurity have worsened in uMkhanyakude [25,27], with only a quarter (26%) of households reporting sufficient food stocks at all times [23]. Jozini local municipality is the most populous municipality in uMkhanyakhude District and comprises of 20 administrative wards. It ranks among the least developed regions of the district with limited access to amenities including tapped water and has high levels of unemployment, poverty, and HIV/AIDS [24].

The Ingwavuma area is located in the north-eastern corner of KZN along the South Africa–Mozambique, and South Africa–Eswatini borders to the north and west, respectively (Figure 1). Like other rural areas in Jozini local municipality, Ingwavuma is inhabited by predominantly isiZulu speaking people under traditional leadership structures. The area is semi-arid and is characterised by three distinct seasons. It has a short winter season between June and August, followed by a hot, dry season until November and a humid summer season until March. The area is hilly with seasonal dams, streams, and rivers that sustain the community's domestic water needs and small-scale gardening. Ingwavuma's sociodemographic indicators are similar to other sub-regions in the district with high levels of poverty, unemployment, and HIV/AIDS. The area also has low levels of literacy and access to basic and social amenities.

Figure 1. The map of South Africa showing Ingwavuma rural community in north-eastern KwaZulu-Natal (KZN). Source: Google Maps.

2.2. Participants and Procedures

We enrolled adult (≥18 years) and non-disabled participants through convenience sampling from randomly selected villages in the study area. Data were collected over two phases. Firstly, the isiZulu questionnaire was administered to participants at their homes through face-to-face interviews by a team of three community research assistants (CRAs) supervised by the investigator. A trained nurse then conducted physical measurements. The CRAs were native isiZulu speakers and had completed at least high school (Metric Level for South Africa). Data were collected electronically using the KoBo-Collect application (Cambridge, MA, USA) on Android mobile devices [28]. This study was approved by the University of KwaZulu-Natal Biomedical Research Ethics Committee (BREC/00000235/2019), and all participants provided written informed consent.

2.3. Measures

We adapted the validated WHO STEPwise approach to surveillance (STEPS) questionnaire [29] to collect data on physical activity. The adaptation of the questionnaire was intended to contextualise questions and examples. The English version questionnaire was translated into isiZulu by a native speaker, revised for linguistic appropriateness with native isiZulu speaking community research assistants (CRAs) during training, and back-translated to English to ensure comparability with initial questions. We pretested the questionnaire in a village within the study area with 30 randomly selected and representative participants. The village was excluded from the main study. Minor revision of terminology and sequencing of questions was done using insights from the pretest. The final questionnaire with open-ended and closed-ended questions had the following five sections: sociodemographic data (age, gender, marital status, formal education level, and occupation), health awareness and knowledge, healthcare utilisation, and modifiable health behaviours.

The following self-reported physical activity data were collected: (a) forms of physical activity (livelihood, travel, and sport and recreational activities); (b) intensity level, whether moderate-intensity physical activity or vigorous physical activity; (c) weekly frequency; (d) estimate of time spent on physical activity; and (e) physical activity-related health awareness and knowledge. Time spent on physical activity was defined in moderate-intensity minutes, where a unit of vigorous-intensity equalled two units of moderate-intensity physical activity. Total physical activity (TPA) was the sum of the product of mean daily duration and weekly frequency of physical activity. The outcome variable TPA was expressed as a: (a) continuous variable to assess the level and predictors of physical activity, and (b) categorical variable to classify TPA against WHO recommendations.

The following categorical predictors were used in the analysis: Age (<30, 30–39, 40–49, 50–59, 60+ years), gender (male, female), formal education level (none, primary, secondary, post-secondary), employment status (unemployed, self-employed, employed, other), and body mass index (BMI) (underweight, normal-weight, overweight, obese). Two health-related predictors were also included, whether participants thought inactivity has adverse effects on health (yes, no, do not know) and whether participants were ever informed to be active (yes, no, do not remember).

2.4. Analyses

Descriptive statistics are presented as means (M) with their standard deviation (SD) for continuous variables, and as frequencies with corresponding percentages for categorical variables. We conducted the one-sample t-test to assess the community's physical activity time against the recommended 150 min, and a two-level multiple linear regression to investigate the relationship between TPA and independent predictors. Model 1 controlled for age and gender as non-modifiable risk factors of physical activity. Formal education level, employment status, BMI, and proxies of health awareness were added into model 2 as modifiable factors. Non-modifiable factors are acquiescent to interventions [30], hence the use of hierarchical regression to study their influence on physical activity. We evaluated linear regression assumptions. Residual scatter plots of standardised predicted values were used to assess homoscedasticity and linearity, and both assumptions were met. Although visual inspection of histograms and residual plots did not suggest non-normal distribution, results of the Shapiro–Wilk test (SW) indicated a violation of the normality assumption (SW = 0.83, df = 392, $p < 0.05$). Nonetheless, multiple linear regression was used in view of the robustness of estimates from large sample data [31–33]. Statistical significance was evaluated at the 0.05 alpha level for all tests. We used IBM SPSS Statistics version 26 (IBM Corp., Armonk, NY, USA) to conduct statistical analyses.

3. Results

Table 1 presents an overview of participant characteristics. A total of 400 consenting participants were recruited into the study, and 392 participants with a mean age of 42.7 years (SD = 17.38) had complete data required for this assessment. The gender distribution was 67% ($n = 265$) female and 32% ($n = 127$) male. Participants in the 18–29 age group ($n = 116$, 30%) and 30–39 ($n = 81$, 21 %) constituted 50% of the sample. About two-thirds of the sample ($n = 244$, 62%) were single. The majority ($n = 327$, 83%) of study participants were unemployed with comparable levels across gender. The mean weight of the participants was 69.1 kg (SD = 18.2). One-fifth were obese, with a notably higher proportion of females (26%) than males (9%).

Results of the one-sample t-test show that the Ingwavuma community's mean TPA (M = 912.2, SD = 870.5) significantly exceeds the WHO recommended 150 min of weekly physical activity, with a statistically significant mean difference of 762.1 (675.8–848.6), t (391) = 17.335, $p < 0.001$. Overall, the Ingwavuma community spent 912.2 (SD = 870.5) minutes on all physical activity (see supplementary file Table S1). However, there were distinct gender differences; males had 37% higher physical activity (1210.6 min, SD = 994.2) than females (769.2, SD = 766.3). Physical activity was also lower among participants aged at least 60 years (M = 714.5, SD = 721.9) and obese participants (M = 657.5, SD = 582.8) who were 33% less active than their normal-weight counterparts. Overall, livelihood activities contributed the most (65%) to TPA with almost even distribution between vigorous- and moderate-intensity activities. Two-thirds (67%) of participants indicated they grew crops in their fields to supplement their food requirements. On the other hand, sport and recreation had the least contribution (10%) to TPA. Time spent on all physical activity was consistently higher among males than females with a six-fold difference between the two genders in sport and recreational activities. Although there was no distinct age-related pattern on livelihood and travel physical activity, there were noteworthy differences in TPA in other predictors. Unmarried participants were most active, while the unemployed and obese ranked lowest. Time on sport and recreation varied inversely with age. Young adults under 30 years spent 201.6 min (SD = 306.9) on sport and recreational activities, while the elderly (60+ years) did not participate in similar activities. Additionally, only a fifth (23%) of participants believed that inactivity could result in adverse health outcomes.

Table 1. Demographic characteristics of the Participants.

Variables		Female		Male		Total	
		$n\left(\overline{X}\right)$	% (SD)	$n\left(\overline{X}\right)$	% (SD)	$n\left(\overline{X}\right)$	% (SD)
Age		42.1	17.5	43.9	17.2	42.7	17.4
Age group	18–29	82	30.9	34	26.8	116	29.6
	30–39	56	21.1	25	19.7	81	20.7
	40–49	41	15.5	18	14.2	59	15.1
	50–59	40	15.1	24	18.9	64	16.3
	60+	46	17.4	26	20.5	72	18.4
Marital status	Single	166	62.6	78	61.4	244	62.2
	Married	42	15.9	20	15.8	62	15.8
	Cohabiting	36	13.6	25	19.7	61	15.6
	Widowed/divorced	21	7.9	4	3.2	25	6.4
Education level	None	88	33.2	35	27.6	123	31.4
	Primary school	67	25.3	38	29.9	105	26.8
	Secondary school	106	40.0	46	36.2	152	38.8
	Post-secondary	4	1.51	8	6.3	12	3.1
Occupational status	Unemployed	229	86.4	98	77.2	327	83.4
	Self employed	19	7.2	9	7.1	28	7.1
	Employed	4	1.5	10	7.9	14	3.6
	Other	13	4.9	10	7.9	23	5.9
Body mass index	Underweight	14	5.3	12	9.5	26	6.6
	Normal-weight	91	34.2	75	59.1	166	42.2
	Overweight	87	32.7	26	205	113	28.8
	Obese	74	27.8	14	11.0	88	22.4
Grow crops in field	No	86	32.5	45	35.4	131	33.4
	Yes	179	67.6	82	64.6	261	66.6
Inactivity may lead to poor health outcomes	No	177	66.8	97	76.4	274	69.9
	Yes	67	25.3	22	17.3	89	22.7
	Don't know	21	7.9	8	6.3	29	7.4
Advised by a health worker to be physically active	No	109	41.0	57	44.9	166	42.2
	Yes	124	46.6	56	44.1	180	45.8
	Don't remember	32	12.4	14	11.0	46	12.0
Total		265	100	127	100	392	100

Notes: $\overline{X}$—mean; SD—Standard deviation.

Table 2 summarizes the classification of physical activities. Almost a tenth (11%) and three quarters (75%) of study participants had low and high TPA levels, respectively. More males (84%) than females (70%) achieved high TPA, and most age groups had comparable proportions of participants in the three TPA classifications except for the 60+ year age group. Two-thirds of the elderly (60+ years) had low TPA, almost three-fold (28%) of the community level. Relative to other sub-categories, low physical activity was higher among participants without formal education (20%), the underweight (27%), and obese (16%). Adjusted for age, the prevalence of low TPA was more prevalent among females for all but those aged 50–59 (Figure 2). Insufficient activity levels also varied with gender. It was higher among females except for 45–59-year-olds and least among males aged 18–29 years (3%), 30–44-year-old males (0%), and females in the 45–59-year-old group (4%).

Table 2. Classification of weekly physical activity.

Variable		Weekly Physical Activity Level (%)		
		Low	**Sufficient**	**High**
Overall		10.9	14.3	74.8
Gender	Female	12.0	17.7	70.3
	Male	8.7	7.1	84.3
Age group	<30	7.7	14.5	77.8
	30–39	4.9	12.4	82.7
	40–49	8.5	6.8	84.8
	50–59	7.8	18.8	73.4
	60+	27.8	18.1	54.2
Marital status	Single	6.5	12.2	81.2
	Cohabiting	16.1	19.4	64.5
	Married	14.8	16.4	68.9
	Other	32.0	16.0	52.0
Education	None	19.5	13.8	66.7
	Primary	6.7	19.1	74.3
	Secondary	7.2	12.4	80.4
	Post-secondary	8.3	0.0	91.7
Occupational status	Unemployed	12.5	14.0	73.5
	Self-employed	3.6	17.9	78.6
	Employed	0.0	7.1	92.9
	Other	4.4	17.4	78.3
Body mass index	Underweight	26.9	7.7	65.4
	Normal-weight	7.8	13.3	78.9
	Overweight	8.6	12.9	78.5
	Obese	16.1	17.3	66.7
Inactivity may lead to poor health outcomes	No	16.7	16.7	66.7
	Yes	8.4	13.1	78.5
	Don't know	17.2	17.2	65.5
Advised to be physically active	No	10.8	17.5	71.7
	Yes	11.7	11.1	77.2
	Don't remember	8.5	14.9	76.6

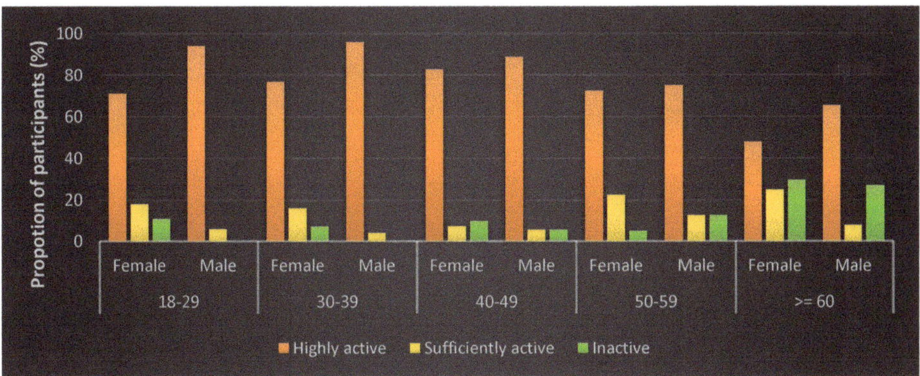

Figure 2. Physical activity classification by age and gender.

Results of the two-step hierarchical linear regression analysis conducted to understand sociodemographic predictors of TPA are presented in Table 3. From the set of dummy predictors entered into model 1 (gender and age group), the model significantly predicted TPA, F (5, 386) = 7.853, $p < 0.001$. A highly significant difference in TPA was predicted between males and females, with males significantly more active by 455.4 min ($\beta = -0.25$, $p < 0.001$). Compared with younger adults (<30 years), elderly participants aged at least 60 years ($\beta = -0.12$, $p < 0.05$) were predicted to attain significantly lower TPA. In the full model, male gender remained a significant predictor of higher TPA ($\beta = 0.22$, $p < 0.05$). From the set of modifiable factors added into model 2, being married ($\beta = -0.12$, $p < 0.05$), cohabiting ($\beta = -0.13$, $p < 0.05$), and lack of awareness on the adverse effects of inactivity were ($\beta = -0.13$, $p < 0.05$) were significant predictors of low TPA. Independent variables entered in model 2 significantly predicted the variability in TPA, F (20, 271) = 3.093, $p < 0.001$

Table 3. Results of hierarchical linear regression on time spent on physical activity in Ingwavuma ($N = 392$).

Predictors		Model 1 B	Model 1 β	Model 1 Sig.	Model 1 95% CI	Model 2 B	Model 2 β	Model 2 Sig.	Model 2 95% CI
	Constant	747.6		0.000	586.5–908.6	873.5		0.000	647.2–1099.7
Gender	(Female) Male	455.4	0.25	0.000	277.9–633.0	408.1	0.22	0.000	214.0–602.2
Age-group	(18–29) 30–39	164.4	0.08	0.175	−73.3–402.0	196.9	0.09	0.126	−55.8–449.6
	40–49	228.6	0.09	0.088	−33.8–491.1	293.7	0.12	0.067	−20.3–607.7
	50–59	−3.3	−0.00	0.980	−259.2–252.7	107.1	0.05	0.548	−243.4–457.7
	≥60	−276.2	−0.12	0.028	−522.8—−29.7	−109.5	−0.05	0.547	−467.2–248.1
Education	(Secondary) None					68.8	0.04	0.653	−231.8–369.4
	Primary					−37.0	−0.02	0.758	−280.0–204.1
	Post-secondary					−84.5	−0.02	0.744	−592.2–423.2
Marital status	(Single) Married					−259.1	−0.12	0.037	−503.0—−15.1
	Cohabiting					−322.3	−0.13	0.013	−577.4—−67.1
	Other					−313.2	−0.09	0.102	−688.6–62.2
Employment status	(Unemployed) Self employed					213.2	0.06	0.207	−118.6–545.1
	Employed					163.3	0.04	0.493	−304.7–631.3
	Other					−175.3	−0.05	0.362	−552.6–202.0
BMI	(Normal-weight) Underweight					−68.0	−0.02	0.704	−419.2–283.2
	Overweight					−6.5	−0.00	0.951	−215.8–202.8
	Obese					−203.7	−0.01	0.087	−437.4–29.9
Inactivity has adverse effects?	(Yes) No					−160.9	−0.08	0.156	−383.2–61.4
	Don't know					−339.7	−0.10	0.040	−663.8—−15.5
Advised to be active	(No) Yes					42.2	0.03	0.530	−89.9–174.4
	R-squared	0.92		0.000		0.14		0.118	
	Adjusted- R-squared	0.81				0.10			

Note: Reference categories of predictors are given in brackets. B—Beta coefficient; CI—Confidence interval.

4. Discussion

This study aimed to assess the levels and correlates of physical activity in rural KZN against the backdrop of the rising burden of CVDs and a trend towards sedentary living in South Africa. To the best of our knowledge, this study is the first to examine physical activity from a large sample of adults in the uMkhanyakude District of KwaZulu-Natal.

Participation in sports and recreational activities was low, with males participating more than females. This finding is consistent with evidence from studies in rural South Africa [34] and other African countries [35]. Low and gendered participation rate in sport and recreation is a concern amidst the rising prevalence of CVDs in rural South Africa. The low participation could be attributed to multiple factors, including a lack of awareness on the significance of sporting and recreational activities, unavailability of enabling infrastructure, and time constraints. The limited access to basic amenities in the area, notably tapped water, may result in time constraints for leisure activities. With such limitations, gendered roles and responsibilities from deep-rooted traditional practices [24] in the area are likely to affect women's participation in leisure activities.

Although physical activity is known to be inversely related to age, the high prevalence of insufficient physical activity among the elderly is worrisome relative to age-related health risks. Substantive evidence shows ageing-induced adverse outcomes, such as elevated CVD risk, reduced skeletal muscle performance, increased risk of falls and loss of bone mass [36,37], and depression [38], can be moderated by physical activity. Physical activity also has an established positive effect on the mental health of the elderly [38].

Several factors potentially explain why the elderly in Ingwavuma are not leveraging the benefits of physical activity. First, the lack of awareness proxied by the high proportion of inactive participants unaware of the adverse health outcomes of inactivity potentially explains the trend among adults. Second, strenuous livelihood and household activities are associated with severe musculoskeletal disorders. The high prevalence of back pain and knee osteoarthritis among the elderly in rural South Africa [39,40] possibly explains the insufficient physical activity. Third, unlike typical rural areas dependent on labour-intensive subsistence farming, agricultural activities have declined in the region over the past few decades due to erratic rainfall [27]. Fourth, the high unemployment in the region [24] likely spares the elderly from livelihood activities and domestic chores as younger adults may assume a significant share of activities. In the absence of leisure activity, this results in physical activity deficits for the elderly.

Health education may help with awareness of the significance and spectrum of sport and recreation activities that do not require extensive hard infrastructure. Our findings indicate the need for health education to showcase the range of minimal cost physical activities services available to rural residents and their potential benefits. Aerobic activities with short bouts of vigorous activities (such as brisk walking) confer significant health benefits for all age-groups and health states [41–43]. Notwithstanding the risk of arrhythmias in older people during physical activity, light intensity activities such as regular walking have been established to be beneficial [10]. As such, the elderly in rural communities should be encouraged to incorporate walking into their leisure activity package and consider it as a health routine than exclusively as part of achieving livelihood and social activities. Importantly, health promotion will also help to influence social constructs that may hinder participation in sport and recreational activities.

Most physical activity in rural South Africa is associated with livelihood activities and is seasonal. Therefore, it is crucial to encourage sport and recreation in rural areas to assist with smoothing seasonal variation in physical activity. It is noteworthy that there was almost no report of sport and recreation among the elderly whose age predisposes them to CVD. Evidence shows that sport and recreational activities have considerable mental, bone and cardiovascular health benefits for the elderly, more so for those with sedentary habits [44–46]. Importantly, leisure-time physical activity also ensures healthy ageing [46]. Gender-based peer activity groups could be explored as a means to promote sport and recreation among the elderly and females of all age groups in rural South Africa.

Our study reports a number of limitations. We only assessed the correlates of physical activity due to the cross-sectional study design hence we could not examine causal relationships that underlie physical activity in Ingwavuma. Furthermore, we relied on a self-reported assessment of physical activity and acknowledge that assessments could have understated or overstated duration, frequency, and intensity, and therefore the extent of physical activity [47]. Notably, estimates were prone to a floor effect as household chores may not have been fully captured in physical activity reports.

Notwithstanding the aforementioned limitations, the study also had noteworthy strengths. The inclusion of adults of both genders allowed for a multilevel assessment of the influence of modifiable factors on physical activity in a rural setting. Furthermore, the large sample size allowed for the assessment of key sociodemographic correlates of physical activity that could be generalised for the community.

5. Conclusions

In conclusion, low participation in sport and recreational activities and the high proportion of the elderly with insufficient activity are worrisome given the rising burden of CVDs. Furthermore, it is essential to encourage rural communities to increase physical activity, given its benefits of enhancing lipoprotein balance [8] and healthy ageing [46]. Physical activity interventions have potential economic gains by reducing the treatment costs of chronic morbidity that may result from the lower prevalence and better control of CVD and its risk factors.

Our policy related recommendation is the promotion of physical activity as a primary health intervention that complements pharmacological treatment of CVDs in rural South Africa, where the burden of lifestyle-related morbidity is rising. We also recommend the provision of rudimentary infrastructure in some areas with rugged terrain to promote physical activity by enhancing neighbourhood walkability [48]. For optimum outcomes, it is needful to develop multilevel physical activity interventions relevant to rural social context. Future research should aim to understand knowledge, attitudes, and perceptions of physical activity; and its social nuances in rural areas. Longitudinal studies with objectively measured physical activity are recommended to provide extant estimates of energy expenditure and assess the determinants of physical activity.

Supplementary Materials: The following are available online at http://www.mdpi.com/1660-4601/17/18/6739/s1, Table S1: Mean weekly minutes (standard deviation; SD) spent on livelihood, travel, and sport and recreational activities in Ingwavuma.

Author Contributions: H.C. conceptualised the study, collected data, conducted statistical analysis, and wrote the first draft. H.C. and M.J.C. reviewed and edited the draft. M.J.C. supervised the process. Both authors read and agreed to the published version of the manuscript. All authors have read and agreed to the published version of the manuscript.

Funding: This research was funded by Tackling Infections To Benefit Africa (TIBA) through grant UoERef:CT-4987(d) and the University of KwaZulu-Natal through a PhD studentship bursary awarded to H.C. by the College of Health Sciences.

Acknowledgments: We thank Ingwavuma community leaders and households for their participation in this research. The researchers are also grateful to Celiwe Tembe, Nozipho Mthembu, and Sinenhlanhla Mthembu for their meticulous work as Community Research Assistants, and TIBASA colleagues for their support.

Conflicts of Interest: The authors declare no conflict of interest.

References

1. WHO. *Global Status Report on Noncommunicable Diseases 2014*; World Health Organization: Geneva, Switzerland, 2014.
2. Abegunde, D.O.; Mathers, C.D.; Adam, T.; Ortegon, M.; Strong, K. The burden and costs of chronic diseases in low-income and middle-income countries. *Lancet* **2007**, *370*, 1929–1938. [CrossRef]
3. Pillay-van Wyk, V.; Msemburi, W.; Laubscher, R.; Dorrington, R.E.; Groenewald, P.; Glass, T.; Nojilana, B.; Joubert, J.D.; Matzopoulos, R.; Prinsloo, M.; et al. Mortality trends and differentials in South Africa from 1997 to 2012: Second national burden of disease study. *Lancet Glob. Health* **2016**, *4*, e642–e653. [CrossRef]

4. Maredza, M.; Bertram, M.Y.; Tollman, S.M. Disease burden of stroke in rural South Africa: An estimate of incidence, mortality and disability adjusted life years. *BMC Neurol.* **2015**, *15*, 54. [CrossRef] [PubMed]
5. Barquera, S.; Pedroza-Tobias, A.; Medina, C. Cardiovascular diseases in mega-countries: The challenges of the nutrition, physical activity and epidemiologic transitions, and the double burden of disease. *Curr. Opin. Lipidol.* **2016**, *27*, 329–344. [CrossRef] [PubMed]
6. Mokdad, A.H.; Marks, J.S.; Stroup, D.F.; Gerberding, J.L. Actual causes of death in the United States, 2000. *JAMA* **2004**, *291*, 1238–1245. [CrossRef]
7. Joubert, J.; Norman, R.; Lambert, E.V.; Groenewald, P.; Schneider, M.; Bull, F.; Bradshaw, D.; South African Comparative Risk Assessment Collaborating Group. Estimating the burden of disease attributable to physical inactivity in South Africa in 2000. *S. Afr. Med. J.* **2007**, *97*, 725–731. [PubMed]
8. Kraus, W.E.; Houmard, J.A.; Duscha, B.D.; Knetzger, K.J.; Wharton, M.B.; McCartney, J.S.; Bales, C.W.; Henes, S.; Samsa, G.P.; Otvos, J.D.; et al. Effects of the amount and intensity of exercise on plasma lipoproteins. *N. Engl. J. Med.* **2002**, *347*, 1483–1492. [CrossRef]
9. LeCheminant, J.D.; Tucker, L.A.; Bailey, B.W.; Peterson, T. The relationship between intensity of physical activity and HDL cholesterol in 272 women. *J. Phys. Act. Health* **2005**, *2*, 333–344. [CrossRef]
10. Wannamethee, S.G.; Shaper, A.G.; Walker, M. Physical activity and mortality in older men with diagnosed coronary heart disease. *Circulation* **2000**, *102*, 1358–1363. [CrossRef]
11. Kikuchi, H.; Inoue, S.; Lee, I.M.; Odagiri, Y.; Sawada, N.; Inoue, M.; Tsugane, S. Impact of moderate-intensity and vigorous-intensity physical activity on mortality. *Med. Sci. Sports Exerc.* **2018**, *50*, 715–721. [CrossRef]
12. Strong, W.B.; Malina, R.M.; Blimkie, C.J.; Daniels, S.R.; Dishman, R.K.; Gutin, B.; Hergenroeder, A.C.; Must, A.; Nixon, P.A.; Pivarnik, J.M.; et al. Evidence based physical activity for school-age youth. *J. Pediatr.* **2005**, *146*, 732–737. [CrossRef] [PubMed]
13. Hallal, P.C.; Victora, C.G.; Azevedo, M.R.; Wells, J.C. Adolescent physical activity and health: A systematic review. *Sports Med.* **2006**, *36*, 1019–1030. [CrossRef] [PubMed]
14. Van Langendonck, L.; Lefevre, J.; Claessens, A.L.; Thomis, M.; Philippaerts, R.; Delvaux, K.; Lysens, R.; Renson, R.; Vanreusel, B.; Vanden Eynde, B.; et al. Influence of participation in high-impact sports during adolescence and adulthood on bone mineral density in middle-aged men: A 27-year follow-up study. *Am. J. Epidemiol.* **2003**, *158*, 525–533. [CrossRef]
15. WHO. *Global Recommendations on Physical Activity for Health*; World Health Organization: Geneva, Switzerland, 2010.
16. US Department of Health and Human Services. *2008 Physical Activity Guidelines for Americans*; US Department of Health and Human Services: Washington, DC, USA, 2008.
17. Manson, J.E.; Greenland, P.; LaCroix, A.Z.; Stefanick, M.L.; Mouton, C.P.; Oberman, A.; Perri, M.G.; Sheps, D.S.; Pettinger, M.B.; Siscovick, D.S. Walking compared with vigorous exercise for the prevention of cardiovascular events in women. *N. Engl. J. Med.* **2002**, *347*, 716–725. [CrossRef]
18. Oyeyemi, A.L.; Moss, S.J.; Monyeki, M.A.; Kruger, H.S. Measurement of physical activity in urban and rural South African adults: A comparison of two self-report methods. *BMC Public Health* **2016**, *16*, 1004. [CrossRef] [PubMed]
19. Cook, I.; Alberts, M.; Brits, J.S.; Choma, S.R.; Mkhonto, S.S. Descriptive epidemiology of ambulatory activity in rural, black South Africans. *Med. Sci. Sports Exerc.* **2010**, *42*, 1261–1268. [CrossRef]
20. Kruger, H.S.; Venter, C.S.; Vorster, H.H.; Margetts, B.M. Physical inactivity is the major determinant of obesity in black women in the North West Province, South Africa: The THUSA study. *Nutrition* **2002**, *18*, 422–427. [CrossRef]
21. Maimela, E.; Alberts, M.; Modjadji, S.E.; Choma, S.S.; Dikotope, S.A.; Ntuli, T.S.; Van Geertruyden, J.P. The prevalence and determinants of chronic non-communicable disease risk factors amongst adults in the dikgale Health Demographic and Surveillance System (HDSS) site, Limpopo province of South Africa. *PLoS ONE* **2016**, *11*, e0147926. [CrossRef]
22. Jones, J.L. Transboundary conservation: Development implications for communities in KwaZulu-Natal, South Africa. *Int. J. Sustain. Dev. World Ecol.* **2009**, *12*, 266–278. [CrossRef]
23. Schoeman, S.; Faber, M.; Adams, V.; Smuts, C.; Ford-Ngomane, N.; Laubscher, J.; Dhansay, M. Adverse social, nutrition and health conditions in rural districts of the KwaZulu-Natal and Eastern Cape provinces, South Africa. *S. Afr. J. Clin. Nutr.* **2016**, *23*, 140–147. [CrossRef]

24. Nell, W.; de Crom, E.; Coetzee, H.; van Eeden, E. The psychosocial well-being of a "forgotten" South African community: The case of Ndumo, KwaZulu-Natal. *J. Psychol. Afr.* **2015**, *25*, 171–181. [CrossRef]
25. Markides, J.; Forsythe, L. *Looking back and Living forward: Indigenous Research Rising Up*; Brill Sense: Leiden, The Netherlands, 2018; Volume 125.
26. Morgenthal, T.L.; Kellner, K.; van Rensburg, L.; Newby, T.S.; van der Merwe, J.P.A. Vegetation and habitat types of the Umkhanyakude Node. *S. Afr. J. Bot.* **2006**, *72*, 1–10. [CrossRef]
27. Hendriks, S.L.; Viljoen, A.; Marais, D.; Wenhold, F.; McIntyre, A.; Ngidi, M.; Van der Merwe, C.; Annandale, J.; Kalaba, M.; Stewart, D. *The Current Rain-fed and Irrigated Production of Food Crops and Its Potential to Meet the Year-round Nutritional Requirements of Rural Poor People in North West, Limpopo, Kwazulu-Natal and the Eastern Cape: Report to the Water Research Commission and Department of Agriculture, Forestry & Fisheries*; Water Research Commission: Pretoria, Soutth Africa, 2016.
28. Initiative, H.H. *Data Collection Tools for Challenging Environments*; KoBoToolbox: Cambridge, MA, USA, 2018.
29. Riley, L.; Guthold, R.; Cowan, M.; Savin, S.; Bhatti, L.; Armstrong, T.; Bonita, R. The World Health Organization STEPwise approach to noncommunicable disease risk-factor surveillance: Methods, challenges, and opportunities. *Am. J. Public Health* **2016**, *106*, 74–78. [CrossRef]
30. Bahr, R.; Holme, I. Risk factors for sports injuries–A methodological approach. *Br. J. Sports Med.* **2003**, *37*, 384–392. [CrossRef] [PubMed]
31. Schmidt, A.F.; Finan, C. Linear regression and the normality assumption. *J. Clin. Epidemiol.* **2018**, *98*, 146–151. [CrossRef]
32. Lumley, T.; Diehr, P.; Emerson, S.; Chen, L. The importance of the normality assumption in large public health data sets. *Annu. Rev. Public Health* **2002**, *23*, 151–169. [CrossRef] [PubMed]
33. Field, A. *Discovering Statistics Using SPSS: (and Sex and Drugs and rock'n'roll)*; Sage: London, UK, 2009.
34. Kruger, H.S.; Venter, C.S.; Vorster, H.H. Physical inactivity as a risk factor for cardiovascular disease in communities undergoing rural to urban transition: The THUSA study: Cardiovascular topics. *Cardiovasc. J. S. Afr.* **2003**, *14*, 16–23.
35. Guthold, R.; Louazani, S.A.; Riley, L.M.; Cowan, M.J.; Bovet, P.; Damasceno, A.; Sambo, B.H.; Tesfaye, F.; Armstrong, T.P. Physical activity in 22 African countries: Results from the World Health Organization STEPwise approach to chronic disease risk factor surveillance. *Am. J. Prev. Med.* **2011**, *41*, 52–60. [CrossRef] [PubMed]
36. Kohrt, W.M.; Bloomfield, S.A.; Little, K.D.; Nelson, M.E.; Yingling, V.R.; American College of Sports Medicine. American College of Sports Medicine position stand: Physical activity and bone health. *Med. Sci. Sports Exerc.* **2004**, *36*, 1985–1996. [CrossRef]
37. Fajemiroye, J.O.; Cunha, L.C.d.; Saavedra-Rodríguez, R.; Rodrigues, K.L.; Naves, L.M.; Mourão, A.A.; Silva, E.F.d.; Williams, N.E.E.; Martins, J.L.R.; Sousa, R.B.; et al. Aging-induced biological changes and cardiovascular diseases. *BioMed Res. Int.* **2018**, *2018*, 7156435. [CrossRef]
38. Schuch, F.B.; Vancampfort, D.; Rosenbaum, S.; Richards, J.; Ward, P.B.; Veronese, N.; Solmi, M.; Cadore, E.L.; Stubbs, B. Exercise for depression in older adults: A meta-analysis of randomized controlled trials adjusting for publication bias. *Braz. J. Psychiatry* **2016**, *38*, 247–254. [CrossRef] [PubMed]
39. Usenbo, A.; Kramer, V.; Young, T.; Musekiwa, A. Prevalence of Arthritis in Africa: A Systematic Review and Meta-Analysis. *PLoS ONE* **2015**, *10*, e0133858. [CrossRef] [PubMed]
40. Geere, J.A.; Hunter, P.R.; Jagals, P. Domestic water carrying and its implications for health: A review and mixed methods pilot study in Limpopo Province, South Africa. *Environ. Health* **2010**, *9*, 52. [CrossRef] [PubMed]
41. Chimen, M.; Kennedy, A.; Nirantharakumar, K.; Pang, T.; Andrews, R.; Narendran, P. What are the health benefits of physical activity in type 1 diabetes mellitus? A literature review. *Diabetologia* **2012**, *55*, 542–551. [CrossRef]
42. Elsawy, B.; Higgins, K.E. Physical activity guidelines for older adults. *Am. Fam. Phys.* **2010**, *81*, 55–59.
43. Janssen, I.; LeBlanc, A.G. Systematic review of the health benefits of physical activity and fitness in school-aged children and youth. *Int. J. Behav. Nutr. Phys. Act.* **2010**, *7*, 40. [CrossRef]
44. Dergance, J.M.; Calmbach, W.L.; Dhanda, R.; Miles, T.P.; Hazuda, H.P.; Mouton, C.P. Barriers to and benefits of leisure time physical activity in the elderly: Differences across cultures. *J. Am. Geriatr. Soc.* **2003**, *51*, 863–868. [CrossRef]

45. Crombie, I.K.; Irvine, L.; Williams, B.; McGinnis, A.R.; Slane, P.W.; Alder, E.M.; McMurdo, M.E. Why older people do not participate in leisure time physical activity: A survey of activity levels, beliefs and deterrents. *Age Ageing* **2004**, *33*, 287–292. [CrossRef]

46. Peel, N.M.; McClure, R.J.; Bartlett, H.P. Behavioral determinants of healthy aging. *Am. J. Prev. Med.* **2005**, *28*, 298–304. [CrossRef]

47. Prince, S.A.; Adamo, K.B.; Hamel, M.E.; Hardt, J.; Connor Gorber, S.; Tremblay, M. A comparison of direct versus self-report measures for assessing physical activity in adults: A systematic review. *Int. J. Behav. Nutr. Phys. Act.* **2008**, *5*, 56. [CrossRef]

48. Oyeyemi, A.L.; Kolo, S.M.; Rufai, A.A.; Oyeyemi, A.Y.; Omotara, B.A.; Sallis, J.F. Associations of neighborhood walkability with sedentary time in nigerian older adults. *Int. J. Environ. Res. Public Health* **2019**, *16*, 1879. [CrossRef] [PubMed]

International Journal of
**Environmental Research
and Public Health**

Article

Urgent Need for Adolescent Physical Activity Policies and Promotion: Lessons from "Jeeluna"

Omar J. Baqal [1], Hassan Saleheen [2] and Fadia S. AlBuhairan [1,3,*

[1] College of Medicine, Alfaisal University, Riyadh 11533, Saudi Arabia; ojaved@alfaisal.edu
[2] National Family Safety Program, King Abdulaziz Medical City-Ministry of National Guard Health Affairs, Riyadh 11426, Saudi Arabia; hassan_nazmus@hotmail.com
[3] Department of Pediatrics & Adolescent Medicine, Aldara Hospital and Medical Center, Riyadh 12714, Saudi Arabia
* Correspondence: falbuhairan@aldaramed.com or fadia.albuhairan@gmail.com

Received: 17 May 2020; Accepted: 18 June 2020; Published: 21 June 2020

Abstract: Physical inactivity is a growing concern in Kingdom of Saudi Arabia (KSA) and globally. Data on physical activity (PA) trends, barriers, and facilitators among adolescents in KSA are scarce. This study aims to identify PA trends amongst adolescents in KSA and associated health and lifestyle behaviors. Data from "Jeeluna", a national study in KSA involving around 12,500 adolescents, were utilized. School students were invited to participate, and a multistage sampling procedure was used. Data collection included a self-administered questionnaire, anthropometric measurements, and blood sampling. Adolescents who performed PA for at least one day per week for >30 min each day were considered to "engage in PA". Mean age of the participants was 15.8 ± 0.8 years, and 51.3% were male. Forty-four percent did not engage in PA regularly. Only 35% engaged in PA at school, while 40% were not offered PA at school. Significantly more 10–14-year old than 15–19-year-old adolescents and more males than females engaged in PA (<0.01). Mental health was better in adolescents who engaged in PA (<0.01). Adolescents who engaged in PA were more likely to eat healthy food and less likely to live a sedentary lifestyle (<0.01). It is imperative that socio-cultural and demographic factors be taken into consideration during program and policy development. This study highlights the urgent need for promoting PA among adolescents in KSA and addressing perceived barriers, while offering a treasure of information to policy and decision makers.

Keywords: adolescent; adolescent health; physical activity; exercise; health surveillance; Kingdom of Saudi Arabia

---

1. Introduction

Physical activity (PA) when performed regularly, promotes well-being and helps in prevention of various health problems. In children and adolescents, PA plays a particularly critical role in growth and development, while promoting their transition into healthier adults, as lifestyle habits and PA patterns are largely established during childhood and adolescence [1]. It can help children and adolescents "see better, feel better and function better" [2]. Essentially, PA achieves this by helping them build strong bones, maintain healthy weight, improve cardiorespiratory fitness, alleviate symptoms of depression and anxiety, and also reduce the risk of developing several illnesses, such as cancer, type 2 diabetes, heart disease and obesity [2]. Studies have shown that physically active students have better grades, school attendance, cognitive performance and classroom behaviors [3,4]. It is recommended that children and adolescents between the ages of 6 and 17 years have at least 60 min of PA each day [5,6].

Despite widely recognized benefits and establishment of guidelines for PA, adolescents' PA behaviors have been far from encouraging. Globally, a significant increase in physical inactivity has been observed [7], which has led to increasing incidence of several health problems, imposing a considerable burden upon healthcare systems of many countries. In the United States (US), it was found that only 21.6% of children and adolescents aged between 6 and 19 years attained 60 or more minutes of moderate-to-vigorous PA on at least 5 days per week [8]. In England, only 21% and 16% of boys and girls, respectively were found to meet minimum PA recommendations [9]. Even though the magnitude of the health issues associated with physical inactivity is well-known, relatively few countries conduct physical inactivity surveillance and monitoring [10].

In the Kingdom of Saudi Arabia (KSA), a consequence of rapid socio-economic growth has been striking lifestyle changes. This has had adverse effects on population health, with an increase in physical inactivity and obesity being particularly observed [11]. Prevalence of physical inactivity in the Saudi population has been found to range between 46% and an astonishing 99% in certain segments of the population [11]. A study involving Saudi school students noted high inactivity levels, with roughly only 44% of males and 20% of females being sufficiently active [12]. Consistent with global PA trends [13], males have been reported to participate more frequently in high-intensity PA than females [10].

In parallel with increasing physical inactivity, KSA has also experienced an increasing prevalence of obesity and activities that promote sedentary lifestyle, with a study noting a negative relationship between PA and obesity [14]. The same study noted a significant inverse relationship between physical activity and computer-usage (a sedentary activity), indicating that the more active the individual, the less time spent on computer-usage, with a higher body mass index (BMI) observed amongst those who reported greater computer-usage time [14]. Geographical factors have been shown to influence PA habits, with a study reporting that youths living in rural desert areas were less physically active than those living in urban or rural farm environments [15].

With the dramatic lifestyle transformation, a change in dietary habits and food consumption patterns has also been noted. Calorie-rich foods such as fast food, soft drinks, and processed snacks have become extremely affordable and accessible to children and adolescents [16]. Such food consumption patterns potentially synergize the negative effect of physical inactivity and sedentary lifestyles on population health, in turn contributing to the growing prevalence of obesity, heart disease and other non-communicable diseases. KSA's hot sun-drenched climate limits physical activity as people prefer staying indoors to avoid exposure to sunlight and heat, though there have been increasing opportunities for indoor places and activities. The widespread vitamin-D deficiency among Saudi children and adolescents has been shown to be influenced by PA and sun exposure [17].

Data on PA trends, determinants, barriers and facilitators among adolescents in Saudi Arabia have been scarce. "Jeeluna" (Arabic for "Our Generation"), a national school-based cross-sectional study involving around 12,500 adolescents in all 13 regions of KSA, found that only 13.7% of adolescents aged 10–19 years engaged in at least 30 min of PA daily [18]. To the best of our knowledge, no other studies on the same have been published exclusively on the adolescent age group (10–19 years), with most published studies focusing solely on older adolescents. This is of limited benefit when attempting to extrapolate research findings to the development of policies, programs and services promoting PA focused particularly on the adolescent age group.

The aim of this study is to identify the sociodemographic characteristics of adolescents who participate in PA in KSA, as well as their associated health and lifestyle behaviors by gender. Such information will support program and policy development that address adolescents' local needs in an evidence-based manner.

2. Materials and Methods

2.1. Study Design and Participants

For this current analysis, we utilized data from the Jeeluna study database. Jeeluna was a cross-sectional, school-based, nationally representative study conducted in Saudi Arabia. Participants included intermediate and secondary school students, aged 10–19 years from all 13 regions of the country.

2.2. Procedures

A stratified, cluster random sampling procedure was used. The detailed sampling methodology has been previously published [18]. Any male/female, intermediate/secondary, public/private school in a Saudi Arabian city/town that functions during the day was eligible to participate in the study. Exclusion criteria included evening schools and schools that served students with special needs. Data collection involved: (1) administration of self-administered questionnaire; (2) anthropometric measurements; and (3) blood sampling for laboratory investigations. Data collection teams operating in all regions received standardized training.

The study received approval from the Institutional Review Board at the King Abdullah International Medical Research Center (KAIMRC) (RC08-092), as well as the Ministry of Education (MOE). It was necessary to seek permission from participating schools' principals, active written parental consent and student assent; participants were given the option to opt out of blood sampling.

For the purpose of this secondary analysis, the focus was on the PA section. It was assessed by asking students how many days in the past week they had engaged in exercise for at least 30 min each day. The cutoff of 30 min was used instead of the recommended 60 min, as PA trends have shown that most children and adolescents do not meet the recommended minimum, thus supporting the use of a lower cutoff in the analysis.

2.3. Measurements

The questionnaire was guided by the Youth Risk Behavior Survey [19] and the Global School-based Student Health Survey [20]. The global definition of health was used, so multiple domains were included and addressed in the questionnaire, including: (1) family; (2) education/schooling; (3) nutrition/dietary behaviors; (4) physical activity; (5) safety; (6) sleep; (7) violence and bullying; (8) tobacco and substance use (including alcohol use); (9) health; (10) health services; and (11) health knowledge. Students were given assurance of the anonymity and confidentiality of their responses.

Adolescents who engaged in PA for 1 or more days in a week were considered to "engage in PA". The section also included questions on sedentary activities (TV, internet, video games and cell phone use). Responses indicating time spent on each activity to be more than 2 h daily were considered to be part of the "sedentary lifestyle" category. With regard to main meals (breakfast, lunch and dinner), a value of "3" meals a day was considered healthy. Everything else, whether lower (none, 1 or 2 meals a day) or higher (4 or >4 meals a day), was put in the "unhealthy" category. We considered guidelines from the American Academy of Pediatrics (AAP) that refer to 3 meals a day as the target for healthy nutrition [21]. The 5 response choices pertaining to absence from school were grouped into 3 categories: "Not absent", "Rarely absent", and the third category including "sometimes", "often" and "always."

The "Mental health" variable was created based on responses to the questionnaire items asking about the frequency of symptoms reflecting depression or anxiety over the past 12 months. "Never", "rarely" and "sometimes" were considered as "No", and "most of the times" and "always" were considered as "Yes." The anthropometric measurement from the study database included in this analysis was BMI, which was based on measured weights and heights. BMI was classified as "Underweight", "Normal", "Overweight" and "Obese", based on Center for Disease Control and Prevention BMI charts for gender and age [22,23].

2.4. Data Analysis

The first step in the analysis was the descriptive analysis. Participants were described in terms of their selected socio-demographic status. Means along with standard deviations and percentages were calculated for continuous and categorical variables, respectively. The relationship between engaging in PA and school performance, health status, and health risk behaviors was calculated using chi-square test. A significance level of less than 0.05 was used for all statistical tests. All data were analyzed using SPSS version 25.0.3 (IBM Corp, Armonk, NY, USA).

3. Results

Participant characteristics are listed in Table 1. Mean age of the participants was 15.8 ± 1.8 years, and 51.3% were male. Most of the participants (97.9%) lived in urban areas. Little over half (53.1%) of the participants fell within the "normal" BMI range; similarly, just over half reported engaging in PA (53.4%). Only about one-third (35.4%) engaged in PA at school.

Table 1. Characteristics of the participants (n = 12,463).

Characteristic	Number (%) [¥]
Age (mean)	
Mean ± SD	15.8 ± 1.8
Gender	
Male	6398 (51.3)
Female	6065 (48.7)
Areas of residence	
Urban	12,205 (97.9)
Rural	258 (2.1)
Nationality	
Saudi	10,318 (82.8)
Non-Saudi	1493 (12.0)
Grade	
Intermediate	6142 (49.3)
Secondary	6321 (50.7)
BMI result	
<5th centile	1858 (14.9)
5–<85th centile	6616 (53.1)
≥85th–<95th centile	1712 (13.7)
≥95th centile	1920 (15.4)
Engage in physical activity	
Yes	6653 (53.4)
No	5450 (43.7)
Engage in physical activity (sports) at school	
Yes	4412 (35.4)
No	2624 (21.1)
No physical activity at school	5052 (40.5)

[¥] Percentages may not add up to 100 due to missing data.

Nationally, 92.7% of males and 7.3% of females were offered PA in schools. Out of the 13 regions, Jizan had the highest mean number of PA days per week in total (2.4 ± 2.8) as well as among males (3.3 ± 2.8). Jizan also boasts the highest proportion of total (44.7%) and male students (99.6%) engaging in PA in school. Interestingly, a stark contrast is observed in PA trends among females in Jizan, which has the lowest mean number of PA days per week in females (0.8 ± 1.7) among all regions, as well as the lowest proportion of females engaging in PA in school (0.4%). Female PA trends were dramatically better in Aljouf, which had the highest proportion of females engaging in PA in school (53.2%), and second highest mean number of PA days (1.8 ± 2.4). Detailed region-wise PA trends are displayed in Table 2.

Table 2. Physical activity relative to region.

Region	Mean No. of Days/Week Spent on >30 min of Exercise			Regional Sample	Physical Activity (Sports) in School Number (%)		
	Total	Male	Female		Total PA	Male	Female
Riyadh	1.7 ± 2.3	2.2 ± 2.4	1.3 ± 2.2	2749	876 (31.8)	853 (97.4)	23 (2.6)
Qasim	1.5 ± 2.2	2.0 ± 2.4	1.0 ± 1.8	734	243 (33.1)	239 (98.4)	4 (1.6)
Makkah	2.0 ± 2.5	2.7 ± 2.7	1.0 ± 1.9	2906	1099 (37.8)	1074 (97.7)	25 (2.3)
Madinah	1.8 ± 2.4	2.5 ± 2.6	1.0 ± 1.9	943	318 (33.7)	312 (98.1)	6 (1.9)
Eastern province	1.8 ± 2.4	2.4 ± 2.6	1.2 ± 2.0	1851	683 (36.8)	577 (84.5)	106 (15.5)
Tabuk	1.7 ± 2.3	2.2 ± 2.5	1.3 ± 2.0	342	111 (32.4)	81 (73.0)	30 (27.0)
Aljouf	1.6 ± 2.2	1.2 ± 2.0	1.8 ± 2.4	361	111 (30.7)	52 (46.8)	59 (53.2)
Hail	1.8 ± 2.3	1.7 ± 2.3	2.0 ± 2.4	358	124 (34.6)	76 (61.3)	48 (38.7)
Northen borders	1.8 ± 2.4	2.7 ± 2.7	1.2 ± 1.9	222	73 (32.8)	69 (94.5)	4 (5.5)
Albaha	1.5 ± 2.3	1.8 ± 2.5	1.3 ± 2.2	314	50 (15.9)	43 (86.0)	7 (14.0)
Aseer	2.2 ± 2.6	3.0 ± 2.8	1.3 ± 2.1	773	324 (41.9)	316 (97.5)	8 (2.5)
Jizan	2.4 ± 2.8	3.3 ± 2.8	0.8 ± 1.7	606	271 (44.7)	270 (99.6)	1 (0.4)
Najran	2.0 ± 2.4	2.3 ± 2.4	1.6 ± 2.3	304	129 (42.4)	127 (98.4)	2 (1.6)
Overall	1.9 ± 2.4	2.4 ± 2.6	1.2 ± 2.1	12,463	4412 (35.4)	4089 (92.7)	323 (7.3)

Figure 1 reflects a decline in PA relative to age, with mean PA days per week of 2.7 at 12 years among males dropping to 2.1 by age 18, before rising to 2.4 by 19 years of age. In females, PA steadily declined between 12 and 19 years of age from 1.6 to 1. Significant differences in age, gender, and areas of residence, were found in terms of engagement in PA, with younger adolescents (59.6% vs. 53.5%, $p < 0.01$), males (68.3% vs. 48.7%, $p < 0.01$), and adolescents living in urban areas (55.1% vs. 47.1%, $p < 0.01$) being more likely to engage in PA. PA habits were not significantly different between Saudi and non-Saudi adolescents, as well as between the four BMI groups, with proportions of adolescents engaging in PA ranging between 54% and 57% across the BMI groups (Table 3).

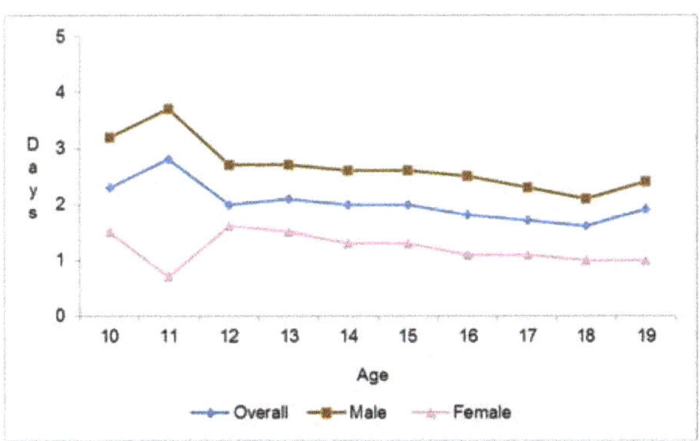

Figure 1. Mean no. of days/week spent on >30 min of physical activity relative to age.

Table 4 represents the relationship between engaging in PA and school performance, health status, and health risk behaviors. Adolescents who engaged in PA were more likely to eat healthy food (99.1% vs. 98.3%, $p < 0.01$), less likely to live a sedentary lifestyle (63.9% vs. 67.3%, $p < 0.01$), and less likely to be absent from school (26.6% vs. 20.3%, $p < 0.01$). With regards to mental health, symptoms reflective of depression (12.2% vs. 17%, $p < 0.01$) and anxiety (5.9% vs. 7.8%, $p < 0.01$) were found to be less prevalent among adolescents who engaged in PA. Prevalence of chronic diseases as well as tobacco/substance use was not significantly different between the two groups. No significant difference was found in number of students with "excellent" academic performance among those who engage in PA and those who do not.

Table 3. Relationship between engaging in physical activity (PA) and demographic characteristics ($n = 12,463$).

Engagement in PA	Age			Gender			Areas of Residence			Nationality			BMI				
	10–14	15–19	p Value	Male	Female	p Value	Urban	Rural	p Value	Saudi	Non-Saudi	p Value	Under-Weight	Normal	Over-Weight	Obese	p Value
									Engage in PA								
Yes	1905 (59.6)	4673 (53.5)	<0.01	4266 (68.3)	2387 (40.7)	<0.01	6532 (55.1)	121 (47.1)	<0.05	5575 (54.9)	835 (56.6)	NS	1010 (55.0)	3541 (54.4)	962 (56.9)	1057 (56.0)	NS
No	1293 (40.4)	4061 (46.5)		1976 (31.7)	3474 (59.3)		5314 (44.9)	136 (52.9)		4579 (45.1)	640 (43.4)		826 (45.0)	2964 (45.6)	729 (43.1)	830 (44.0)	

Table 4. Relationship between engaging in PA and school performance, health status, and health risk behaviors among adolescents in Saudi Arabia ($n = 12,463$).

Variable	Total			Male			Female		
	PA	No PA	p-Value	PA	No PA	p-Value	PA	No PA	p-Value
School performance									
Excellent academic performance	2845 (43.3)	2383 (44.2)	NS	1709 (40.5)	766 (39.1)	NS	1136 (48.2)	1617 (47.0)	NS
Not absent from school	1745 (26.6)	1095 (20.3)	<0.01	1282 (30.4)	533 (27.3)	<0.05	463 (19.7)	562 (16.3)	<0.01
Chronic diseases	566 (8.8)	438 (8.2)	NS	386 (9.4)	174 (9.1)	NS	180 (7.7)	264 (7.7)	NS
Mental Health									
Sadness/frustration	800 (12.2)	915 (17.0)	<0.01	388 (9.2)	242 (12.4)	<0.01	412 (17.5)	673 (19.6)	<0.05
Anxiety	383 (5.9)	416 (7.8)	<0.01	171 (4.1)	117 (6.0)	<0.01	212 (9.1)	299 (8.8)	NS
Healthy food	6584 (99.1)	5344 (98.3)	<0.01	4231 (99.3)	1958 (99.2)	NS	2353 (98.7)	3386 (97.7)	<0.05
Sedentary lifestyle	4222 (63.9)	3643 (67.3)	<0.01	2653 (62.7)	1246 (63.5)	NS	1569 (66.1)	2397 (69.4)	<0.01
Tobacco/substance use	1658 (25.6)	1350 (25.3)	NS	1041 (25.1)	521 (27.0)	NS	617 (26.4)	829 (24.3)	NS

4. Discussion

Research activity on PA and lifestyle habits has not matched the rapid pace at which lifestyle changes have been occurring in the KSA. This study, based on the Jeeluna national school-based study, sheds light on the status quo of PA among the adolescents of KSA and related health and lifestyle habits. Over 40% of adolescents in our study did not engage in PA at all. The remaining adolescents were part of a wide-ranging category that included those who engaged in PA once a week as well as those who engaged in PA daily. Twenty percent of males and a worrying 59% of females did not engage in PA regularly. Locally reported studies that utilized objective PA measurement techniques showed that 60% of Saudi children and 71% of youths did not engage in PA of sufficient frequency and duration [24,25].

To put local estimates including ours in perspective, it is suitable to compare them with PA estimates from other countries. A systematic assessment of literature on PA in Arab countries revealed that physical inactivity was alarmingly high in children and adolescents, reaching about 80% in all national surveys except Tunisia [26]. A Finnish study reported that 59% of boys and 50% of girls aged 15–16 years reported engaging in 60 min or more of PA per day; these estimates dropped to 23% (boys) and 10% (girls) when moderate-to-high physical activity was considered [27]. In the US, the Youth Risk Behavior Surveillance revealed that only 18.4% of adolescents met the PA guidelines [28]. The Global School-based Student Health Survey (GSHS), which included over 72,000 young people across 34 countries, showed that only 24% of boys and 15% of girls engaged in sufficient PA to meet recommendations [29]. Recently, WHO published a pooled analysis of survey data reported by 1.6 million 11 to 17-year-old adolescents across 146 countries. It found that more than 80% of school-going adolescents globally did not meet current daily PA recommendations, including 85% of girls and 78% of boys [30].

These worrying estimates of inactivity are of a great concern as inactivity has been associated with increased risk of development of cardiovascular and metabolic health conditions [2]. Modern life has paralleled technological advances, reflected in increased screen time for everyone, whether for work or leisure. This lifestyle has aimed to minimize efforts and systematically reduce energy expenditure and, in the process, to discourage physical exertion. We found that 67% of adolescents who did not exercise pursued a sedentary lifestyle. Spending time on the previously mentioned sedentary activities soaks up valuable time that could otherwise be utilized in health-promoting activities such as regular exercise. Amusingly, a lack of time has been regularly reported by individuals for their lack of PA [31,32]. Other factors that contribute to the reported adolescent inactivity rates in KSA may include lack of PA programs in schools, hot weather conditions, over-dependence on cars rather than walking even for short distances, the surrounding built environment which is not pedestrian friendly, poor peer and family support and socio-cultural barriers, which particularly impact females.

We found that males were significantly more likely to engage in PA than females. This finding is consistent with findings of several other studies [16,33,34]. This could be attributed to cultural beliefs and lifestyle patterns in KSA that possibly discourage participation of females in recreational activities outside the house. It is perhaps due to these socio-cultural differences that experts have suggested a specific gender-based consideration when making PA promotion recommendations [35]. Although males have been generally considered to have better opportunities for recreational activities including exercise than females, rapid changes have fortunately been witnessed in the past 1–2 years with more sports opportunities for females being made available in the country. The opening of female-only gyms, sports clubs, and female national sports teams have encouraged more women to engage in regular exercise. Interestingly, our study found that females who exercise were significantly more likely to eat heathy food and less likely to live a sedentary lifestyle than females who did not exercise, while there was no such significant difference between the two male groups regarding healthy eating and sedentary lifestyle. This could reflect active consciousness and awareness regarding healthy eating and lifestyle habits among females who exercise regularly. On the other hand, males seem to manage to get sufficient PA, possibly through school and recreational activities while still personally not living an active healthy lifestyle.

Our study found a stark contrast between PA offered to males and females in schools. For example, in Jizan, 99.6% of male adolescents reported being offered PA at school, while only a shocking 0.4% of females reported the same. A local study showed that males from public schools were more active than males at private schools, whereas the opposite was true for females; they were more active in private schools [12]. While PA programs in female public schools are nearly nonexistent, private schools provide students with opportunities for regular exercise, encouraging an active and healthier lifestyle. Our findings are in agreement with this observation as most of the schools in Jizan belong to the public education system. We also reported Jizan to have the highest mean number of PA days per week among males in the whole country, while the lowest among females. Jizan is considered to be relatively more conservative when compared with the bigger cities of Jeddah and Riyadh, and thus, the cultural and lifestyle barriers to female participation in PA mentioned earlier may be augmented in this region.

We also found that adolescents aged 10–14 years were more likely to engage in PA than older adolescents aged 15–19 years. PA decline in adolescence has been previously observed [36], a trend that seems to continue into adulthood [37,38]. Increasing academic workload in secondary school could contribute to this decline, and factors such as more time spent on sedentary activities [39] as well as taking up part-time jobs [40] could also be related. This highlights the need for qualitative studies to clearly understand the reasons for this decline and to promote PA and other healthy habits in adolescence, a stage of building habits, many of which last a lifetime.

There was no significant difference in PA habits between Saudi and non-Saudi adolescents, as well as between different BMI groups. These interesting findings counter expectations, as it would generally be convenient to believe that expats from more open societies with lesser socio-cultural barriers would be living a more active lifestyle. This highlights the impact of lifestyle changes an individual adopts when living in a foreign country.

Among all of the benefits of PA and its positive effects on general health, its impact on mental health is well recognized. Exercise is known to boost self-esteem, body image and overall mood, in line with our findings of adolescents who engage in PA feeling sad or hopeless significantly less frequently than those who do not engage in PA [41]. Exercise is widely recommended by health authorities for the prevention and treatment of various non-communicable diseases, although the above mentioned benefits may depend on type, timing, and intensity of PA. Physical activity has been well-studied in the area of depression, with a study reporting 20–33% lower odds of depression in the active groups in prospective cohort studies [42]. For tasks involving more complex executive functioning, exercise was found to be associated with enhanced cognitive functioning in a meta-analysis of randomized controlled trials looking into exercise training studies in adults aged 55–80 [43,44].

We found that students who engaged in PA were significantly more likely to not be absent from school. Based on 2005–2008 NHANES (National Health and Nutrition Examination Survey) data, it was found that excessive TV watching and inactivity as well as high activity levels (>7 times per week) were independently associated with severe school absenteeism [45]. As discussed previously, excessive screen time could be taking away valuable time from adolescents that they could have spent being more physically active. Moreover, excessive TV watching can have various negative effects on a child's cognitive and socio-emotional development. Having discussed the benefits of PA on mental health, low PA levels could hamper an adolescent's participation in school, such that school is something they do not look forward to due to negative self-esteem, body image, feelings of sadness and worry, or lack of motivation.

Consequently, it is indeed worrying that despite the myriad of known health benefits, PA trends among our adolescents remain poor. PA, even in small amounts can considerably enhance physical and psychological well-being. School-based PA programs can address the enormous discrepancy between female and male PA habits in KSA by improving availability and accessibility to PA opportunities. Health care professionals, starting from the primary health care system and school health system, must convey the benefits of regular physical activity to individuals and their families and do everything in their capacity to facilitate their participation in PA. The American Heart Association emphasizes that "the advice from healthcare professionals significantly influences adoption of healthy lifestyle behaviors, including regular PA, and can increase satisfaction with medical care" [46]. Several medical institutions, particularly in the US have begun assessing PA as a vital sign tool (indicator of general physical condition) to identity individuals who could benefit from prompt counseling or appropriate referral. Parents and guardians play a critical role in ensuring adolescents at least meet the minimum recommended amount of PA as per latest guidelines. The remarkable technological literacy of this young generation can be positively utilized by promoting active e-sports video games (e.g., Wii Sports) and health and fitness tracking applications. Above all, national health authorities, particularly the Ministry of Health, Ministry of Education, and Sports Authority must strategize an efficient multi-level action plan by taking into consideration existing data on local PA trends and lifestyle habits among adolescents, including results of this landmark study. This will also significantly contribute towards lowering the currently estimated total costs attributable to physical inactivity in KSA.

The Jeeluna study included over 12,000 adolescents, making it the largest study on adolescent health in the region to date. Additionally, the study was conducted in all 13 regions of KSA; the national representativeness of the study population enhances its generalizability. However, the study may be viewed in light of a few limitations. The study questionnaire was self-administered and may be associated with recall bias. Physical activity was not measured objectively, nor was the type of PA reported taken into consideration. In search of an explanation for the distortion of the graph line at 11 years of age, we found that the study population aged 11 years was small in number. The questionnaire did not take into consideration food portions when asking about frequency of meals. Nonetheless, the questionnaire was put through several rounds of expert review and was pilot tested for clarity and comprehension amongst the target respondent group. The study thus sets a benchmark for future adolescent health research in the country and the wider region.

5. Conclusions

The Jeeluna study highlights the urgent need for promotion of PA among adolescents in KSA and addressing perceived barriers. It is imperative that socio-cultural and demographic factors be taken into consideration during program and policy development. Schools play a critical role in improving PA habits among adolescents and remain an important focus of action, particularly for female PA promotion. Lack of PA, unhealthy food and sedentary activities are not isolated, independent health risk factors and are often found clustered among adolescents, synergistically increasing risk of disease and poor health several-fold. In order to be effective, these issues must be addressed in a cross-sectoral approach, involving multiple ministries and other stakeholders.

The study sheds light on the urgent and important issue of physical inactivity, which should be of utmost national and international interest. Our study offers valuable information to policymakers, educators and health professionals involved in the progression of adolescent health to understand young people's health in their social context. Further studies are required to be able to assess trends in PA and to evaluate impact of more recently implemented programs in the country and hence support strategies for further promotion of PA and the reduction of sedentary behavior among adolescents.

Author Contributions: Conceptualization, O.J.B. and F.S.A.; Data curation, F.S.A.; Formal analysis, H.S.; Funding acquisition, F.S.A.; Investigation, F.S.A.; Methodology, F.S.A.; Project administration, F.S.A.; Resources, F.S.A.; Software, H.S.; Supervision, F.S.A.; Validation, H.S.; Visualization, O.J.B., F.S.A. and H.S.; Writing—original draft, O.J.B.; Writing—review and editing, O.J.B., F.S.A. and H.S. All authors have read and agree to the published version of the manuscript.

Funding: JEELUNA study was supported and funded by King Abdullah International Medical Research Center (Protocol RC08-092).

Conflicts of Interest: The authors declare no conflict of interest.

References

1. Azevedo, M.R.; Araújo, C.L.; Cozzensa da Silva, M.; Hallal, P.C. Tracking of physical activity from adolescence to adulthood: A population-based study. *Rev. Saude Publica* **2007**, *41*, 69–75. [CrossRef] [PubMed]
2. 2018 Physical Activity Guidelines Advisory Committee. *2018 Physical Activity Guidelines Advisory Committee Scientific Report*; U.S. Department of Health and Human Services: Washington, DC, USA, 2018. Available online: https://health.gov/paguidelines/second-edition/report/pdf/PAG_Advisory_Committee_Report.pdf (accessed on 26 July 2018).
3. Centers for Disease Control and Prevention. *The Association between School-Based Physical Activity, Including Physical Education, and Academic Performance*; Centers for Disease Control and Prevention, US Department of Health and Human Services: Atlanta, GA, USA, 2010. Available online: https://www.cdc.gov/healthyyouth/health_and_academics/pdf/pa-pe_paper.pdf (accessed on 26 July 2018).
4. Michael, S.L.; Merlo, C.L.; Basch, C.E.; Wentzel, K.R.; Wechsler, H. Critical connections: Health and academics. *J. Sch. Health* **2015**, *85*, 740–758. [CrossRef] [PubMed]
5. US Department of Health and Human Services. *2008 Physical Activity Guidelines for Americans*; US Department of Health and Human Services: Washington, DC, USA, 2008. Available online: https://health.gov/paguidelines/guidelines/summary.aspx (accessed on 26 July 2018).
6. World Health Organization. Physical Activity and Young People. Available online: http://www.who.int/dietphysicalactivity/factsheet_young_people/en/ (accessed on 26 July 2018).
7. Kohl, H.W., 3rd; Craig, C.L.; Lambert, E.V.; Inoue, S.; Alkandari, J.R.; Leetongin, G.; Kahlmeier, S. The pandemic of physical inactivity: Global action for public health. *Lancet* **2012**, *380*, 294–305. [CrossRef]
8. National Physical Activity Plan Alliance. 2016 US Report Card on Physical Activity for Children and Youth. 2016. Available online: http://www.physicalactivityplan.org/reportcard/2016FINAL_USReportCard.pdf (accessed on 26 July 2018).
9. Department of Health. *Start Active, Stay Active: A Report on Physical Activity from the Four Home Countries*; DH: London, UK, 2011. Available online: https://assets.publishing.service.gov.uk/government/uploads/system/uploads/attachment_data/file/216370/dh_128210.pdf (accessed on 26 July 2018).
10. Bauman, A.; Bull, F.; Chey, T.; Craig, C.L.; Ainsworth, B.E.; Sallis, J.F.; Bowles, H.R.; Hagstromer, M.; Sjostrom, M.; Pratt, M.; et al. The International Prevalence Study on Physical Activity: Results from 20 countries. *Int. J. Behav. Nutr. Phys. Act.* **2009**, *6*, 21. [CrossRef] [PubMed]
11. Al-Hazzaa, H.M. Physical inactivity in Saudi Arabia. An under served public health issue. *Saudi Med. J.* **2010**, *31*, 1278–1279. [PubMed]
12. Al-Hazzaa, H.M.; Alahmadi, M.A.; Al-Sobayel, H.I.; Abahussain, N.A.; Qahwaji, D.M.; Musaiger, A.O. Patterns and determinants of physical activity among Saudi adolescents. *J. Phys. Act. Health* **2014**, *11*, 1202–1211. [CrossRef] [PubMed]
13. Hallal, P.C.; Andersen, L.B.; Bull, F.C.; Guthold, R.; Haskell, W.; Ekelund, U.; Lancet Physical Activity Series Working Group. Global physical activity levels: Surveillance progress, pitfalls, and prospects. *Lancet* **2012**, *380*, 247–257. [CrossRef]
14. Al-Nakeeb, Y.; Lyons, M.; Collins, P.; Al-Nuaim, A.; Al-Hazzaa, H.; Duncan, M.J.; Nevill, A. Obesity, Physical Activity and Sedentary Behavior amongst British and Saudi Youth: A Cross-Cultural Study. *Int. J. Environ. Res. Public Health* **2012**, *9*, 1490–1506. [CrossRef]
15. Al-Nuaim, A.A.; Al-Nakeeb, Y.; Lyons, M.; Al-Hazzaa, H.M.; Nevill, A.; Collins, P.; Duncan, M.J. The Prevalence of Physical Activity and Sedentary Behaviours Relative to Obesity among Adolescents from Al-Ahsa, Saudi Arabia: Rural versus Urban Variations. *J. Nutr. Metab.* **2012**, *2012*, 417589. [CrossRef]
16. Collison, K.S.; Zaidi, M.Z.; Subhani, S.N.; Al-Rubeaan, K.; Shoukri, M.; Al-Mohanna, F.A. Sugar-sweetened carbonated beverage consumption correlates with BMI, waist circumference, and poor dietary choices in school children. *BMC Public Health* **2010**, *10*, 234. [CrossRef]
17. Al-Othman, A.; Al-Musharaf, S.; Al-Daghri, N.M.; Krishnaswamy, S.; Yusuf, D.S.; Alkharfy, K.M.; Al-Saleh, Y.; Al-Attas, O.S.; Alokail, M.S.; Moharram, O.; et al. Effect of physical activity and sun exposure on vitamin D status of Saudi children and adolescents. *BMC Pediatr.* **2012**, *12*, 92. [CrossRef] [PubMed]
18. AlBuhairan, F.S.; Tamim, H.; Al Dubayee, M.; AlDhukair, S.; Al Shehri, S.; Tamimi, W.; El Bcheraoui, C.; Magzoub, M.E.; de Vries, N.; Al Alwan, I. Time for an Adolescent Health Surveillance System in Saudi Arabia: Findings From "Jeeluna". *J. Adolesc. Health* **2015**, *57*, 263–269. [CrossRef] [PubMed]

19. Kann, L.; Kinchen, S.; Shanklin, S.L.; Flint, K.H.; Kawkins, J.; Harris, W.A.; Lowry, R.; Olsen, E.O.; McManus, T.; Chyen, D.; et al. Youth risk behavior surveillance—United States, 2013. *Surveill. Summ.* **2014**, *63*, 1–168.

20. World Health Organization. Global School-Based Student Health Survey. Available online: http://www.who.int/chp/gshs/en/ (accessed on 9 June 2018).

21. Hagan, J.F.; Shaw, J.S.; Duncan, P.M. (Eds.) *Bright Futures: Guidelines for Health Supervision of Infants, Children, and Adolescents*, 4th ed.; American Academy of Pediatrics: Elk Grove Village, IL, USA, 2017.

22. Centers for Disease Control and Prevention. 2 to 20 Years: Boys. Body Mass Index-for-Age Percentiles. Available online: https://www.cdc.gov/growthcharts/data/set1clinical/cj41l023.pdf (accessed on 26 January 2020).

23. Centers for Disease Control and Prevention. 2 to 20 Years: Girls. Body Mass Index-for-Age Percentiles. Available online: https://www.cdc.gov/growthcharts/data/set1clinical/cj41l024.pdf (accessed on 26 January 2020).

24. Al-Hazzaa, H.M. Physical activity, fitness and fatness among Saudi children and adolescents: Implications for cardiovascular health. *Saudi Med. J.* **2002**, *23*, 144–150. [PubMed]

25. Al-Hazzaa, H. Prevalence of physical inactivity in Saudi Arabia: A brief review. *EMHJ-East. Mediterr. Health J.* **2004**, *10*, 663–670.

26. Sharara, E.; Akik, C.; Ghattas, H.; Makhlouf Obermeyer, C. Physical inactivity, gender and culture in Arab countries: A systematic assessment of the literature. *BMC Public Health* **2018**, *18*, 639. [CrossRef]

27. Tammelin, T.; Ekelund, U.; Remes, J.; Näyhä, S. Physical activity and sedentary behaviors among Finnish youth. *Med. Sci. Sports Exerc.* **2007**, *39*, 1067–1074. [CrossRef]

28. Eaton, D.K.; Kann, L.; Kinchen, S.; Shanklin, S.; Ross, J.; Hawkins, J.; Harris, W.A.; Lowry, R.; McManus, T.; Chyen, D.; et al. Centers for Disease Control and Prevention (CDC): Youth risk behavior surveillance—United States, 2009. *MMWR Surveill. Summ.* **2010**, *59*, 1–142. [PubMed]

29. Guthold, R.; Cowan, M.J.; Autenrieth, C.S.; Kann, L.; Riley, L.M. Physical activity and sedentary behavior among schoolchildren: A 34-country comparison. *J. Pediatr.* **2010**, *157*, 43–49. [CrossRef] [PubMed]

30. Guthold, R.; Stevens, G.A.; Riley, L.M.; Bull, F.C. Global trends in insufficient physical activity among adolescents: A pooled analysis of 298 population-based surveys with 1·6 million participants. *Lancet Child Adolesc. Health* **2019**, S2352–S4642. [CrossRef]

31. Santos, M.S.; Hino, A.A.; Reis, R.S.; Rodriguez-Añez, C.R. Prevalência de barreiras para a prática de atividade física em adolescentes. *Rev. Bras. Epidemiol.* **2010**, *13*, 94–104. [CrossRef] [PubMed]

32. Hammerschmidt, P.; Tackett, W.; Golzynski, M.; Golzynski, D. Barriers to and Facilitators of Healthful Eating and Physical Activity in Low-income Schools. *J. Nutr. Educ. Behav.* **2011**, *43*, 63–68. [CrossRef] [PubMed]

33. Henry, C.J.; Lightowler, H.J.; Al-Hourani, H.M. Physical activity and levels of inactivity in adolescent females ages 11-16 years in the United Arab Emirates. *Am. J. Hum. Biol. Off. J. Hum. Biol. Assoc.* **2004**, *16*, 346–353. [CrossRef] [PubMed]

34. Fernandes, R.A.; Christofaro, D.G.; Casonatto, J.; Kawaguti, S.S.; Ronque, E.R.; Cardoso, J.R.; Freitas Junior, I.F.; Oliveira, A.R. Cross-sectional association between healthy and unhealthy food habits and leisure physical activity in adolescents. *J. Pediatr.* **2011**, *87*, 252–256. [CrossRef]

35. White, J.L.; Ransdell, L.B.; Vener, J.; Flohr, J.A. Factors related to physical activity adherence in women: Review and suggestions for future research. *Women Health* **2005**, *41*, 123–148. [CrossRef]

36. Dumith, S.C.; Gigante, D.P.; Domingues, M.R.; Kohl, H.W., 3rd. Physical activity change during adolescence: A systematic review and a pooled analysis. *Int. J. Epidemiol.* **2011**, *40*, 685–698. [CrossRef]

37. Mäkinen, T.E.; Borodulin, K.; Tammelin, T.H.; Rahkonen, O.; Laatikainen, T.; Prättälä, R. The effects of adolescence sports and exercise on adulthood leisure-time physical activity in educational groups. *Int. J. Behav. Nutr. Phys. Act.* **2010**, *7*, 27. [CrossRef]

38. Milanović, Z.; Pantelić, S.; Trajković, N.; Sporiš, G.; Kostić, R.; James, N. Age-related decrease in physical activity and functional fitness among elderly men and women. *Clin. Interv. Aging* **2013**, *8*, 549–556. [CrossRef]

39. Sharif, I.; Sargent, J.D. Association between television, movie, and video game exposure and school performance. *Pediatrics* **2006**, *118*, E1061–E1070. [CrossRef]

40. Malina, R.; Bouchard, C.; Bar-Or, O. Risk factors during growth and adult health. In *Growth, Maturation, and Physical Activity*; Malina, R., Bouchard, C., Bar-Or, O., Eds.; Human Kinetics: Champaign, FL, USA, 2004; pp. 587–608.

41. Zamani Sani, S.H.; Fathirezaie, Z.; Brand, S.; Pühse, U.; Holsboer-Trachsler, E.; Gerber, M.; & Talepasand, S. Physical activity and self-esteem: Testing direct and indirect relationships associated with psychological and physical mechanisms. *Neuropsychiatr. Dis. Treat.* **2016**, *12*, 2617–2625. [CrossRef]

42. Dishman, R.K.; Heath, G.W.; Lee, I.-M. *Physical Activity Epidemiology*, 2nd ed.; Human Kinetics: Champaign, FL, USA, 2013.

43. Colcombe, S.; Kramer, A.F. Fitness effects on the cognitive function of older adults: A meta-analytic study. *Psychol. Sci.* **2003**, *14*, 125–130. [CrossRef] [PubMed]

44. Verburgh, L.; Königs, M.; Scherder, E.J.; Oosterlaan, J. Physical exercise and executive functions in preadolescent children, adolescents and young adults: A meta-analysis. *Br. J. Sports Med.* **2014**, *48*, 973–979. [CrossRef] [PubMed]

45. Hansen, A.; Pritchard, T.; Melnic, I.; Zhang, J. Physical activity, screen time, and school absenteeism: Self-reports from NHANES 2005–2008. *Curr. Med. Res. Opin.* **2016**, *32*, 651–659. [CrossRef] [PubMed]

46. Berra, K.; Rippe, J.; Manson, J. Making physical activity counseling a priority in clinical practice: The time for action is now. *JAMA* **2015**, *314*, 2617–2618. [CrossRef] [PubMed]

International Journal of
*Environmental Research
and Public Health*

Article

Examining Obedience Training as a Physical Activity Intervention for Dog Owners: Findings from the Stealth Pet Obedience Training (SPOT) Pilot Study

Katie Potter [1,*], Brittany Masteller [2] and Laura B. Balzer [3]

[1] Department of Kinesiology, University of Massachusetts Amherst, Amherst, MA 01003, USA
[2] Department of Exercise & Sport Studies, Smith College, Northampton, MA 01063, USA; bmasteller@smith.edu
[3] Department of Biostatistics and Epidemiology, University of Massachusetts Amherst, Amherst, MA 01003, USA; lbalzer@umass.edu
* Correspondence: katie.potter@umass.edu; Tel.: +1-413-545-6058

Abstract: Dog training may strengthen the dog–owner bond, a consistent predictor of dog walking behavior. The Stealth Pet Obedience Training (SPOT) study piloted dog training as a stealth physical activity (PA) intervention. In this study, 41 dog owners who reported dog walking ≤3 days/week were randomized to a six-week basic obedience training class or waitlist control. Participants wore accelerometers and logged dog walking at baseline, 6- and 12-weeks. Changes in PA and dog walking were compared between arms with targeted maximum likelihood estimation. At baseline, participants (39 ± 12 years; females = 85%) walked their dog 1.9 days/week and took 5838 steps/day, on average. At week 6, intervention participants walked their dog 0.7 more days/week and took 480 more steps/day, on average, than at baseline, while control participants walked their dog, on average, 0.6 fewer days/week and took 300 fewer steps/day (difference between arms: 1.3 dog walking days/week; 95% CI = 0.2, 2.5; 780 steps/day, 95% CI = −746, 2307). Changes from baseline were similar at week 12 (difference between arms: 1.7 dog walking days/week; 95% CI = 0.6, 2.9; 1084 steps/day, 95% CI = −203, 2370). Given high rates of dog ownership and low rates of dog walking in the United States, this novel PA promotion strategy warrants further investigation.

Keywords: dog walking; exercise; health behavior change; stealth health; pet ownership; human–animal interaction; animal-assisted intervention; targeted learning

Citation: Potter, K.; Masteller, B.; Balzer, L.B. Examining Obedience Training as a Physical Activity Intervention for Dog Owners: Findings from the Stealth Pet Obedience Training (SPOT) Pilot Study. *Int. J. Environ. Res. Public Health* **2021**, *18*, 902. https://doi.org/10.3390/ijerph18030902

Academic Editors: Luciana Zaccagni and Emanuela Gualdi-Russo

Received: 4 December 2020
Accepted: 14 January 2021
Published: 21 January 2021

1. Introduction

Fewer than one in four American adults meet federal physical activity (PA) guidelines [1]. Part of the problem may be messaging around PA. Traditional messaging frames PA as a critically important health behavior that people should engage in to improve health and prevent disease. While this may motivate PA adoption short-term, this framing provides a controlled source of motivation for PA (in the form of external pressure to be healthy), and therefore is unlikely to motivate consistent participation over time [2,3]. Sustainable PA might be more realistically achieved by encouraging people to engage in activities they enjoy, care about, or find purposeful and that naturally involve PA (i.e., that increase PA as a side-effect) [4,5]. This approach preserves autonomy and therefore better aligns with the science of motivation and decision-making [2]. Interventions that take this approach are called "stealth interventions" [4,5]. Stealth interventions have been successful in promoting healthy eating and weight control [6–8].

A relevant aspect of American culture that lends itself well to stealth PA interventions is dog ownership. Almost half of households in the United States (46%) own at least one dog [9]. Research has demonstrated that owners who walk their dog(s) are more likely to meet PA guidelines than those who do not [10]. The strength of the dog–owner bond is a

173

key correlate of dog walking behavior [11,12], as owners who have a strong relationship with their dog(s) feel a greater sense of responsibility to walk and perceive more motivation and support from their dog(s) for walking [11,13]. A recent national survey study found that only 42% of American dog owners walk their dog [14], suggesting a potential target for stealth PA interventions.

Interventions designed to strengthen the dog–owner bond are already offered in communities across the United States. These interventions are obedience training classes. In addition to strengthening the dog–owner bond [15,16], these classes teach basic manners, including loose leash walking, which may increase dog-walking self-efficacy [17] and reduce behavior problems that can interfere with dog walking [13]. To the best of our knowledge, dog obedience training classes have never been examined in the context of PA promotion.

The purpose of the Stealth Pet Obedience Training (SPOT) randomized trial was to pilot a six-week, basic obedience training class as a stealth PA intervention for inactive dog owners (NCT04329741). Again, the logic behind this approach is that some dog owners who take the course may naturally develop a new, personal source of motivation for PA (in the form of dog walking) as a result of becoming more attached to their dog through the dog training experience. We consider this a stealth approach not because dog owners will be unaware that they are engaging in more PA, but because the intervention is a course focused on improving dog obedience, not increasing PA. We hypothesized that the intervention would lead to greater increases in average steps/day and daily minutes of moderate–vigorous PA (MVPA) at 6 and 12 weeks, as compared to the waitlist control. Secondary outcomes included dog walking days per week, daily sedentary minutes, and psychosocial variables plausibly mediating PA changes: strength of dog–owner bond (emotional closeness with dog), dog walking self-efficacy, and social support (from the dog) for walking.

2. Materials and Methods

2.1. Study Design and Population

The SPOT study was an individually randomized trial with a waitlist control group. Adult dog owners (21+ years) who reported walking their dog(s) ≤3 times/week (for no more than 20 min/walk) were eligible for participation. Exclusion criteria were regular exercise (defined as ≥3 times per week for ≥20 min), previously attending obedience training with their current dog, presence of any condition that limits walking ability, or presence of uncontrolled diabetes or hypertension. These criteria were set to identify a sample of inactive dog owners who could safely walk for exercise. All participants provided informed consent. This study was approved by the Institutional Review Board (protocol ID: 2017-3945) and the Institutional Animal Care and Use Committee (protocol ID: 2017-0018) at the University of Massachusetts-Amherst. Randomization was conducted by an external researcher.

2.2. Procedures

Participants were recruited through university-affiliated social media outlets. The study was advertised as an investigation of how attending a dog obedience training course affects the dog–owner bond and the health and quality of life of dog owners. Participants randomized to the intervention group enrolled in a six-week basic obedience training class led by a certified behavior adjustment trainer. The class covered basic commands (e.g., sit, down, watch), loose leash walking, and polite greetings, among other skills. The importance of dog walking was implied, but not specifically emphasized. Classes were held once per week for 45 min, with 5–8 students per class. Participants in the control group were asked not to enroll in an obedience training class or train their dog at home until after completing 12-week assessments. After completing these assessments, control group participants received a voucher to take the same obedience training class for free. Baseline measures were taken in August, six-week measures in October, and 12-week measures in December 2017 in Massachusetts, United States.

2.3. Measures

Participants reported their demographics (e.g., age, sex), characteristics of their dog (e.g., age, size), and whether they had a yard ("Do you have a yard where your dog can run free?" (yes/no)). To estimate dog size, participants were given four options (small, medium, large, giant) and provided example dog breeds for each option. Research staff measured each participant's height and weight at orientation for body mass index calculation.

2.3.1. Process Evaluation

The study process was evaluated through the following metrics. Retention was assessed as the percentage of randomized participants who completed 12-week assessments. Class attendance, defined as the average number of classes attended across the 6 weeks, served as the indicator of intervention engagement. Intervention fidelity was indicated by the proportion of intervention participants who agreed or strongly agreed with the prompt "I am happy with the behavior of my dog" at week 6 compared to control participants.

2.3.2. Physical Activity and Sedentary Behavior

For seven consecutive days at baseline, post-program (at 6 weeks), and 6 weeks post-program (at 12 weeks), participants wore an ActiGraph wGT3X-BT monitor (ActiGraph LLC, Pensacola, FL, USA) on their right hip [18–20] and logged all leisure-time PA, including dog walking, in a paper log booklet. The ActiGraph, a research-grade triaxial accelerometer deemed valid [21] and reliable [22] in free-living conditions, was used to assess changes in steps, MVPA, and sedentary behavior. Participants wore the device during all waking hours, except when showering/swimming.

ActiGraph data were processed using Actilife Version 6.13.3 (ActiGraph LLC, Pensacola, FL, USA); validated cut-points for adults [23] were used to evaluate minutes spent in different PA intensity categories. To be included in analyses, participants had to wear the device ≥8 h/day for ≥4 days, including one weekend day; otherwise, data were considered missing [24]. Participant data were averaged across valid wear-days to produce daily estimates.

2.3.3. Psychosocial Outcomes

The 10-item perceived emotional closeness subscale from the Cat/Dog–Owner Relationship Scale (C/DORS) was used to assess the dog–owner bond [25]. Each item is scored on a five-point scale (from 1–5). Sample items include "My pet provides me with constant companionship" and "My pet is there whenever I need to be comforted". The 10 item scores were summed and divided by 10 to yield a subscale score ranging from 1–5, with higher scores indicating better perceived relationship quality.

Subscales from the Dogs and Walking Survey (DAWGS) [26] were used to assess self-efficacy beliefs about dog walking and support provided by the dog for walking. The self-efficacy subscale consists of nine items from the Exercise Confidence Survey [27] modified for dog walking. Participants are asked to rate how confident they are that they would consistently walk their dog under a number of circumstances if they really wanted to (1 = very unconfident to 5 = very confident). Scores on the nine items were summed to produce a dog walking self-efficacy score ranging from 9–45. Three five-point items (1 = strongly disagree to 5 = strongly agree) were used to assess dog support for walking. These items were "Having my dog makes me walk more", "My dog provides encouragement for me to go on walk"', and "My dog provides social support for me to go on walks". Scores were summed to produce a total dog support for walking score ranging from 3–15.

2.4. Statistical Analyses

A sample size of 40 adults was selected to provide at least 80% power to detect differences between randomized groups of at least 1500 average steps/day and at least 76 min/week of MVPA. Such differences are consistent with meeting 50% of aerobic physical activity guidelines, which have been associated with improvements in cardiorespiratory fitness among previously sedentary individuals [28,29].

We compared 6- and 12-week outcomes between randomized arms using targeted maximum likelihood estimation (TMLE), which provides precision and power gains over an unadjusted approach (e.g., the Student's *t*-test) in randomized trials [30,31]. We used a pre-specified adjustment strategy and excluded participants whose outcome assessments were missing at the timepoint of interest [32]. All PA outcomes were parameterized in terms of the change from baseline. Statistical inference was based on the *t*-distribution and with a two-sided hypothesis test at the 5% significance level. All analyses were completed with R v3.5.1. 9. (The R Foundation, Vienna, Austria). Additional details about the statistical approach are provided in the Supplementary Materials.

3. Results

The median age of participants was 37 years, and most were non-Hispanic White (87%) and female (87%) (Table 1). The median age of the study dogs was 3 years. About half the dogs were large (50–90 lbs; *n* = 18), followed by medium-sized (20–49 lbs; *n* = 10), small/toy-sized (<20 lbs; *n* = 7), and giant-sized (>90 lbs; *n* = 4). At baseline, participants averaged 5838 steps/day, 22 MVPA minutes/day, and reported dog walking 1.9 days/week (Table 1).

Table 1. Baseline characteristics of Stealth Pet Obedience Training (SPOT) study participants and their dogs.[1]

	Overall (*n* = 39)	Intervention (*n* = 19)	Control (*n* = 20)
Age, median (min–max) in years	37 (21–72)	39 (27–72)	34 (21–54)
Sex, *n* (%) female	34 (87%)	16 (84%)	18 (90%)
Race, *n* (%) non-Hispanic White	34 (87%)	18 (95%)	16 (80%)
Annual income, *n* (%)			
<$40,000	5 (13%)	4 (21%)	1 (5%)
$40,000—$80,000	16 (41%)	5 (26%)	11 (55%)
>$80,000	18 (46%)	10 (53%)	8 (40%)
Education, *n* (%)			
High school or GED	8 (21%)	3 (16%)	5 (25%)
College degree	15 (38%)	8 (42%)	7 (35%)
Graduate or professional degree	16 (41%)	8 (42%)	8 (40%)
Body mass index, median (min–max) in kg/m^2	30 (20.5–44.8)	30 (20.5–44.8)	29.8 (22.6–42.3)
Dog's age, median (min–max) in years	3 (0–11)	4 (0–10)	2.8 (0–11)
Dog's size, *n* (%)			
Giant (>90 lbs)	4 (10%)	2 (11%)	2 (10%)
Large (50–90 lbs	18 (46%)	11 (58%)	7 (35%)
Medium (20–49 lbs)	10 (26%)	5 (26%)	5 (25%)
Small/toy (<20 lbs)	7 (18%)	1 (5%)	6 (30%)
Yard where dog can run freely, *n* (%)	25 (64%)	15 (79%)	10 (50%)
Agree or strongly agree with prompt "I am happy with the behavior of my dog", *n* (%)	8 (21%)	3 (16%)	5 (25%)
Days/week with at least 1 dog walk, mean (SD)	1.9 (2.1)	1.7 (1.9)	2.2 (2.4)
Steps/day, mean (SD)	5838 (2141)	5840 (2132)	5836 (2208)
Moderate-to-vigorous physical activity (MVPA) minutes/day, mean (SD)	22 (14)	22 (16)	21 (13)
Sedentary minutes/day, mean (SD)	542 (87)	544 (68)	540 (104)
Emotional closeness with dog, median (min–max)[2]	3.9 (2.4–5)	3.7 (2.5–4.9)	4.2 (2.4–5)
Social support from dog for walking, median (min–max)[3]	11 (3–15)	11 (3–15)	11 (7–15)
Self-efficacy for dog walking, median (min–max)[3]	29 (9–45)	29 (9–45)	30 (18–45)

[1] Excludes two intervention participants who dropped out prior to study start. Missing ActiGraph data (steps, MVPA, sedentary minutes) for one participant. [2] Emotional closeness score from cat/dog–owner relationship scale (C/DORS)—scale 1–5. [3] Social support from dog for walking (scale 3–15) and self-efficacy for dog walking (scale 9–45) from dogs and walking scale (DAWGS).

Four of the 21 participants randomized to intervention dropped out before or during the six-week class. All of the remaining 17 intervention participants completed the 6-week assessments, and 16 of 17 completed 12-week assessments. All 20 control participants completed both 6- and 12-week assessments. Altogether, 36 of 41 randomized participants completed 12-week assessments for an overall study retention rate of 88%. Intervention participants attended an average 5.6 out of 6 classes and were 33% more likely to agree or strongly agree with the prompt "I am happy with the behavior of my dog" than control participants at week 6 (95% CI = 6%, 60%).

ActiGraph wear time criteria were met by 38 participants (19 per group) at baseline, 34 participants (16 intervention, 18 control) at 6 weeks, and 33 participants (16 intervention, 17 control) at 12 weeks. At 6 weeks, intervention participants took 480 more steps/day than they had at baseline, while control participants took 300 fewer steps/day than at baseline (Figure 1). Thus, the difference in the average change in steps/day between randomized arms was 780 steps/day (95% CI = −746, 2307). These differences persisted at 12 weeks, when intervention participants had essentially no change in their steps/day from baseline, and control participants decreased by 1084 steps/day for an average difference of 1084 steps/day (95% CI = −203, 2370).

These changes in daily steps were echoed in changes in daily MVPA minutes (Figure 1). At 6 weeks, intervention participants averaged 4.7 more MVPA minutes than they had at baseline, while control participants averaged <1 more MVPA minute than at baseline, for a difference of 4.3 min (95% CI = −7.8, 16.4). At 12 weeks, both intervention and control participants had decreased their MVPA minutes by 1.9 and 3.1 min from baseline, respectively (difference = 1.2; 95% CI = −6.4, 8.8).

Self-reported days with at least one dog walk are reported in Table 2. At 6 weeks, intervention participants walked their dog, on average, 0.7 more days/week than they had at baseline, while control participants walked their dog, on average, 0.6 fewer days/week (difference = 1.3; 95% CI = 0.2, 2.5). At 12 weeks, intervention participants increased their dog walking by 0.9 more days/week compared to baseline, and control participants reduced their dog walking by 0.8 days/week (difference = 1.7; 95% CI = 0.6, 2.9).

Trends were also observed in daily sedentary minutes (Table 2). At 6 weeks, intervention participants had increased their sedentary time by 2.1 min on average, while control participants had increased their sedentary time by 16 min on average (difference = −13.9; 95% CI = −41.2, 13.4). By 12 weeks, intervention participants had increased their daily sedentary time by <1 min from baseline, whereas control participants increased their sedentary time by a half-hour (30.2 min), for an average difference of −29.3 min/day (95% CI = −69.8, 11.2).

Intervention effects on psychosocial outcomes are shown in Table 2. At 6 and 12 weeks, both groups reported similar levels of emotional closeness with their dog. Intervention participants averaged 3.8–3.9 out of 5 possible points on the C/DORS emotional closeness subscale, whereas control participants averaged 4.0–4.1 points. Both groups also reported similar social support from the dog for walking; at 6 weeks, intervention participants averaged 11.2 out of a maximum 15 points compared to an average 10.4 points in the control arm. Intervention participants did, however, report higher confidence in their ability to walk the dog in the face of barriers; at 6 weeks, their average score was 31.9 (maximum score of 45), compared to an average score of 29.0 in the control group (difference = 2.9; 95% CI = −0.7, 6.4). Similar trends were observed at 12 weeks (difference = 7.4; 95% CI = −1.2, 15.9).

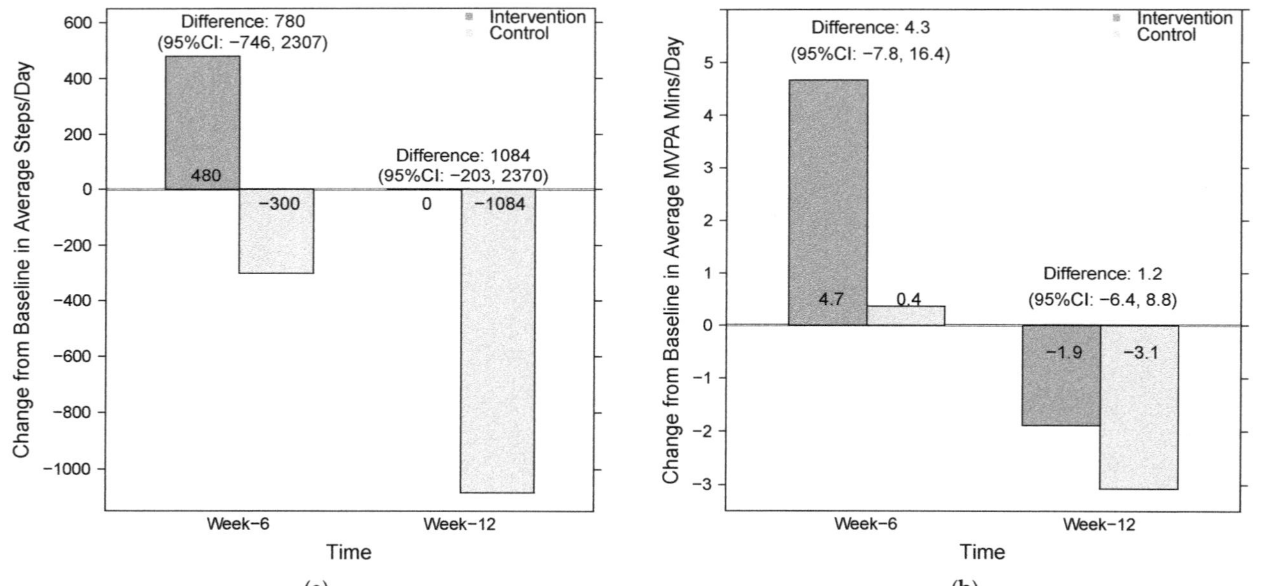

Figure 1. Change in average steps/day (**a**) and moderate-to-vigorous physical activity (MVPA) minutes/day (**b**) among participants in the Stealth Pet Obedience Training (SPOT) pilot study from baseline to 6 and 12 weeks. Analysis restricted to participants with valid ActiGraph data at both baseline and timepoint of interest: 16 intervention participants and 18 control participants at week 6, and 16 intervention participants and 17 control participants at week 12.

Table 2. Intervention effects on secondary outcomes in the Stealth Pet Obedience Training (SPOT) pilot study.

	At Week 6			At Week 12		
	Intervention Average	Control Average	Difference (95% CI)	Intervention Average	Control Average	Difference (95% CI)
Change in dog walking days/week [1]	0.73	−0.62	1.35 (0.21, 2.49)	0.92	−0.83	1.75 (0.58, 2.91)
Change in sedentary minutes/day [1]	2.10	16.00	−13.90 (−41.20, 13.40)	0.90	30.20	−29.30 (−69.80, 11.20)
Emotional closeness with dog [2]	3.90	4.00	0 (−0.30, 0.20)	3.80	4.10	−0.30 (−0.50, 0)
Social support from the dog for walking [3]	11.20	10.40	0.70 (−0.90, 2.30)			
Self-efficacy for dog walking [3]	31.90	29.00	2.90 (−0.70, 6.40)	23.30	15.90	7.40 (−1.20, 15.90)

[1] Changes at both week 6 and week 12 are compared to baseline. [2] Emotional closeness from cat/dog owner–relationship scale (C/DORS); scores can range from 1–5. [3] Social support from dog for walking (score range 3–15) and self-efficacy for dog walking (score range 9–45) from dogs and walking scale (DAWGS). Social support for dog walking data not available at 12 weeks.

4. Discussion

The SPOT study piloted a stealth health approach to increasing PA among inactive dog owners. Immediately post-intervention (at 6 weeks), the intervention group averaged more steps and MVPA minutes per day than the waitlist control group, although differences were modest (780 steps/day and 4.7 MVPA minutes/day, on average). The modest group difference in daily steps was maintained at the follow-up assessment (1080 steps/day, on average), and was driven by a decrease in steps by control participants. Given the timeline of the study (i.e., baseline measures taken in summer, 12-week measures taken in winter), a plausible explanation for these findings is that obedience training buffered against a decrease in leisure-time PA with the onset of winter [33], which was observed in the control group.

Most existing interventions to increase dog walking focus on health benefits for the dog and/or owner [34–37] and promote dog walking as something owners should do. In contrast, the approach tested in this study aimed to strengthen the dog–owner bond and foster a sustainable PA habit as a side effect (small sample sizes prohibited formal mediation analyses). Importantly, basic dog obedience training is already available in communities across the United States, and to ensure the potential for population-level dissemination, the program was not modified in any way. The success of this pilot was reflected in the high rate of class attendance and low study attrition. Furthermore, intervention participants were more likely to be happy with their dogs' behavior at 6 weeks compared to control participants, demonstrating that the course was effective in achieving its primary purpose.

Increases in steps in the range of 1000–2000 steps/day have been associated with reduced risk of type II diabetes [38], cardiovascular disease [39,40], and all-cause mortality [41–43]. Immediately post-intervention, there were average daily increases of nearly 500 steps and 5 MVPA minutes (the equivalent of 500 steps taken at a rate of ≥100 steps/min [44]) among intervention participants. Notably, nearly a third (31%) of intervention participants increased their steps/day by at least 1000 steps at both 6 and 12 weeks. In comparison, PA interventions that use pedometers to increase walking behavior have led to short-term increases of about 2000 steps/day [45]. Altogether, the modest changes in PA shown in the SPOT pilot are promising given that the intervention was not advertised for PA and did not specifically focus on dog walking. This stealth approach may reach inactive adults who are uninterested in or unmotivated by traditional PA promotion efforts, and therefore modest intervention efficacy may translate to large public health impact [46].

Intervention participants also accumulated fewer daily sedentary minutes than control participants immediately post-intervention and at the 12-week follow-up assessment. The group differences in sedentary behavior were driven by decreases from baseline among control participants, which may have been due to seasonal changes. Since intervention participants were asked to work on new skills outside of class, it is plausible that more time spent training or playing with one's dog (irrespective of walking) buffered against an increase in sedentary behavior with the onset of winter. Although the best sedentary behavior interventions to date primarily aim to change sedentary behavior (not PA) [47,48], this finding suggests that obedience training could positively impact both PA and sedentary behavior among dog owners. A few studies have examined dog ownership in relation to sedentary behavior [49–51], but no intervention studies have attempted to leverage the dog–owner bond to reduce sedentary behavior.

We anticipated that changes in PA would be driven by differences in the dog–owner bond, but did not observe meaningful differences in the dog–owner relationship. All participants were highly attached to their dog at baseline, and therefore a ceiling effect may have occurred. It is also possible that intervention participants did experience increased feelings of attachment toward their dog, but that the sample size was too small or the questionnaire not nuanced enough to detect changes. Qualitative data collection may be helpful for measuring changes in the human–animal bond in this study population.

Intervention participants did, however, report higher dog walking self-efficacy than control participants at 6 and 12 weeks. The self-efficacy construct is actively debated, with some experts suggesting it is confounded with motivation [52,53]. In SPOT, participants were asked to rate their confidence in their ability to consistently walk their dog under a number of circumstances (e.g., after a long day at work, when undergoing a stressful life change) if they really wanted to. The "if you really wanted to" qualifier attempts to assess perceived capability independent of motivational factors [54]. We attribute higher self-efficacy for dog walking among intervention participants to the mastery of program skills, including loose leash walking and the "leave it" command. However, it is also possible that the acquisition of these skills made dog walking more enjoyable, and this greater enjoyment led to greater motivation to walk.

The major strength of this study is the innovative stealth health approach tested. Stealth approaches may engage individuals who are uninterested in lifestyle change for better health and disease prevention (and therefore would not participate in traditional interventions), and may promote sustainable change given that the change is personally meaningful or enjoyable. The stealth approach tested in this study has potential for high reach as, by some estimates, nearly 50% of American households have a dog [9]. Furthermore, dog walking may be a particularly sustainable form of PA to promote, as it serves a purpose and lends itself well to habit formation [55]. Other strengths of this study include the rigorous randomized design and use of objective measures of PA.

Limitations of this study include its small sample size, which led to large variability in PA outcomes. Baseline assessments were held in summer, and follow-up assessments were held in winter; therefore, the study timeline corresponded with major changes in weather and lifestyle (i.e., back to school) that may affect PA levels. The small sample size in concert with these seasonal changes likely made intervention-driven changes in PA difficult to detect. Finally, although the researchers aimed to determine whether changes in PA naturally occurred as a side-effect of attending a dog training course, participants were aware that the research team was interested in changes in PA and health. This transparency was required for both ethical and practical reasons (i.e., participants wore an accelerometer to provide PA data), but may have led to an expectancy effect. Therefore, changes in PA may not have been driven entirely by autonomous motivation developed as a side-effect of dog training, as theorized.

5. Conclusions

The results of this pilot, randomized trial suggest that attending a basic dog obedience training course may lead dog owners to walk more and sit less. If the positive effects observed in this trial are replicated in larger trials with longer follow-up periods, interdisciplinary partnerships can work to normalize and incentive obedience training among new and current dog owners, and to increase accessibility for low-income owners. Given the high rate of dog ownership [9] and low rate of dog walking in the United States [14], this novel approach to PA promotion has the potential for considerable public health impact and warrants further investigation.

Supplementary Materials: The following are available online at https://www.mdpi.com/1660-460 1/18/3/902/s1, Appendix S1: Further details on the statistical analysis, Figure S1: Study CONSORT diagram.

Author Contributions: Conceptualization, K.P.; methodology, K.P. and L.B.B.; formal analysis, L.B.B.; investigation, K.P. and B.M.; validation, K.P. and L.B.B.; resources, K.P.; writing—original draft preparation, K.P. and B.M.; writing—review and editing, K.P., B.M., and L.B.B.; visualization, L.B.B.; supervision, K.P.; project administration, K.P.; funding acquisition, K.P. All authors have read and agreed to the published version of the manuscript.

Funding: This research was funded by the American College of Sports Medicine Foundation, grant number 16-00293.

Institutional Review Board Statement: The study was conducted according to the guidelines of the Declaration of Helsinki, and approved by the Institutional Review Board (protocol ID: 2017-3945; date of approval 07/06/2017) and the Institutional Animal Care and Use Committee (protocol ID: 2017-0018; date of approval 07/11/2017) of the University of Massachusetts-Amherst.

Informed Consent Statement: Informed consent was obtained from all subjects involved in the study.

Acknowledgments: The authors wish to thank Caryl-Rose Pofcher of My Dog, LLC (Amherst, MA, United States) and all undergraduate and graduate students who assisted with data collection for the SPOT study.

Conflicts of Interest: All the authors are dog owners. The authors declare no other conflict of interest.

References

1. Blackwell, D.L.; Clarke, T.C. State variation in meeting the 2008 federal guidelines for both aerobic and muscle-strengthening activities through leisure-time physical activity among adults aged 18–64: United States, 2010–2015. *Natl. Health Stat. Rep.* **2018**, *112*, 1–22.
2. Segar, M.L.; Richardson, C.R. Prescribing pleasure and meaning cultivating walking motivation and maintenance. *Am. J. Prev. Med.* **2014**, *47*, 838–841. [CrossRef]
3. Teixeira, P.J.; Carraça, E.V.; Markland, D.; Silva, M.N.; Ryan, R.M. Exercise, physical activity, and Self-Determination Theory: A systematic review. *Int. J. Behav. Nutr. Phys. Act.* **2012**, *9*, 78. [CrossRef]
4. Robinson, T.N. Stealth interventions for obesity prevention and control: Motivating behavior change. In *Obesity Prevention*; Dubé, L., Bechara, A., Dagher, A., Drewnowski, A., Lebel, J., James, P., Yada, R.Y., Eds.; Academic Press: San Diego, CA, USA, 2010; Chapter 25; pp. 319–327.
5. Robinson, T.N. Save the world, prevent obesity: Piggybacking on existing social and ideological movements. *Obesity* **2010**, *18*, S17–S22. [CrossRef]
6. Hekler, E.B.; Gardner, C.D.; Robinson, T.N. Effects of a college course about food and society on students' eating behaviors. *Am. J. Prev. Med.* **2010**, *38*, 543–547. [CrossRef]
7. Robinson, T.N.; Killen, J.D.; Kraemer, H.C.; Wilson, D.M.; Matheson, D.M.; Haskell, W.L.; Pruitt, L.A.; Powell, T.M.; Owens, A.S.; Thompson, N.S.; et al. Dance and reducing television viewing to prevent weight gain in African-American girls: The Stanford GEMS pilot study. *Ethn. Dis.* **2003**, *13*, S65–S77.
8. Weintraub, D.L.; Tirumalai, E.C.; Haydel, K.F.; Fujimoto, M.; Fulton, J.E.; Robinson, T.N. Team sports for overweight children: The Stanford Sports to Prevent Obesity Randomized Trial (SPORT). *Arch. Pediatr. Adolesc. Med.* **2008**, *162*, 232–237. [CrossRef]
9. Applebaum, J.W.; Peek, C.W.; Zsembik, B.A. Examining U.S. pet ownership using the General Social Survey. *Soc. Sci. J.* **2020**, 1–10. [CrossRef]
10. Soares, J.; Epping, J.N.; Owens, C.J.; Brown, D.R.; Lankford, T.J.; Simoes, E.J.; Caspersen, C.J. Odds of getting adequate physical activity by dog walking. *J. Phys. Act. Health* **2015**, *12*, S102–S109. [CrossRef]
11. Westgarth, C.; Christley, R.M.; Christian, H.E. How might we increase physical activity through dog walking? A comprehensive review of dog walking correlates. *Int. J. Behav. Nutr. Phys. Act.* **2014**, *11*, 83. [CrossRef]
12. Westgarth, C.; Christian, H.E.; Christley, R.M. Factors associated with daily walking of dogs. *BMC Vet. Res.* **2015**, *11*, 116. [CrossRef] [PubMed]
13. Westgarth, C.; Knuiman, M.; Christian, H.E. Understanding how dogs encourage and motivate walking: Cross-sectional findings from RESIDE. *BMC Public Health* **2016**, *16*, 1019. [CrossRef] [PubMed]

14. Richards, E.A. Does dog walking predict physical activity participation: Results from a national survey. *Am. J. Health Promot.* **2016**, *30*, 323–330. [CrossRef]

15. Clark, G.I.; Boyer, W.N. The effects of dog obedience training and behavioural counselling upon the human-canine relationship. *Appl. Anim. Behav. Sci.* **1993**, *37*, 147–159. [CrossRef]

16. Bennett, P.C.; Rohlf, V.I. Owner-companion dog interactions: Relationships between demographic variables, potentially problematic behaviours, training engagement and shared activities. *Appl. Anim. Behav. Sci.* **2007**, *102*, 65–84. [CrossRef]

17. Hoerster, K.D.; Mayer, J.A.; Sallis, J.F.; Pizzi, N.; Talley, S.; Pichon, L.C.; Butler, D.A. Dog walking: Its association with physical activity guideline adherence and its correlates. *Prev. Med.* **2011**, *52*, 33–38. [CrossRef]

18. Tudor-Locke, C.; Barreira, T.V.; Schuna, J.M. Comparison of step outputs for waist and wrist accelerometer attachment sites. *Med. Sci. Sports Exerc.* **2015**, *47*, 839–842. [CrossRef]

19. Ellis, K.; Kerr, J.; Godbole, S.; Lanckriet, G.; Wing, D.; Marshall, S. A random forest classifier for the prediction of energy expenditure and type of physical activity from wrist and hip accelerometers. *Physiol. Meas.* **2014**, *35*, 2191–2203. [CrossRef]

20. Migueles, J.H.; Cadenas-Sanchez, C.; Ekelund, U.; Delisle Nyström, C.; Mora-Gonzalez, J.; Löf, M.; Labayen, I.; Ruiz, J.R.; Ortega, F.B. Accelerometer data collection and processing criteria to assess physical activity and other outcomes: A systematic review and practical considerations. *Sports. Med.* **2017**, *47*, 1821–1845. [CrossRef]

21. Lyden, K.; Keadle, S.K.; Staudenmayer, J.; Freedson, P.S. A method to estimate free-living active and sedentary behavior from an accelerometer. *Med. Sci. Sports Exerc.* **2014**, *46*, 386–397. [CrossRef]

22. Aadland, E.; Ylvisåker, E. Reliability of the Actigraph GT3X+ accelerometer in adults under free-living conditions. *PLoS ONE* **2015**, *10*, e0134606. [CrossRef] [PubMed]

23. Freedson, P.S.; Melanson, E.; Sirard, J. Calibration of the Computer Science and Applications, Inc. accelerometer. *Med. Sci. Sports Exerc.* **1998**, *30*, 777–781. [CrossRef] [PubMed]

24. Troiano, R.P.; Berrigan, D.; Dodd, K.W.; Mâsse, L.C.; Tilert, T.; McDowell, M. Physical activity in the United States measured by accelerometer. *Med. Sci. Sports Exerc.* **2008**, *40*, 181–188. [CrossRef] [PubMed]

25. Howell, T.J.; Bowen, J.; Fatjó, J.; Calvo, P.; Holloway, A.; Bennett, P.C. Development of the Cat-Owner Relationship Scale (CORS). *Behav. Processes* **2017**, *141*, 305–315. [CrossRef] [PubMed]

26. Richards, E.A.; McDonough, M.H.; Edwards, N.E.; Lyle, R.M.; Troped, P.J. Development and Psychometric Testing of the Dogs and WalkinG Survey (DAWGS). *Res. Q. Exerc. Sport* **2013**, *84*, 492–502. [CrossRef]

27. Sallis, J.F.; Pinski, R.B.; Grossman, R.M.; Patterson, T.L.; Nader, P.R. The Development of Self-Efficacy Scales for Health-Related Diet and Exercise Behaviors. *Health Educ. Res.* **1988**, *3*, 283–292. [CrossRef]

28. Church, T.S.; Earnest, C.P.; Skinner, J.S.; Blair, S.N. Effects of different doses of physical activity on cardiorespiratory fitness among sedentary, overweight or obese postmenopausal women with elevated blood pressure: A randomized controlled trial. *JAMA* **2007**, *297*, 2081–2091. [CrossRef]

29. Jordan, A.N.; Jurca, G.M.; Locke, C.T.; Church, T.S.; Blair, S.N. Pedometer indices for weekly physical activity recommendations in postmenopausal women. *Med. Sci. Sports Exerc.* **2005**, *37*, 1627–1632. [CrossRef]

30. van der Laan, M.J.; Rose, S. *Targeted Learning: Causal Inference for Observational and Experimental Data*; Springer: New York, NY, USA, 2011.

31. Rosenblum, M.; van der Laan, M.J. Simple, efficient estimators of treatment effects in randomized trials using generalized linear models to leverage baseline variables. *Int. J. Biostat.* **2010**, *6*. [CrossRef]

32. Balzer, L.B.; van der Laan, M.J.; Petersen, M.L.; SEARCH Collaboration. Adaptive pre-specification in randomized trials with and without pair-matching. *Stat. Med.* **2016**, *35*, 4528–4545. [CrossRef]

33. Pivarnik, J.M.; Reeves, M.J.; Rafferty, A.P. Seasonal variation in adult leisure-time physical activity. *Med. Sci. Sports Exerc.* **2003**, *35*, 1004–1008. [CrossRef] [PubMed]

34. Kushner, R.F.; Blatner, D.J.; Jewell, D.E.; Rudloff, K. The PPET study: People and pets exercising together. *Obesity* **2006**, *14*, 1762–1770. [CrossRef] [PubMed]

35. Rhodes, R.E.; Murray, H.; Temple, V.A.; Tuokko, H.; Higgins, J.W. Pilot study of a dog walking randomized intervention: Effects of a focus on canine exercise. *Prev. Med.* **2012**, *54*, 309–312. [CrossRef]

36. Byers, C.G.; Wilson, C.C.; Stephens, M.B.; Goodie, J.L.; Netting, F.E.; Olsen, C.H. Owners and pets exercising together: Canine response to veterinarian-prescribed physical activity. *Anthrozoös* **2014**, *27*, 325–333. [CrossRef]

37. Richards, E.A.; Ogata, N.; Cheng, C.-W. Evaluation of the Dogs, Physical Activity, and Walking (Dogs PAW) intervention: A randomized controlled trial. *Nurs. Res.* **2016**, *65*, 191–201. [CrossRef]

38. Kraus, W.E.; Yates, T.; Tuomilehto, J.; Sun, J.-L.; Thomas, L.; McMurray, J.J.V.; Bethel, M.A.; Holman, R.R. Relationship between baseline physical activity assessed by pedometer count and new-onset diabetes in the NAVIGATOR trial. *BMJ Open Diabetes Res. Care* **2018**, *6*, e000523. [CrossRef]

39. Yates, T.; Haffner, S.M.; Schulte, P.J.; Thomas, L.; Huffman, K.M.; Bales, C.W.; Califf, R.M.; Holman, R.R.; McMurray, J.J.V.; Bethel, M.A.; et al. Association between change in daily ambulatory activity and cardiovascular events in people with impaired glucose tolerance (NAVIGATOR Trial): A cohort analysis. *Lancet* **2014**, *383*, 1059–1066. [CrossRef]

40. Jefferis, B.J.; Parsons, T.J.; Sartini, C.; Ash, S.; Lennon, L.T.; Papacosta, O.; Morris, R.W.; Wannamethee, S.G.; Lee, I.-M.; Whincup, P.H. Does total volume of physical activity matter more than pattern for onset of CVD? A prospective cohort study of older British men. *Int. J. Cardiol.* **2019**, *278*, 267–272. [CrossRef]

41. Dwyer, T.; Pezic, A.; Sun, C.; Cochrane, J.; Venn, A.; Srikanth, V.; Jones, G.; Shook, R.P.; Shook, R.; Sui, X.; et al. Objectively measured daily steps and subsequent long term all-cause mortality: The Tasped prospective cohort study. *PLoS ONE* **2015**, *10*, e0141274.
42. Jefferis, B.J.; Parsons, T.J.; Sartini, C.; Ash, S.; Lennon, L.T.; Papacosta, O.; Morris, R.W.; Wannamethee, S.G.; Lee, I.-M.; Whincup, P.H. Objectively measured physical activity, sedentary behaviour and all-cause mortality in older men: Does volume of activity matter more than pattern of accumulation? *Br. J. Sports Med.* **2019**, *53*, 1013–1020. [CrossRef]
43. Lee, I.-M.; Shiroma, E.J.; Kamada, M.; Bassett, D.R.; Matthews, C.E.; Buring, J.E. Association of step volume and intensity with all-cause mortality in older women. *JAMA Intern. Med.* **2019**, *179*, 1105–1112. [CrossRef]
44. Tudor-Locke, C.; Han, H.; Aguiar, E.J.; Barreira, T.V.; Schuna, J.M., Jr.; Kang, M.; Rowe, D.A. How fast is fast enough? Walking cadence (steps/min) as a practical estimate of intensity in adults: A narrative Rreview. *Br. J. Sports Med.* **2018**, *52*, 776–788. [CrossRef]
45. Bravata, D.M.; Smith-Spangler, C.; Sundaram, V.; Gienger, A.L.; Lin, N.; Lewis, R.; Stave, C.D.; Olkin, I.; Sirard, J.R. Using pedometers to increase physical activity and improve health: A systematic review. *JAMA* **2007**, *298*, 2296–2304. [CrossRef]
46. Glasgow, R.E.; Vogt, T.M.; Boles, S.M. Evaluating the public health impact of health promotion interventions: The RE-AIM framework. *Am. J. Public Health* **1999**, *89*, 1322–1327. [CrossRef]
47. Prince, S.A.; Saunders, T.J.; Gresty, K.; Reid, R.D. A comparison of the effectiveness of physical activity and sedentary behaviour interventions in reducing sedentary time in adults: A systematic review and meta-analysis of controlled trials: Interventions and sedentary behaviours. *Obes. Rev.* **2014**, *15*, 905–919. [CrossRef]
48. Gardner, B.; Smith, L.; Lorencatto, F.; Hamer, M.; Biddle, S.J.H. How to reduce sitting time? A review of behaviour change strategies used in sedentary behaviour reduction interventions among adults. *Health Psychol. Rev.* **2016**, *10*, 89–112. [CrossRef]
49. Garcia, D.O.; Wertheim, B.C.; Manson, J.E.; Chlebowski, R.T.; Volpe, S.L.; Howard, B.V.; Stefanick, M.L.; Thomson, C.A. Relationships between dog ownership and physical activity in postmenopausal women. *Prev. Med.* **2015**, *70*, 33–38. [CrossRef]
50. White, M.N.; King, A.C.; Sallis, J.F.; Frank, L.D.; Saelens, B.E.; Conway, T.L.; Cain, K.L.; Kerr, J. Caregiving, transport-related, and demographic correlates of sedentary behavior in older adults: The Senior Neighborhood Quality of Life Study. *J. Aging Health* **2016**, *28*, 812–833. [CrossRef]
51. Dall, P.M.; Ellis, S.L.H.; Ellis, B.M.; Grant, P.M.; Colyer, A.; Gee, N.R.; Granat, M.H.; Mills, D.S. The influence of dog ownership on objective measures of free-living physical activity and sedentary behaviour in community-dwelling older adults: A longitudinal case-controlled study. *BMC Public Health* **2017**, *17*. [CrossRef]
52. Williams, D.M.; Rhodes, R.E. The confounded self-efficacy construct: Conceptual analysis and recommendations for future research. *Health Psychol. Rev.* **2016**, *10*, 113–128. [CrossRef]
53. Rhodes, R.E.; Williams, D.M.; Mistry, C.D. Using short vignettes to disentangle perceived capability from motivation: A test using walking and resistance training behaviors. *Psychol. Health Med.* **2016**, *21*, 639–651. [CrossRef] [PubMed]
54. Rhodes, R.E.; Courneya, K.S. Differentiating motivation and control in the Theory of Planned Behavior. *Psychol. Health Med.* **2004**, *9*, 205–215. [CrossRef]
55. Potter, K.; Sartore-Baldwin, M. Dogs as support and motivation for physical activity. *Curr. Sports Med. Rep.* **2019**, *18*, 275–280. [CrossRef]

International Journal of
Environmental Research and Public Health

Article

The Use of Fear versus Hope in Health Advertisements: The Moderating Role of Individual Characteristics on Subsequent Health Decisions in Chile

Pablo Farías

Departamento de Administración, Facultad de Economía y Negocios, Universidad de Chile, Santiago 8330015, Chile; pfarias@fen.uchile.cl

Received: 3 November 2020; Accepted: 4 December 2020; Published: 7 December 2020

Abstract: No studies have addressed the way the effectiveness of fear and hope advertisements differs across differently characterized individuals. The present study aims to find out in which situations related to different individual characteristics do fear and hope advertisements work as tools in generating healthy eating intention and physical activity intention. This study conducted an experiment using 283 adults from Chile. The results suggest that fear versus hope appeals in health advertisements have a more positive influence on healthy eating intention. The results suggest that the effect of fear advertisements on healthy eating intention is positively moderated by the frequency of fast food consumption and is negatively moderated by self-efficacy. The results suggest that fear versus hope appeals in health advertisements have no main effect on physical activity intention. However, the results suggest that the effect of fear advertisements on physical activity intention is positively moderated by perceived body weight and past healthy eating behavior and is negatively moderated by subjective norms. The results indicate that when making health advertising, homogenous messages are not persuasive for heterogeneous audiences. The present study results suggest that fear and hope advertisements should be delivered considering the individual characteristics identified in the present study.

Keywords: healthy eating intention; physical activity intention; emotions; aversive state; reinforcement; social support; self-concept

---

1. Introduction

Physical activity and healthy eating are part of the solution for the ongoing obesity epidemic [1,2]. In the last decade, national policy initiatives focused on physical activity and healthy eating have spread in Latin American countries. Programs aimed at promoting healthy lifestyles by addressing eating behaviors and physical activity are at the center of every government's agenda. Despite these efforts, in Latin America, the prevalence of overweight and obese adults has increased markedly in the last three decades [3–7].

Emotions play a significant role in advertising [8,9]. The emotions of fear and hope could stimulate individuals to make health-related decisions [10–12]. In the U.S., Krishen and Bui [13] reported that fear-based framing of health messages could lead to positive decision intentions and, consequently, help people make better health-related decisions. Fear is a powerful emotion in getting individuals to change their behavior. Advertising, especially health advertising, is meant to be effective and aims to change individuals' behaviors towards their health. Thus, fear advertising may be considered a useful tool for changing people's attitudes towards healthy eating and participating in physical activity [10,14]. The frequency of fear appeals in advertisements varies across countries and cultures [15,16]. Unfortunately, in Latin America, not enough is known about the potential that fear-mongering could have in health advertisements. There is no research analyzing its effectiveness in the region, and fear-mongering advertising has seldom been employed. Consequently, this research study's first objective is to examine the effects of fear-based framing of health advertisements on healthy eating intention and physical activity intention among adults in Chile.

A major limitation of the Krishen and Bui [13] study was that they treated subjects as a homogenous group. The way recipients process messages from fear appeals could be affected by their particular characteristics [17,18]. No studies have addressed the way the effectiveness of fear-based advertisements differs across differently characterized individuals, or the circumstances under which fear-based advertisements could potentially influence health-related choices. In fact, Krishen and Bui [13] recommend that future research includes the moderating roles of self-esteem, self-control, and other individual characteristics on subsequent health decisions. For these reasons, this study's second objective is to incorporate the individual characteristics (frequency of fast food consumption, body weight, past behavior, subjective norms, social influences regarding physical activity, self-efficacy, self-esteem, and self-control) as moderators. This research aims to identify the situations related to different individual attributes in which fear-mongering advertisements work as a useful tool to elicit behavior changes. Table 1 summarizes the identified hypotheses to achieve both research objectives.

Table 1. Proposed hypotheses.

Variable	Hypothesis	Healthy Eating Intention (a)	Physical Activity Intention (b)
Advertisement			
Fear	H1	+	+
Moderating role of individual characteristics			
Fear × Frequency of fast food consumption	H2	+	+
Fear × Perceived body weight	H3	+	+
Fear × Past behavior	H4	+	+
Fear × Subjective norm	H5	−	−
Fear × Social influences regarding physical activity	H6	−	−
Fear × Self-efficacy	H7	−	−
Fear × Self-esteem	H8	−	−
Fear × Self-control	H9	−	−

Note: (+) indicates that the variable has a positive effect on the dependent variable and (−) indicates that the variable has a negative effect on the dependent variable.

1.1. Fear Versus Hope Appeals in Health Advertisements

The frequency of fear appeals in advertisements varies across countries and cultures [15,16]. Fear appeals are more frequently used in Canada and China than in France [15]. This resonates with the cultural value of uncertainty avoidance, a characteristic more present in France than in Canada or China [15,19]. Previous studies have found evidence that fear works [20]. Krishen and Bui [13] reveal that, in the U.S., fear-based framing of health messages can direct individuals to positive decision intentions, thus helping them make better-planned health-related decisions. The question then arises: do cultural aspects influence the effects of fear versus hope appeals in health advertisements? Latin America has higher uncertainty avoidance levels than the U.S. [19,21]. The uncertainty avoidance score expresses the extent to which the members of a culture feel threatened by ambiguous or unknown situations and have created beliefs and institutions to try avoiding these circumstances [19]. Given that cultures with high uncertainty avoidance levels have beliefs and institutions to deal with fear, Matsumoto [22] suggests that people may tend not to recognize fear or may attenuate attributions of its intensity when sensed or expressed. As there is no information on the potential of fear appeals in health advertisements in Latin America, to examine the fear-priming effects, the first hypothesis is proposed:

Hypothesis 1. *When subjects view a fear advertisement in Chile, they will be more likely to report (a) greater healthy eating intention and (b) greater physical activity intention, than when they view a hope advertisement.*

1.2. The Moderating Role of Individual Characteristics

A major limitation of Krishen and Bui's [13] study was that they treated subjects as a homogenous group. The way recipients process messages from fear appeals could be affected by their particular characteristics [17]. No research has studied the circumstances under which fear advertisements can potentially influence health-related decisions and how their effectiveness varies across individuals. This study aims at figuring out if different reactions from people with different characteristics will affect the results of fear advertising. Krishen and Bui [13] suggest that traits such as self-esteem and self-control, among others, should be considered in future research.

1.2.1. Fear Appeals and Avoiding an Aversive State

Fast food is generally low cost and the promotion is active, referring to consumers' social surroundings and behaviors [23,24]. Frequent fast food consumption has been linked to obesity, diabetes, hypertension, and heart disease [25,26]. In a hope environment, a favorable outcome could occur (e.g., healthy living), whereas, in a fear environment, an unfavorable outcome could be avoided or even resolved (e.g., death) [27]. The emotion of fear could stimulate individuals to deal with the causes of that emotion (fast food consumption) [10]. Fear engenders a desire to escape from or avoid an aversive state [28]. As a primary aversive emotion, fear arises in situations of menace to the organism (i.e., the individual) and enables them to respond to them adaptively [29]. Through fear, individuals with frequent fast food consumption may have a greater intention to eat healthily and do physical activity in order to escape from this aversive state [28,29], that is, the negative effects (e.g., obesity, diabetes, hypertension, heart disease) associated with frequent consumption of fast food. [25,26]. Consequently, the effects of fear (vs. hope) appeals in health advertisements on healthy eating intention and physical activity intention could be stronger among individuals having higher (e.g., once a day, more than once a day) rather than lower (e.g., occasionally, never) frequency of fast food consumption. Hence:

Hypothesis 2. *The effects of fear (vs. hope) appeals in health advertisements on (a) healthy eating intention and (b) physical activity intention are stronger among individuals who consume fast food more frequently than others.*

Researchers have shown that overweight individuals are more likely to be present-biased [30]. Present-biased individuals tend to value more immediate rewards than rewards that would happen over time [31]. In the case of overweight individuals, the immediate action of eating healthy or doing physical activity would be relatively more appealing if they feel rewarded by doing so. The reward could be psychological or material; it is highly valued as long as it is immediate. Fear-based health advertising could thus have a stronger effect on overweight individuals as the promoted effects would be immediate (e.g., avoid hypertension and heart diseases). Hence:

Hypothesis 3. *The effects of fear (vs. hope) appeals in health advertisements on (a) healthy eating intention and (b) physical activity intention are stronger among individuals who are overweight.*

1.2.2. Fear Appeals and Reinforcement

Fear appeals elicit increased brain processing and more robust recall [32]. The effects of fear in terms of past behavior have been researched, especially in individual decision-making leading to some sort of pain. If an individual remembers a painful time in their life, they are likely to act against similar events in the future [33]. Fear appeals could be used in a way that allows fear of reinforcing appropriate past behavior [34]. Fear could thus work as reinforcement in making healthy decisions. Hence:

Hypothesis 4. *The effects of fear (vs. hope) appeals in health advertisements on (a) healthy eating intention and (b) physical activity intention are stronger among individuals who have higher past healthy behaviors.*

1.2.3. Hope Appeals and Social Support

Social support is the individual's support behaviors from people in their social network, which enhance their mood, performance, and activities. Social support is a significant factor affecting a person's eating and physical activity intentions and behaviors [35–38]. The effects of subjective norms and social influences regarding healthy eating intentions and physical activity intentions could be enhanced if the individual is motivated more by hope (positive goals are expected) rather than fear (negative outcomes should be avoided). Therefore, hope (vs. fear) appeals could be more effective in cases where individuals have high subjective norms and social influences regarding healthy eating intentions and physical activity intentions. Hence:

Hypothesis 5. *The effects of hope (vs. fear) appeals in health advertisements on (a) healthy eating intention and (b) physical activity intention are stronger among individuals who have higher subjective norms.*

Hypothesis 6. *The effects of hope (vs. fear) appeals in health advertisements on (a) healthy eating intention and (b) physical activity intention are stronger among individuals who have more social influences regarding physical activity.*

1.2.4. Hope Appeals and Self-Concept

Hope as an emotion emerges when a concrete positive goal is envisioned. It comprises a cognitive element of expecting and an affective feel-good element about the expected events or outcomes [39,40]. Hope is a positive motivational state. Hopeful individuals are more intrinsically motivated and enjoy pursuing goals [41]. In order to reach such goals and feel hope, an individual must be rather determined and goal-oriented and possess some self-efficacy, self-esteem, and self-control.

As concluded by Snyder [42], hope messaging can contribute to goal-oriented behaviors geared towards improving states already in satisfactory levels rather than in other unsatisfactory ones. Hope requires mental representations of positively valued abstract future situations, and, more specifically, it requires setting goals, planning how to achieve them, using imagination, creativity, cognitive flexibility, and mental exploration of novel situations [29]. An individual's beliefs about their capabilities to perform well in situations that affect their lives is denoted by self-efficacy. It rules how individuals feel, think, motivate themselves, and behave in such cases. People with high self-efficacy are determined and motivated, whereas people with low self-efficacy doubt their capabilities, are not committed to pursuing their tasks and goals, and see difficulties as obstacles and challenging tasks as personal threats [43,44]. For example, Mowen et al. [45] comment that individuals with higher self-efficacy experience lower fear levels.

For intrinsically motivated people (e.g., high in self-esteem or self-control), hope appeals in health advertisements may work as a better motivator than fear appeals. People who are more likely to be externally motivated (e.g., individuals low in self-esteem or self-control) may find fear more appealing than hope, due to the stronger emotional effects fear has on an individual. Hence:

Hypothesis 7. *The effects of hope (vs. fear) appeals in health advertisements on (a) healthy eating intention and (b) physical activity intention are stronger among individuals who have higher levels of self-efficacy.*

Hypothesis 8. *The effects of hope (vs. fear) appeals in health advertisements on (a) healthy eating intention and (b) physical activity intention are stronger among individuals who have higher levels of self-esteem.*

Hypothesis 9. *The effects of hope (vs. fear) appeals in health advertisements on (a) healthy eating intention and (b) physical activity intention are stronger among individuals who have high levels of self-control.*

2. Materials and Methods

2.1. Procedures

Data were collected using an online self-administered questionnaire sent to a convenience sample of adults from Chile. Some measures included in the questionnaire may represent a personally sensitive issue. Respondents may not have wanted to face an interviewer and would like to have completed the survey alone. In order to obtain sensitive information (e.g., perceived body weight, self-esteem), to offer perceived respondent anonymity (i.e., respondents' perceptions that the interviewer or researcher will not discern their identities), and to avoid social desirability (i.e., for respondents to answer what they feel to be acceptable in front of others), an invitation to participate in an anonymous online survey on health advertisements was emailed to a database of people over 18 years of age living in Chile. The final sample size was 283 adults (response rate = 23.2%). All response data were anonymous. Participants were randomly assigned one of two events (fear or hope advertisement). Subjects were told to watch a video (fear or hope advertisement) for about two minutes. Then, they had to complete an online questionnaire.

2.2. Stimuli

Fear and hope advertisements had the same modality (audiovisual) and a similar length (102 and 113 s, respectively). The fear advertisement is an edited version of a video released by the Children's Healthcare of Atlanta. Through Jim's life, the video starts with him lying on a hospital bed after suffering a heart attack at the age of 32, then rewinds to expose the patterns of an unhealthy and sedentary lifestyle that lead him to that point. In different stages of his life, Jim is shown eating unhealthy foods, sitting around watching television, and playing computer games. In the edited version, sentences were modified to communicate that adults should change their habits to improve their own health.

The hope advertisement is an edited version of a video released by Coke. The video shows that they published advertisements in a newspaper to promote weight loss pills (magic pills). Without mentioning the brand, the advertisement said the first individuals to call would win a free sample of magic pills. From all the calls, three individuals were chosen to receive the magic pills. Nonetheless, they were not expecting the many obstacles on their way there such as staircases, barking dogs, damsels in distress, and grannies with packages to carry. In the end, the participants find out that it was all a well-staged, provocative, and inspiring prank to encourage them to lose weight. They were shown the tapings of themselves to learn that every bit of effort and activity in their routine contributes more to their weight and fitness than a magical pill. The video was shortened by removing some obstacles. The mention of Coke was also removed from the video to avoid brand effects. In the edited version, sentences were modified to communicate that adults should change their habits to improve their own health. Both original videos had not yet been exhibited in Chile.

2.3. Measures

In the interest of having a valid and reliable instrument, all scales used in this study were adopted from previous research [13,46–53]. Subjects were asked to fill in the questionnaire based on the scenario (fear or hope advertisement) presented a moment ago. Appendix A contains the measures included in the questionnaire. The dependent variables were healthy eating intention [46] and physical activity intention [47].

The manipulation check was an advertising message (fear vs. hope) [13]. Respondents rated the advertisement message on a bipolar anchored scale with "fear about changes in your weight for the future" and "hope for changes in your weight for the future" as endpoints. They responded to the following statement: "The advertisement message referenced what you would … " on a 7-point scale (−3 to +3). Higher numbers signal the hope advertisement message, while lower numbers indicate the fear advertisement message.

The independent variables were subjective norm (food) [48], subjective norm (exercise) [49], social influences regarding physical activity [47], self-efficacy [47], number of perceived barriers to a healthy lifestyle [47], past behavior (exercise) [50], past behavior (food) [46], frequency of fast food consumption [51], self-esteem [52], and self-control [53]. Finally, participants self-reported gender, age, marital status, height, weight, and perceived body weight (underweight, average weight, and overweight). The body mass index (BMI) was calculated with the reported height and weight data.

2.4. Hierarchical Linear Regression Analyses

The data were utilized in a series of hierarchical linear regression analyses to estimate the path coefficients for the hypothesized relationships. To minimize multicollinearity, the independent variables employed were mean-centered before creating the interaction terms. Regressions were estimated using the following equations to explain healthy eating intention and physical activity intention (the dependent variables, Y_i) for each individual i:

$$Y_i = \beta_o + \sum_{k=1}^{15} \beta_k \times X_{k,i} \text{ (Model 1)} \tag{1}$$

$$Y_i = \beta_o + \sum_{k=1}^{15} \beta_k \times X_{k,i} + \beta_{16} \times F_i \text{ (Model 2)} \tag{2}$$

$$Y_i = \beta_o + \sum_{k=1}^{15} \beta_k \times X_{k,i} + \beta_{16} \times F_i + \sum_{k=1}^{15} \beta_{16+k} \times X_{k,i} \times F_i \text{ (Model 3)} \tag{3}$$

Model 1 includes in the regression the constant (β_o) and k = 15 individual characteristics (frequency of fast food consumption, perceived body weight, past behavior (food), past behavior (exercise), subjective norm (food), subjective norm (exercise), social influences regarding physical activity, self-efficacy, self-esteem, self-control, number of perceived barriers to healthy lifestyle, female, age, married, and body mass index) for each individual i ($X_{k,i}$) as independent variables. Model 2 incorporates into the regression the binary variable that measures whether individual i was exposed to fear advertisement ($F_i = 1$) or hope advertisement ($F_i = 0$). Model 3 adds 15 interaction terms ($F_i \times X_{k,i}$).

3. Results

The specifics of the sample are detailed in Table 2. Women account for 48% of the sample and 29% of the sample is married. Tables 3 and 4 exhibit the hypotheses test results. First of all, no multicollinearity existed among the employed constructs in this study, which can be confirmed through the variance inflation factors (VIFs) for each regression coefficient. They range from a low of 1.012 to a high of 3.731 in each regression, suggesting that they are at acceptable levels.

Table 2. Descriptive statistics.

Variable	Mean	Standard Deviation	Minimum	Maximum
Dependent variables				
Healthy eating intention	4.00	0.88	1.00	5.00
Physical activity intention	3.25	0.79	1.00	4.00
Individual characteristics				
Frequency of fast food consumption	2.70	1.65	0.00	8.00
Perceived body weight	0.36	0.48	0.00	1.00
Past behavior—food	3.61	1.02	1.00	5.00
Past behavior—exercise	4.14	3.22	0.00	14.00
Subjective norm—food	5.22	1.69	1.00	7.00
Subjective norm—exercise	5.32	1.43	1.00	7.00
Social influences regarding physical activity	0.67	0.53	−1.00	1.00
Self-efficacy	7.67	2.19	0.00	10.00
Self-esteem	6.33	1.74	1.00	9.00
Self-control	4.71	1.46	1.00	7.00
Number of perceived barriers to healthy lifestyle	0.97	0.81	0.00	5.00
Female	0.48	0.50	0.00	1.00
Age	2.44	2.40	1.00	10.00
Married	0.29	0.45	0.00	1.00
Body mass index	23.83	3.33	15.78	33.84
Advertisement and manipulation check				
Fear	0.53	0.50	0.00	1.00
Advertising message (fear vs. hope)	−0.18	2.72	−3.00	3.00

Table 3. Hierarchical linear regression predicting healthy eating intention.

Variable	Hypothesis	Model 1	Model 2	Model 3
Intercept		4.000 **	4.000 **	4.002 **
Individual characteristics (method = enter)				
Frequency of fast food consumption		−0.043	−0.035	−0.031
Perceived body weight		0.028	0.014	0.058
Past behavior—food		0.349 **	0.353 **	0.337 **
Past behavior—exercise		0.100 *	0.102 *	0.121 *
Subjective norm—food		0.053	0.053	0.077
Subjective norm—exercise		0.089	0.084	0.068
Social influences regarding physical activity		0.004	0.008	0.010
Self-efficacy		0.192 **	0.191 **	0.177 **
Self-esteem		−0.092	−0.087	−0.089
Self-control		0.028	0.024	0.025
Number of perceived barriers to healthy lifestyle		−0.089	−0.101 *	−0.109 *
Female		0.166 **	0.173 **	0.169 **
Age		0.033	0.041	0.049
Married		0.084	0.078	0.068
Body mass index		−0.003	0.007	−0.042
Advertisement (method = enter)				
Fear	H1a: +		0.083 *	0.083 *
Interaction effects (method = stepwise)				
Fear × Frequency of fast food consumption	H2a: +			0.083 *
Fear × Self-efficacy	H7a: −			−0.123 **

Table 3. *Cont.*

Variable	Hypothesis	Model 1	Model 2	Model 3
Maximum VIF value		3.596	3.613	3.731
R^2		0.437	0.445	0.472
Adjusted R^2		0.405	0.411	0.436
R^2 change		0.437	0.008	0.027
Partial F value		13.790 **	3.997 *	6.807 **
N		283	283	283

Note: Unstandardized regression coefficients are reported. * significant at $p = 0.05$, ** significant at $p = 0.01$.

Table 4. Hierarchical linear regression predicting physical activity intention.

Variable	Hypothesis	Model 1	Model 2	Model 3
Intercept		3.254 **	3.254 **	3.257 **
Individual characteristics (method = enter)				
Frequency of fast food consumption		−0.008	−0.004	0.011
Perceived body weight		0.047	0.040	0.034
Past behavior—food		0.016	0.018	0.048
Past behavior—exercise		0.068	0.069	0.050
Subjective norm—food		0.065	0.065	0.032
Subjective norm—exercise		0.061	0.058	0.057
Social influences regarding physical activity		0.079	0.081	0.082
Self-efficacy		0.310 **	0.309 **	0.297 **
Self-esteem		−0.021	−0.018	−0.006
Self-control		0.026	0.023	0.026
Number of perceived barriers to healthy lifestyle		−0.017	−0.024	−0.045
Female		0.044	0.048	0.061
Age		−0.143 **	−0.139 **	−0.128 *
Married		0.018	0.015	0.013
Body mass index		−0.029	−0.023	−0.006
Advertisement (method = enter)				
Fear	H1b: +		0.045	0.050
Interaction effects (method = stepwise)				
Fear × Perceived body weight	H3b: +			0.107 *
Fear × Past behavior—food	H4b: +			0.122 **
Fear × Subjective norm—food	H5b: −			−0.154 **
Fear × Subjective norm—exercise	H5b: −			−0.187 **
Maximum VIF value		3.596	3.613	3.662
R^2		0.292	0.295	0.357
Adjusted R^2		0.252	0.252	0.308
R^2 change		0.292	0.003	0.063
Partial F value		7.331 **	1.175	6.381 **
N		283	283	283

Note: Unstandardized regression coefficients are reported. * significant at $p = 0.05$, ** Significant at $p = 0.01$.

3.1. Manipulation Check for the Advertising Messages

An analysis of variance (ANOVA) was performed to ensure that the manipulation of the advertisement message was successful. As expected, there was a significant difference ($F(1.281) = 23.641$, $p < 0.01$) between the hope ($M = 0.62$) and the fear cases ($M = −0.90$) with means in the appropriate direction.

3.2. Regressions Predicting Healthy Eating Intention

As Table 3 summarizes, the Model 1 regression analysis results indicate that individual characteristics explain 43.7% of the variance in healthy eating intention. Adding the fear versus hope appeals in health advertisements in Model 2 increased the R^2 value to 44.5% ($\Delta F = 3.997$, $p < 0.05$).

Consistent with Krishen and Bui [13], the results suggest that fear versus hope appeals in health advertisements have a more positive influence on healthy eating intention ($\beta = 0.083$, $p < 0.05$). That is, when subjects view a fear advertisement, they will be more likely to report greater healthy eating intention than when they view a hope-conditioned advertisement. Therefore, H1a is supported.

Adding the interaction terms in Model 3, using stepwise regression (15 interactions terms; fear advertisement × 15 individual characteristics) increased the R^2 value to 47.2% ($\Delta F = 6.807$, $p < 0.01$). The results suggest that the effect of fear advertisements on healthy eating intention is positively moderated by the frequency of fast food consumption ($\beta = 0.083$, $p < 0.05$). Therefore, H2a is supported.

The results also suggest that the effect of fear advertisements on healthy eating intention is negatively moderated by self-efficacy ($\beta = -0.123$, $p < 0.01$). Therefore, H7a is supported.

The results suggest that the effect of fear advertisements on healthy eating intention is not moderated by perceived body weight, past behavior, subjective norm, social influences, self-esteem, and self-control (ps > 0.05). Therefore, H3a, H4a, H5a, H6a, H8a, and H9a are not supported.

3.3. Regressions Predicting Physical Activity Intention

As summarized in Table 4, the Model 1 regression analysis results indicate that individual characteristics explain 29.2% of the variance in physical activity intention. Adding the fear versus hope appeals in health advertisements in Model 2 increased the R^2 value by only 0.3% ($\Delta F = 1.175$, $p > 0.10$). The results suggest that fear versus hope appeals in health advertisements have no main effect on physical activity intention ($p > 0.10$). Therefore, H1b is not supported.

Adding the interaction terms in Model 3, using stepwise regression (15 interactions terms; fear advertisement × 15 individual characteristics) increased the R^2 value to 35.7% ($\Delta F = 6.381$, $p < 0.01$). The results suggest that the effect of fear advertisements on physical activity intention is positively moderated by perceived body weight ($\beta = 0.107$, $p < 0.05$). Therefore, H3b is supported. The results suggest that the effect of fear advertisements on physical activity intention is positively moderated by past healthy eating behavior ($\beta = 0.122$, $p < 0.01$). Therefore, H4b is supported. The results also suggest that the effect of fear advertisements on physical activity intention is negatively moderated by subjective norm, food ($\beta = -0.154$, $p < 0.01$) and subjective norm, exercise ($\beta = -0.187$, $p < 0.01$). Therefore, H5b is supported.

The results suggest that the effect of fear advertisements on physical activity intention is not moderated by the frequency of fast food consumption, social influences, self-efficacy, self-esteem, and self-control ($p > 0.05$). Therefore, H2b, H6b, H7b, H8b, and H9b are not supported.

4. Discussion

Table 5 presents a summary of the results of the present study. The results in Chile are consistent with what was observed by Krishen and Bui [13] in the United States, showing that fear has a greater impact than hope in generating healthy eating intention. Consistent with previous research [17,18], the results also suggest that, when making health advertising, homogenous messages are not persuasive for heterogeneous audiences. Based on the new evidence found, it would be better to employ fear in cases where the audience is expected to have high past healthy eating behavior, high fast food consumption, or are perceived as overweight. Hope would be more effective in situations where the audience is expected to be high in subjective norms or self-efficacy.

Table 5. Summary of the study's results

Variable	Hypothesis	Healthy Eating Intention (a)	Physical Activity Intention (b)
Advertisement			
Fear	H1	+	
Moderating role of individual characteristics			
Fear × Frequency of fast food consumption	H2	+	
Fear × Perceived body weight	H3		+
Fear × Past behavior	H4		+
Fear × Subjective norm	H5		−
Fear × Social influences regarding physical activity	H6		
Fear × Self-efficacy	H7	−	
Fear × Self-esteem	H8		
Fear × Self-control	H9		

Note: (+) indicates that the variable has a positive effect on the dependent variable and (−) indicates that the variable has a negative effect on the dependent variable. Blank spaces indicate that the hypotheses were not supported by the study's results.

The present study results suggest that fear and hope advertisements should be delivered considering the individual characteristics identified in the present study. Online channels (websites, social networks) easily allow audience segmentation using users' interests and behaviors (e.g., fast food consumption, perceived body weight, past healthy eating behavior, self-efficacy). Therefore, policymakers and organizations concerned about people's healthy eating and physical activity can use these online channels to send segmented messages (fear or hope appeals) to people using their own individual characteristics.

The frequency of buying fast food is an easy variable to measure and, for this reason, is useful for both companies and regulators to segment consumers [54,55]. The results suggest that fear messages should be directed towards frequent fast food consumers and hope messages towards less frequent fast food consumers in order to increase healthy eating intention. The results suggest that to increase healthy eating intention, fear can be communicated in warning messages accompanying the promotion or delivery of fast food [3,56]. In contrast, for people with a high level of self-efficacy, it is suggested to use hope appeals to increase healthy eating intention. This finding is consistent with Mowen et al. [45] who suggest that people with higher self-efficacy experience lower levels of fear.

Consistent with social support being a significant factor affecting a person's eating and physical activity intentions and behaviors [35–38], the results also suggest that to increase people's physical activity, hope messages should include the people who are important to the individual (subjective norm). In contrast, it is suggested to use fear appeals when the individual has a high level of perceived body weight (aversive state [28,29]) and past behavior (reinforcement [34]).

Several other topics are worth exploring in the future. Future research may analyze other advertisements (e.g., using fear and hope advertisements with the same theme, include a greater number of ads in varying themes), media (e.g., radio, print advertising, social media), emotions (e.g., love, guilt, pride, sadness, gratitude, shame), countries (e.g., European and Asian countries), and segments (e.g., children, adolescents). Such studies could increase the generalizability of the results as well as their applicability to health advertising. Considering that the observed effects may only be short-term, future research can also analyze the long-term effects (e.g., through panel data [57,58]) of the relationships proposed in this research. It would be interesting to analyze if, when repeating the stimuli over weeks, months, and years (e.g., repeating messages with fear appeals every week for a year), the effects on healthy eating intention and physical activity intention are maintained over time. Future research may also investigate the effects of mixing such stimuli (e.g., mixing, swapping ads with fear appeals with ads with hope appeals). Future research may also include other interesting and observable response variables such as the intention to recommend or share the health advertisement and behavioral variables (e.g., physical activity performed by the individual) or outcome (e.g., the weight of the individual).

5. Conclusions

Consistent with Krishen and Bui [13], the results suggest that in Chile, when comparing fear versus hope appeals in health advertisements, fear-conditioned advertisements have a greater positive impact on healthy eating intentions. In other words, when subjects view a fear advertisement, they will be more likely to report greater healthy eating intention than when they view a hope advertisement. This study also aimed to determine when, according to different individual characteristics, fear advertising works as a powerful tool for generating healthy eating and physical activity intention. The results suggest that the effect of fear advertisements on healthy eating intention is positively moderated by the frequency of fast food consumption and is negatively moderated by self-efficacy. The results suggest that fear versus hope appeals in health advertisements have no main effect on physical activity intention. However, the results suggest that the effect of fear advertisements on physical activity intention is positively moderated by perceived body weight and past healthy eating behavior and is negatively moderated by the subjective norm of food and subjective norm of exercise.

The results suggest that when making health advertising, standardized messages are not persuasive for diverse audiences. The present study results suggest that fear and hope appeals should be delivered considering the individual characteristics identified in the present study. Firms, governments, regulators, and other entities must consider these individual characteristics for an effective use of fear and hope appeals in the messages used to increase healthy eating intention and physical activity intention. Future research should also include these individual characteristics to analyze the effects of fear and hope appeals on healthy eating intention and physical activity intention.

Funding: This research received no external funding.

Conflicts of Interest: The authors declare no conflict of interest.

Appendix A

Questionnaire (in order of appearance)
Healthy eating intention (Chan et al. [46]; Cronbach's alpha = 0.80)
Do you intend to engage in healthy eating over the next week?
How likely is it that you will engage in healthy eating over the next week?
(5-point scale, 1 = definitely no, 5 = definitely yes)
Physical activity intention (Hopman-Rock et al. [47])
Do you plan to be more physically active in the short term? The answer categories were
1 = no, absolutely not, to 4 = yes, definitely.
Advertising message (fear vs. hope) (adapted from Krishen and Bui [13])
Respondents rated the advertisement message on a bipolar anchored scale with "fear about changes in your weight for the future" and "hope for changes in your weight for the future" as endpoints responding to the following statement: "The advertisement message referenced what you would … " on a 7-point scale (−3 to +3); higher numbers signal the hope-conditioned advertisement message while lower numbers indicate the fear advertisement message.
Subjective norm-food (Conner et al. [48])
People who are important to me think I should eat a healthy diet (unlikely–likely; scored 1 to 7)
Subjective norm-exercise (Courneya [49])
Most people who are important to me think I should engage in regular physical activity (7-point Likert scale, 1 = strongly disagree, 7 = strongly agree)
Social influences regarding physical activity (Hopman-Rock et al. [47])
How do you think people in your environment will react if you exercise more? Answer categories were: positive +1, neutral 0, negative-1.

Self-efficacy (Hopman-Rock et al. [47])

Do you think you will be able to be more physically active? The answer categories ranged from 0 = I am sure I cannot to 10 = I am sure I can.

Number of perceived barriers to healthy lifestyle (adapted from Hopman-Rock et al. [47])

A number of questions looked at several potential barriers: no time, no interest, not used to it, too expensive, no progress, feeling unhealthy, other problems. Answer categories were 1 = agree and 0 = disagree. Sum score for 7 questions.

Past behavior-exercise (Abraham and Sheeran [50])

How many days did you exercise in the last two weeks?

Past behavior-food (Chan et al. [46])

How often did you engage in healthy eating in the past month? (5-point scale, 1 = never, 5 = very often).

Frequency of fast food consumption (Dunn et al. [51])

Participants were asked to report the frequency of fast food consumption on a scale including responses: 0 = never, 1 = occasionally, 2 = once a month, 3 = once a fortnight, 4 = once a week, 5 = 2–3 times a week, 6 = 4–6 times a week, 7 = once a day, and 8 = more than once a day.

Self-esteem (Robins et al. [52])

I have high self-esteem (9-point Likert scale, 1 = strongly disagree, 9 = strongly agree)

Self-control (Daly et al. [53])

Self-control was measured using a single item where participants rated their level of self-control on a scale from 1 (little self-control) to 7 (disciplined).

Female

Male (0), Female (1)

Age

18–25 (1), 26–30 (2), 31–35 (3), 36–40 (4), 41–45 (5), 46–50 (6), 51–55 (7), 56–60 (8), 61–65 (9), +65 (10)

Married

Married (1), not married (0)

Perceived body weight

Underweight (0), average weight (0), overweight (1)

Height

Height in centimeters

Weight

Weight in kilograms

Body mass index (BMI)

The height and weight data were used to calculate the body mass index (BMI).

References

1. Toomey, E.C.; Matvienko-Sikar, K.; Doherty, E.; Harrington, J.; Hayes, C.B.; Heary, C.; Hennessy, M.; Kelly, C.; McHugh, S.; McSharry, J.; et al. A collaborative approach to developing sustainable behaviour change interventions for childhood obesity prevention: Development of the Choosing Healthy Eating for Infant Health (CHErIsH) intervention and implementation strategy. *Br. J. Health Psychol.* **2020**, *25*, 275–304. [CrossRef] [PubMed]

2. Polechoński, J.; Nierwińska, K.; Kalita, B.; Wodarski, P. Can Physical Activity in Immersive Virtual Reality Be Attractive and Have Sufficient Intensity to Meet Health Recommendations for Obese Children? A Pilot Study. *Int. J. Environ. Res. Public Health* **2020**, *17*, 8051. [CrossRef] [PubMed]

3. Alonso-Dos-Santos, M.; Ulloa, R.Q.; Quintana, Á.S.; Quijada, D.V.; Farías, P. Nutrition Labeling Schemes and the Time and Effort of Consumer Processing. *Sustainability* **2019**, *11*, 1079. [CrossRef]

4. Cuevas, A.; Barquera, S. COVID-19, Obesity, and Undernutrition: A Major Challenge for Latin American Countries. *Obesity* **2020**, *28*, 1791–1792. [CrossRef] [PubMed]

5.	Martorell, M.; Ulloa, N.; González, M.E.; Martínez-Sanguinetti, M.A.; Celis-Morales, C. Obesidad, desnutrición y cambio climático: Una sindemia que Chile deberá enfrentar. *Revista Médica de Chile* **2020**, *148*, 882–884. [CrossRef]

6.	Pérez-Ferrer, C.; Auchincloss, A.H.; De Menezes, M.C.; Kroker-Lobos, M.F.; Cardoso, L.D.O.; Barrientos-Gutierrez, T. The food environment in Latin America: A systematic review with a focus on environments relevant to obesity and related chronic diseases. *Public Health Nutr.* **2019**, *22*, 3447–3464. [CrossRef]

7.	Wolfenden, L.; Ezzati, M.; Larijani, B.; Dietz, W. The challenge for global health systems in preventing and managing obesity. *Obes. Rev.* **2019**, *20*, 185–193. [CrossRef]

8.	Chen, M.-F. Impact of fear appeals on pro-environmental behavior and crucial determinants. *Int. J. Advert.* **2016**, *35*, 74–92. [CrossRef]

9.	Poels, K.; Dewitte, S. The Role of Emotions in Advertising: A Call to Action. *J. Advert.* **2019**, *48*, 81–90. [CrossRef]

10.	Campo, M.; Champely, S.; Lane, A.M.; Rosnet, E.; Ferrand, C.; Louvet, B. Emotions and performance in rugby. *J. Sport Health Sci.* **2019**, *8*, 595–600. [CrossRef]

11.	Kemp, E.; Bui, M.; Krishen, A.; Homer, P.M.; Latour, M.S. Understanding the power of hope and empathy in healthcare marketing. *J. Consum. Mark.* **2017**, *34*, 85–95. [CrossRef]

12.	Scarpina, F.; Varallo, G.; Castelnuovo, G.; Capodaglio, P.; Molinari, E.; Alessandro, M. Implicit facial emotion recognition of fear and anger in obesity. *Eat. Weight. Disord. Stud. Anorexia Bulim. Obes.* **2020**, 1–9. [CrossRef] [PubMed]

13.	Krishen, A.S.; Bui, M. Fear advertisements: Influencing consumers to make better health decisions. *Int. J. Advert.* **2015**, *34*, 533–548. [CrossRef]

14.	Kemp, E.; Cowart, K.; Bui, M. (Myla) Promoting consumer well-being: Examining emotion regulation strategies in social advertising messages. *J. Bus. Res.* **2020**, *112*, 200–209. [CrossRef]

15.	Bartikowski, B.; Laroche, M.; Richard, M.-O. A content analysis of fear appeal advertising in Canada, China, and France. *J. Bus. Res.* **2019**, *103*, 232–239. [CrossRef]

16.	Wu, E.C.; Cutright, K.M. In God's Hands: How Reminders of God Dampen the Effectiveness of Fear Appeals. *J. Mark. Res.* **2018**, *55*, 119–131. [CrossRef]

17.	De Vos, S.; Crouch, R.; Quester, P.; Ilicic, J. Examining the Effectiveness of Fear Appeals in Prompting Help-Seeking: The Case of At-Risk Gamblers. *Psychol. Mark.* **2017**, *34*, 648–660. [CrossRef]

18.	Teng, L.; Zhao, G.; Li, F.; Liu, L.; Shen, L. Increasing the persuasiveness of anti-drunk driving appeals: The effect of negative and positive message framing. *J. Bus. Res.* **2019**, *103*, 240–249. [CrossRef]

19.	Arrindell, W. Culture's consequences: Comparing values, behaviors, institutions, and organizations across nations. *Behav. Res. Ther.* **2003**, *41*, 861–862. [CrossRef]

20.	Pechmann, C.; Catlin, J.R. The effects of advertising and other marketing communications on health-related consumer behaviors. *Curr. Opin. Psychol.* **2016**, *10*, 44–49. [CrossRef]

21.	Farías, P. Medición y representación gráfica de las distancias culturales entre países latinoamericanos. *Converg. Rev. de Cienc. Sociales* **2016**. [CrossRef]

22.	Matsumoto, D. Face, culture, and judgments of anger and fear: Do the eyes have it? *J. Nonverbal Behav.* **1989**, *13*, 171–188. [CrossRef]

23.	Bloom, P.N.; Novelli, W.D. Problems and Challenges in Social Marketing. *J. Mark.* **1981**, *45*, 79. [CrossRef] [PubMed]

24.	Patel, P.C.; Struckell, E.M.; Ojha, D. Calorie labeling law and fast food chain performance: The value of capital responsiveness under sales volatility. *J. Bus. Res.* **2020**, *117*, 346–356. [CrossRef]

25.	Grier, S.A.; Kumanyika, S.K. The Context for Choice: Health Implications of Targeted Food and Beverage Marketing to African Americans. *Am. J. Public Health* **2008**, *98*, 1616–1629. [CrossRef] [PubMed]

26.	Nkosi, V.; Rathogwa-Takalani, F.; Voyi, K. The Frequency of Fast Food Consumption in Relation to Wheeze and Asthma Among Adolescents in Gauteng and North West Provinces, South Africa. *Int. J. Environ. Res. Public Health* **2020**, *17*, 1994. [CrossRef]

27.	MacInnis, D.J.; De Mello, G.E. The Concept of Hope and its Relevance to Product Evaluation and Choice. *J. Mark.* **2005**, *69*, 1–14. [CrossRef]

28.	Baumgartner, H.; Pieters, R.; Bagozzi, R.P. Future-oriented emotions: Conceptualization and behavioral effects. *Eur. J. Soc. Psychol.* **2008**, *38*, 685–696. [CrossRef]

29. Jarymowicz, M.; Bar-Tal, D. The dominance of fear over hope in the life of individuals and collectives. *Eur. J. Soc. Psychol.* **2006**, *36*, 367–392. [CrossRef]

30. Borghans, L.; Golsteyn, B.H. Time discounting and the body mass index. *Econ. Hum. Biol.* **2006**, *4*, 39–61. [CrossRef]

31. Hunter, R.F.; Tang, J.; Hutchinson, G.; Chilton, S.; Holmes, D.; Kee, F. Association between time preference, present-bias and physical activity: Implications for designing behavior change interventions. *BMC Public Health* **2018**, *18*, 1–12. [CrossRef] [PubMed]

32. Mostafa, M.M. Neural correlates of fear appeal in advertising: An fMRI analysis. *J. Mark. Commun.* **2018**, *26*, 40–64. [CrossRef]

33. Mowrer, O.H. A stimulus-response analysis of anxiety and its role as a reinforcing agent. *Psychol. Rev.* **1939**, *46*, 553–565. [CrossRef]

34. Job, R.F.S. Effective and ineffective use of fear in health promotion campaigns. *Am. J. Public Health* **1988**, *78*, 163–167. [CrossRef]

35. Ajzen, I. From intentions to actions: A theory of Planned Behavior. In *Action Control*; Kuhl, J., Beckmann, J., Eds.; Springer: Berlin, Germany, 1985; pp. 11–39.

36. Farías, P. Identifying the factors that influence eWOM in SNSs: The case of Chile. *Int. J. Advert.* **2017**, *36*, 852–869. [CrossRef]

37. Latimer-Cheung, A.E.; Ginis, K.A.M. The importance of subjective norms for people who care what others think of them. *Psychol. Health* **2005**, *20*, 53–62. [CrossRef]

38. Wu, F.; Sheng, Y. Social support network, social support, self-efficacy, health-promoting behavior and healthy aging among older adults: A pathway analysis. *Arch. Gerontol. Geriatr.* **2019**, *85*, 103934. [CrossRef]

39. Bar-Tal, D. Why Does Fear Override Hope in Societies Engulfed by Intractable Conflict, as It Does in the Israeli Society? *Politi Psychol.* **2001**, *22*, 601–627. [CrossRef]

40. Choi, T.R.; Choi, J.H.; Sung, Y. I hope to protect myself from the threat: The impact of self-threat on prevention-versus promotion-focused hope. *J. Bus. Res.* **2019**, *99*, 481–489. [CrossRef]

41. Rego, A.; Sousa, F.; Marques, C.; e Cunha, M.P. Hope and positive affect mediating the authentic leadership and creativity relationship. *J. Bus. Res.* **2014**, *67*, 200–210. [CrossRef]

42. Snyder, C.R. TARGET ARTICLE: Hope Theory: Rainbows in the Mind. *Psychol. Inq.* **2002**, *13*, 249–275. [CrossRef]

43. Bandura, A. Recycling misconceptions of perceived self-efficacy. *Cogn. Ther. Res.* **1984**, *8*, 231–255. [CrossRef]

44. Chan, D.K.C.; Zhang, L.; Lee, A.S.Y.; Hagger, M.S. Reciprocal relations between autonomous motivation from self-determination theory and social cognition constructs from the theory of planned behavior: A cross-lagged panel design in sport injury prevention. *Psychol. Sport Exerc.* **2020**, *48*, 101660. [CrossRef]

45. Mowen, J.C.; Harris, E.G.; Bone, S.A. Personality traits and fear response to print advertisements: Theory and an empirical study. *Psychol. Mark.* **2004**, *21*, 927–943. [CrossRef]

46. Chan, K.; Prendergast, G.P.; Ng, Y.-L. Using an expanded Theory of Planned Behavior to predict adolescents' intention to engage in healthy eating. *J. Int. Consum. Mark.* **2016**, *28*, 16–27. [CrossRef]

47. Hopman-Rock, M.; Borghouts, J.A.; Leurs, M.T. Determinants of participation in a health education and exercise program on television. *Prev. Med.* **2005**, *41*, 232–239. [CrossRef]

48. Conner, M.; Norman, P.; Bell, R. The theory of planned behavior and healthy eating. *Health Psychol.* **2002**, *21*, 194–201. [CrossRef]

49. Courneya, K.S. Understanding readiness for regular physical activity in older individuals: An application of the theory of planned behavior. *Health Psychol.* **1995**, *14*, 80–87. [CrossRef]

50. Abraham, C.; Sheeran, P. Deciding to exercise: The role of anticipated regret. *Br. J. Health Psychol.* **2004**, *9*, 269–278. [CrossRef]

51. Dunn, K.I.; Mohr, P.; Wilson, C.J.; Wittert, G.A. Determinants of fast-food consumption. An application of the Theory of Planned Behaviour. *Appetite* **2011**, *57*, 349–357. [CrossRef]

52. Robins, R.W.; Hendin, H.M.; Trzesniewski, K.H. Measuring Global Self-Esteem: Construct Validation of a Single-Item Measure and the Rosenberg Self-Esteem Scale. *Pers. Soc. Psychol. Bull.* **2001**, *27*, 151–161. [CrossRef]

53. Daly, M.; Delaney, L.; Baumeister, R.F. Self-control, future orientation, smoking, and the impact of Dutch tobacco control measures. *Addict. Behav. Rep.* **2015**, *1*, 89–96. [CrossRef] [PubMed]

54. Farías, P. Determinants of knowledge of personal loans' total costs: How price consciousness, financial literacy, purchase recency and frequency work together. *J. Bus. Res.* **2019**, *102*, 212–219. [CrossRef]
55. Olavarrieta, S.; Hidalgo, P.; Manzur, E.; Farías, P. Determinants of in-store price knowledge for packaged products: An empirical study in a Chilean hypermarket. *J. Bus. Res.* **2012**, *65*, 1759–1766. [CrossRef]
56. Farías, P. Promoting the Absence of Pesticides through Product Labels: The Role of Showing a Specific Description of the Harmful Effects, Environmental Attitude, and Familiarity with Pesticides. *Sustainability* **2020**, *12*, 8912. [CrossRef]
57. Noh, J.-W.; Kim, J.; Cheon, J.; Lee, Y.; Kwon, Y.D. Relationships between Extra-School Tutoring Time, Somatic Symptoms, and Sleep Duration of Adolescent Students: A Panel Analysis Using Data from the Korean Children and Youth Panel Survey. *Int. J. Environ. Res. Public Health* **2020**, *17*, 8037. [CrossRef]
58. Lacko, A.; Ng, S.W.; Popkin, B.M. Urban vs. Rural Socioeconomic Differences in the Nutritional Quality of Household Packaged Food Purchases by Store Type. *Int. J. Environ. Res. Public Health* **2020**, *17*, 7637. [CrossRef]

Publisher's Note: MDPI stays neutral with regard to jurisdictional claims in published maps and institutional affiliations.

International Journal of
Environmental Research and Public Health

Article

The Relationship between Physical Activity, Mobile Phone Addiction, and Irrational Procrastination in Chinese College Students

Mengyao Shi [1], Xiangyu Zhai [2], Shiyuan Li [1], Yuqing Shi [1] and Xiang Fan [1,3,*]

[1] Department of Physical Education, Shanghai Jiao Tong University, Shanghai 200240, China; smy0214@sjtu.edu.cn (M.S.); lishiyuan2019@sjtu.edu.cn (S.L.); stangyuan@sjtu.edu.cn (Y.S.)
[2] Graduate School of Sport Sciences, Waseda University, Saitama 359-1192, Japan; xiangyu.zhai@akane.waseda.jp
[3] Shanghai Research Center for Physical Fitness and Health of Children and Adolescents, Shanghai University of Sport, Shanghai 200438, China
* Correspondence: fansheva@sjtu.edu.cn

Abstract: The aim of the current study was to examine the associations between physical activity, mobile phone addiction, and irrational procrastination after adjustment for potential confounding variables. The participants were 6294 first- and second-year students recruited as a cluster sample from three public universities in Shanghai, China. Physical activity, mobile phone use, and irrational procrastination were assessed using the International Physical Activity Questionnaire-Short Form (IPAQ-SF), the mobile phone addiction index scale (MPAI), and the irrational procrastination scale (IPS). The participants were divided into four groups according to their mobile phone usage status and physical activity level. The binary logistic regression model was used to predict the probability of serious irrational procrastination among different groups. The emergence of serious of irrational procrastination under physical activity of different intensity and different mobile phone addiction statuses was predicted by a multiple linear regression model. In this study, the combination of insufficient physical activity and mobile phone addiction is positively associated with high levels of irrational procrastination. Furthermore, students who exhibited both mobile phone addiction behaviors and insufficient physical activity tended to have significantly higher odds of reporting high levels of irrational procrastination than those students who exhibited one behavior or neither behavior. After adjusting for the effects of age, BMI, tobacco, alcohol use, and sedentary time, the result is consistent with previous outcomes. These findings suggest that intervention efforts should focus on the promotion of physical activity and reduction of mobile phone addiction.

Keywords: physical activity; mobile phone addiction; procrastination; college students; China

Citation: Shi, M.; Zhai, X.; Li, S.; Shi, Y.; Fan, X. The Relationship between Physical Activity, Mobile Phone Addiction, and Irrational Procrastination in Chinese College Students. *Int. J. Environ. Res. Public Health* **2021**, *18*, 5325. https://doi.org/10.3390/ijerph18105325

Academic Editors: Luciana Zaccagni and Emanuela Gualdi-Russo

Received: 14 April 2021
Accepted: 14 May 2021
Published: 17 May 2021

Publisher's Note: MDPI stays neutral with regard to jurisdictional claims in published maps and institutional affiliations.

1. Introduction

Irrational procrastination is defined as voluntary postponement of an intended action even though the individual knows they will be worse off as a result of the postponement [1]. In the late 1970s, irrational procrastination was officially demarcated in the field of psychology, and procrastination behavior was considered an explicit manifestation of some mental health conditions [2]. Previous evidence has shown that serious procrastination can cause serious problems in people's studies, work, and life, including strong feelings of remorse, guilt, and constant self-denial [3], and lead to increased pressure, depression, anxiety, and fatigue [4]. A British study has indicated that the more urban people procrastinate, the greater their health problems [5]. Procrastination is a common behavior among both teenagers and adults [6]. Globally, people of all ages, especially college students, are negatively impacted by irrational procrastination [7]. While one study reports approximately 70–80% of college students experience various degrees of procrastination [8],

Int. J. Environ. Res. Public Health **2021**, *18*, 5325. https://doi.org/10.3390/ijerph18105325　　　　https://www.mdpi.com/journal/ijerph

another indicates that the prevalence of procrastination among college students is as high as 85.9%, and it is even higher in China [9,10]. Previous studies have shown that the interaction between a series of behavioral, cognitive, and psychological elements may lead to procrastination [11], including addition behavior, stress, anxiety, and depression [12,13].

Healthy lifestyle behaviors are closely related to the physical and mental health of college students. In particular, physical activity has been found to help reduce various mental illnesses [14], and it is a readily available and low-cost treatment. Following the latest 2020 report by the World Health Organization (WHO), increasing physical activity can improve both physical and mental health [15]. However, in 2016, approximately 28% of adults, including college students around the world, were insufficiently physically active [16]. Previous studies have found a correlation between physical activity and procrastination among college students: physical activity has a positive effect on irrational procrastination [9]. However, the effects of different intensities of physical activity have on procrastination have not yet been investigated. Therefore, we used the irrational procrastination scale (IPS) and the International Physical Activity Questionnaire-Short Form (IPAQ-SF) to measure more participants than the previous studies have measured, aiming to find a more accurate relationship between physical activity and irrational procrastination.

Young people aged 18 to 22 are the largest and fastest growing group of users of mobile phones [17]. Contemporary college students tend to study in a paperless and electronic manner, and mobile phones are widely used in life and study [18]. Normal cell phone use can improve efficiency and have a positive impact on mental health, but the consequences of excessive use of mobile phones prove detrimental to mental health, including but not limited to the overload, sleep disorders, physical problems such as neck pain, role conflict, and being unable to respond to all calls and text messages and the subsequent feeling of guilt [19,20].

Previous evidence has shown that problematic smartphone usage negatively affects procrastination [21,22]. Besides, female college students exhibited more serious mobile phone addiction than male college students [23]. However, few existing studies have discussed the impact of physical activity and mobile phone addiction on irrational procrastination among college students. Since engaging in physical activity and using mobile phones have become part of people's daily routines, considering these behaviors in tandem is necessary in order to develop interventions aimed at reducing procrastination.

The purpose of this study was to investigate the association between physical activity and mobile phone addition combined with irrational procrastination among Chinese college students. We monitored both the physical activity and mobile phone addiction status of our participants. In this paper, we also discussed the effects of different intensities of physical activity on the participants' procrastination behaviors. Due to factors such as physical activity, mobile phone addiction, and irrational procrastination having different effects on different genders, we specifically discuss men and women separately.

2. Participants and Procedure

2.1. Participants and Ethical Considerations

This cross-sectional research study was conducted in October 2020. The sample population included all the healthy first- and second-year students enrolled in three universities in Shanghai, China. We distributed 8546 questionnaires and collected 6786 completed questionnaires on a voluntary basis. A total of 492 incomplete questionnaires (7.25% of those collected) were excluded from our sample. Ultimately, a total of 6294 participants with an average age of 18.57 ± 1.82 years were voluntarily recruited in this survey. There were more male participants (n = 4,310, 68.48%) than female participants (n = 1984, 31.52%), and it included a total of 3463 first-year students and 2831 second-year students.

According to the standardized survey administration protocol, the survey was managed by trained research assistants in laboratories of the participating universities between 11 September and 11 October 2020. The students completed the questionnaire online after having the methods, process benefits, and possible inconveniences explained to them by the research assistants. The participants were provided sufficient time to answer the questions and were instructed how to fill out the survey, including detailed instructions on how to answer the questions. All the students provided informed consent and expressed their willingness to participate. The questionnaire contained data including sociodemographic and anthropometric characteristics, levels of physical activity, mobile phone addiction, and procrastination. Finally, the participants completed a physical fitness test under the supervision of trained testers. This study was approved by the ethics committee of Shanghai Jiao Tong University (NO.H2020043I).

2.2. Methods

2.2.1. Sociodemographic and Anthropometric Characteristics

The sociodemographic characteristics of the participants, including gender, age, and grades, were self-reported by the participants. Their anthropometric characteristics, including height and weight, were surveyed after questionnaire completion using an objective measuring instrument (HK6800-ST, Hengkang, Shenzhen, China). BMI was calculated using an internationally accepted method: a person's weight (kilograms) divided by their height in meters squared (kg/m^2). The participants were required to be barefoot when measured for height, and height measurements were accurate to 0.1 cm; weight measurements were accurate to 0.1 kg.

2.2.2. Physical Activity

The physical activity of the participants was evaluated by issuing the IPAQ-SF [24,25], which is a reliable and effective physical fitness measurement tool [26]. The participants were required to provide information about the frequency and duration of their exercise, as well as the quality of the activity (vigorous or moderate) and the amount of light physical activity and time spent sitting over the preceding seven days. Among their responses, vigorous-intensity physical activity (VPA) was defined as performing at six or more METs (metabolic equivalents of task). On a scale relative to the individuals' personal capacities, VPA is usually ranked at 7 or 8 on a scale of 0–10. Moderate intensity physical activity (MPA) refers to physical activity performed between three to six times the intensity of rest. Light-intensity physical activity (LPA) refers to activities with an energy cost of less than three times the energy expenditure at rest for that person, including slow walking, bathing, or other incidental activities that do not result in a substantial increase in heart rate or breathing rate [15]. The different intensities of the participants' physical activities are analyzed in Sections 3 and 4.

The data reporting vigorous exercise and moderate exercise exceeding 180 min/day [27] were excluded from this study. According to the latest recommendations of the WHO, adequate physical activity for adults is defined as 300 min of moderate activity, 150 min of vigorous activity, or an equivalent combination of both types of physical activity per week [15]. Therefore, the participants who conformed to the above recommendations were classified under the sufficient physical activity group. Otherwise, they were classified as having insufficient physical activity.

2.2.3. Mobile Phone Addiction

The mobile phone addiction index scale (MPAI) was used to investigate the participants' levels of mobile phone addiction [28], which has good reliability and validity among Chinese college students (Cronbach's α: 0.86) [29]. The MPAI, revised by Leung et al., is a survey regarding mobile phone use with a total of 17 questions. Each question was assigned a score based on a five-point scale. The higher the score, the more addicted the participants are to their mobile phones. In addition, questions 3, 4, 5, 6, 8, 9, 14, and 15 were mobile phone addiction screening questions. If the participants responded to five or more questions with a rating of three or above, they were considered mobile phone addicts. The others were treated as non-phone addicts.

2.2.4. Irrational Procrastination

Procrastination can be conceptualized in terms of general procrastinating traits (measured, for example, by the general procrastination scale and the irrational procrastination scale) and task-specific procrastinating propensity (measured, for example, by the Procrastination Assessment Scale for Students and the bedtime procrastination scale [30,31]. Data regarding the irrational procrastination of the participants were collected using the IPS. This version of the IPS has been verified to have good reliability and validity among Chinese college students [32], and the Cronbach's alpha for the current study was 0.79. The scale consisted of nine items, each using five-point Likert scoring, wherein "1" and "5" meant "strongly disagree" and "strongly agree", respectively. Reverse scoring was used for three of the items, and the total score ranged from 0 to 36. The higher the score, the more serious was the participant's irrational procrastination behavior. This scale has been tested and is applicable to the survey and research of college students in China [32]. Irrational procrastination behavior was divided into serious irrational procrastination (i.e., score > 18), slight irrational procrastination (i.e., $0 < \text{score} \leq 18$), and no irrational procrastination (i.e., score = 0) according to the theoretical median number [33].

2.2.5. Lifestyle Behaviors

Smoking and drinking status were determined by the participants' self-reported responses to the questionnaires' questions about the participants' tobacco and alcohol use "Have you ever smoked?" and "Have you ever had alcohol?". The participants who answered "always" were segmented within the "smoker group" or "drinker group", and everybody else were segmented within the "no tobacco users" or "no alcohol users".

2.3. Statistical Analyses

The demographic information of the participating students was described using means, standard deviations, and percentages. Furthermore, age, BMI, physical activity time, and MPAI and IPS scores were determined using t-tests. The use of tobacco and alcohol, mobile phone addiction conditions, physical activity level, and degree of irrational procrastination between the genders were examined using χ^2 tests. Normality of the data was checked using the Kolmogorov–Smirnov test; if the skewness coefficient (SC) and the kurtosis coefficient (KC) were each between −1.96 and +1.96, approximate normal distribution could be established [34]. PA_{male}, SC = 0.99, KC = 0.63; PA_{female}, SC = 1.11, KC = 0.98; MPAI $score_{male}$, SC = −0.03, KC = 0.08; MPAI $score_{female}$, SC = −0.19, KC = 0.06; IPS $score_{male}$, SC = −0.05, KC = 0.15; IPS $score_{female}$, SC = −0.05, KC = −0.01; all of them were normally distributed. Before the t-test, homogeneity of variance was checked by the Levene's test ($p > 0.05$ was variance homogeneity). If the result shows homogeneity of variance, read the t-test result in the "Equal variances assumed" row. If the result shows variance non-homogeneity, read the t-test result in the "Equal variances not assumed" row. The Pearson correlation test was used to examine the correlation between the time spent engaging in physical activity and irrational procrastination scores and the relationship between MPAIs and IPSs. The binary logistic regression model was used to predict the probability of serious irrational procrastination among different groups. The emergence of serious of irrational procrastination under physical activity of different intensity and different mobile phone addiction statuses was predicted by a multiple linear regression model. Depending on whether physical activity was sufficient or not and the two conditions of mobile phone addiction (addicts and non-addicts), the participants were divided into four different combination groups (i.e., Group I: non-mobile phone addicts + sufficient physical activity; Group II: mobile phone addicts + sufficient physical activity; Group III: non-mobile phone addicts + insufficient physical activity; Group IV: mobile phone addicts + insufficient physical activity). Each combination group (Group I, ref.) was formed after adjusting for age, BMI, smoking and drinking status, and sedentary time per day. The acceptable threshold of statistical significance was specified as $p < 0.05$, and statistical significance was set at $p < 0.01$. The collected data were statistically analyzed using version 25 of the IBM Statistical Package for the Social Sciences (SPSS) software (International Business Machines), and normality and homoscedasticity were also checked by SPSS 25.0 (IBM, Armonk, NY, USA).

3. Results

The total of 6294 student participants joined this study. The average BMI of those who conformed to the standard was 22.21 ± 3.61. The participants' average daily time spent engaging in moderate to vigorous physical activity was 59.48 ± 36.98 min, and 74.2% had sufficient physical activity levels. The physical activity times and levels (rate of sufficiency) of men were significantly higher than those of women ($F = 4.51$, $p = 0.034$). The average sedentary time of the participants was 486.20 ± 154.46 min, and in this regard, there was no significant difference between male and female students ($F = 1.21$, $p = 0.271$). The average score of the MPAI was 29.16 ± 12.66 points. According to the mobile phone addiction screening questions, 41.2% of participants were deemed mobile phone addicts, and the degree of mobile phone addiction among male students was significantly lower than that among female students ($F = 6{,}83$, $p = 0.018$). The average score of the participants' IPS was 17.32 ± 5.85 points. Of the college students, 53% displayed serious irrational procrastination behavior, and only 0.2% displayed no procrastination behavior at all. The proportion of serious procrastination behavior among male students was slightly higher than among female students. There was no statistical difference between the IPS scores of male and female students ($F = 1.65$, $p = 0.200$). Table 1 shows demographic information and other basic statistical information about the participants.

Table 1. Participants' demographics and characteristics ($n = 6294$).

Participant Characteristics	Total (n = 6294)	Men (n = 4310)	Women (n = 1984)	p-Value
	n (%) or Mean $\pm$ SD			
Age (years)	18.57 ± 1.82	18.60 ± 1.84	18.50 ± 1.75	<0.01 *
BMI (kg/m^2)	22.21 ± 3.61	22.83 ± 3.77	20.89 ± 2.81	<0.01 *
Physical activity				
Physical activity time	59.48 ± 36.98	61.31 ± 37.37	55.51 ± 35.79	<0.01 *
Sufficient	4668 (74.2)	3328 (77.2)	1340 (67.5)	<0.01 [†]
Insufficient	1626 (25.8)	982 (22.8)	644 (32.5)	
Sedentary time	486.20 ± 154.46	486.23 ± 157.53	486.14 ± 147.62	<0.01 *
Mobile phone addiction				
MPAI score	29.16 ± 12.66	28.47 ± 12.81	30.66 ± 12.20	<0.01 *
Addicts	2591 (41.2)	1721 (39.9)	870 (43.9)	<0.01 [†]
Non-addicts	37035 (8.8)	2589 (60.1)	1114 (56.1)	
Irrational procrastination				
IPS score	17.32 ± 5.85	17.35 ± 5.82	17.26 ± 5.89	<0.01 *
Serious procrastination	3336 (53.0)	2328 (54.0)	1008 (50.8)	<0.01 [†]
Slight procrastination	2944 (46.8)	1974 (45.8)	970 (48.9)	
No procrastination	14 (0.2)	8 (0.2)	6 (0.3)	
Tobacco use				<0.01 [†]
Never	6208 (98.6)	4241 (98.4)	1967 (99.1)	
Yes	86 (1.4)	69 (1.6)	17 (0.9)	
Alcohol use				<0.01 [†]
Never	4298 (68.3)	2755 (63.9)	1543 (77.8)	
Yes	1996 (31.0)	1555 (36.1)	441 (22.2)	

Note: p *, t-tests; p [†], χ^2 test; BMI, body mass index; MPAI, mobile phone addiction index scale; IPS, irrational procrastination scale. All values represent raw non-standardized scores.

Table 2 showed a significant positive correlation between the participants' MPAI and IPS scores ($p < 0.01$). The higher the degree of mobile phone addiction, the higher the IPS score. The IPS scores of mobile phone addicts were significantly higher than those of non-mobile phone addicts, and the results showed a significant difference in the IPS scores of mobile phone addicts and non-mobile phone addicts ($p < 0.01$) among both men and women.

Table 2. The Pearson correlation coefficient for the degree of irrational procrastination exhibited by the participants (IPS score) and across different groups.

Participant Characteristics	Total (*n* = 6294)	Men (*n* = 4310)	Women (*n* = 1984)
	Mean ± SD		
Physical activity			
Sufficient	16.97 ± 5.87 **	17.05 ± 5.86 **	16.76 ± 5.90 **
Insufficient	18.34 ± 5.64 **	18.36 ± 5.57 **	18.32 ± 5.74 **
Mobile phone addiction			
Addicts	19.97 ± 5.10 **	20.04 ± 5.06 **	19.83 ± 5.19 **
Non-addicts	15.47 ± 5.61 **	15.56 ± 5.61 **	15.26 ± 5.61 **

Note: **, $p < 0.01$; p-value for significant IPS score differences between the different groups determined using a *t*-test.

Furthermore, we contrasted the linear correlations between different intensities of physical activity and irrational procrastination among college students. The duration of VPA was significantly negatively correlated with the participants' irrational procrastination scores ($\beta = -0.107$). The absolute value of r was the highest among all the intensities. Similarly, MPA and LPA were also negatively associated with irrational procrastination (MPV: $\beta = -0.083$; LPA: $\beta = -0.069$). In Figure 1, a linear correlation slope can be seen for the three intensities of physical activity and irrational procrastination scores. The conclusion to be drawn is as follows: the higher the intensity of physical activity, the greater the degree of its correlation with irrational procrastination. In addition, we found a significant positive correlation between the amount of sedentary time on workdays and irrational procrastination ($r = 0.086$).

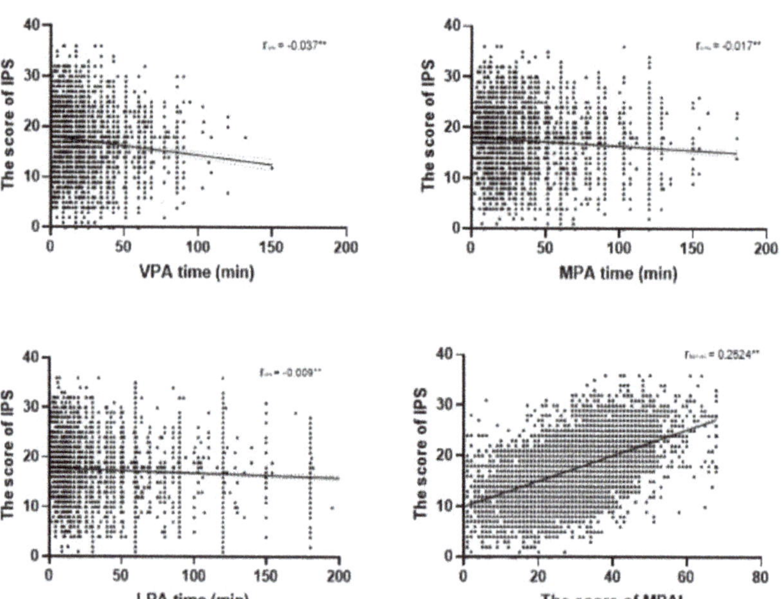

Figure 1. The correlation between the degree of irrational procrastination and the different intensities of physical activity and mobile phone addiction status (** $p < 0.01$).

Table 3 shows results of the multiple linear regression analysis which was used to test the predicted relationship between the IPS scores and the different intensities and durations of physical activity and the MPAI score. VPA ($\beta = -0.043$; 95% CI = $-0.069 \sim -0.017$) and MPA ($\beta = -0.033$; 95% CI = $-0.059 \sim -0.006$) were significantly negatively correlated with irrational procrastination in male students, while MPA ($\beta = -0.053$; 95% CI = $-0.092 \sim -0.014$)

was negatively associated with the IPS score in both genders, and the MPAI score showed a significant positive correlation with the IPS score for both male and female students (βmale = 0.526; 95% CI = 0.500~0.552; βfemale = 0.554; 95% CI = 0.516~0.592) even after adjusting for age, BMI, tobacco use, alcohol use, and sedentary time. The adjusted scores explained that the variance (R2) of the model for male and female students was 0.299 and 0.333, respectively, and the regression models were statistically significant for both genders.

Table 3. The Pearson correlation coefficient for the degree of irrational procrastination exhibited by participants (IPS score) and across different groups.

Variable	Male (Adjusted R^2 = 0.299 **)		Female (Adjusted R^2 = 0.333 **)	
	β (95% CI)	*p*-Value	β (95% CI)	*p*-Value
LPA	−0.026 (−0.052~−0.001)	0.057	−0.014 (−0.053~0.025)	0.491
MPA	−0.033 (−0.059~−0.006)	0.017	−0.053 (−0.092~−0.014)	0.007
VPA	−0.043 (−0.069~−0.017)	0.001	−0.034 (−0.073~0.004)	0.080
MPAI	0.526 (0.500~0.552)	0.000	0.554 (0.516~0.592)	0.000

Note: adjusted for age, BMI, tobacco use, alcohol use, and sedentary time. ** $p < 0.01$.

Table 4 presents the binary logistic regression models, which show the odds of serious irrational procrastination under different combinations of physical activity conditions and mobile phone addiction statuses. All the students were classified into four groups according to their mobile phone addiction statuses and physical activity levels. Group I was specified as the reference group (non-mobile phone addicts × sufficient physical activity). Studies have shown that lack of physical activity or being addicted to the phone increases the odds of serious irrational procrastination. When we further controlled for the effects of age, BMI, tobacco use, alcohol use, and sedentary time, male students with excessive mobile phone usage and insufficient physical activity respectively saw 3.25 (95% CI = 2.793~3.771) times and 1.35 (95% CI = 1.094~1.659) times increased odds of reporting serious irrational procrastination. Compared to the control groups, serious irrational procrastination in men with both mobile phone addiction and insufficient physical activity was 4.26 (95% CI = 3.418~5.312) times higher.

Table 4. The odds of serious irrational procrastination under different combinations of physical activity conditions and mobile phone addiction statuses.

Group	Men		Women	
	OR (95% CI)	aOR[a] (95% CI)	OR (95% CI)	aOR [a] (95% CI)
Group I	1 (ref.)	1 (ref.)	1 (ref.)	1 (ref.)
Group II	3.325 ** (2.870~3.853)	3.246 ** (2.793~3.771)	3.113 ** (2.469~3.853)	3.223 ** (2.537~4.094)
Group III	1.350 ** (1.101~1.655)	1.347 ** (1.094~1.659)	1.229 (0.920~1.640)	1.229 (0.967~1.746)
Group IV	4.510 ** (3.632~5.600)	4.261 ** (3.418~5.312)	4.024 ** (3.062~5.288)	4.183 ** (3.105~5.555)

Note: [a] adjusted for age, BMI, tobacco use, alcohol use, and sedentary time; ** $p < 0.01$.

Similar to male students, insufficient physical activity and mobile phone addiction also had a significant positive correlation with serious irrational procrastination among female students even after adjusting for age, BMI, tobacco use, alcohol use, and sedentary time. The presence of both insufficient physical activity and mobile phone addiction increased the odds of serious irrational procrastination by 4.18 times (95% CI = 3.105~5.555) compared with the reference group. Female students with mobile phone addiction only had 3.22 (95% CI = 2.537~4.094) times higher odds of serious irrational procrastination as compared to non-mobile phone addicts and those who engaged in sufficient physical activity. However, among the female students who were not mobile phone addicts, insufficient physical activity was not significantly correlated with serious irrational procrastination when compared to those who engaged in sufficient physical activity.

4. Discussion

The purpose of this study was to investigate the prevalence and correlation between insufficient physical activity and mobile phone addiction, as well as the presence of irrational procrastination among Chinese college students. In this study, the IPS scores of the participants were 17.32 ± 5.85, and only 0.2% of the participants exhibited no irrational procrastination. This is consistent with the assertions of previous studies which stated that the occurrence rate of procrastination among college students reached over 95% [10]. In addition, we found that the prevalence of serious irrational procrastination (53.0%) was similar to that reported in previous studies on college students (over 50%) [4] and that the odds of irrational procrastination in this group were higher than those among teenagers (over 40%) [35]. We believe that those variances have to do with the educational stage and lifestyles of the students who participated in the study. All the participants in the present study were first- or second-year students from universities in China; they had just left high school where their learning was passive and moved into new independent lives and learning styles. Numerous changes in living conditions, various academic pressures, and being away from parents or others who could supervise them, they had to face many issues themselves and deal with their emotional burden, which was all completely different from their previous experience. In view of this, procrastination could become an ineffective coping strategy to counteract their problems. This may have led to an increased procrastination behavior among them. The incidence of insufficient physical activity in our investigation (25.8%) is also higher than the national Chinese average (14.1%) [16]; heavy college workload could be a potential reason for this finding as well. The revised WHO guidelines on physical activity and sedentary behavior may also have led to these differences.

The MPAI scores of the study participants were 29.16 ± 12.66. According to the MPAI's mobile phone addiction screening questions, 41.2% were considered mobile phone addicts; this finding is significantly higher than those of previous studies conducted in China [36] (21.3%) and other countries [37]. The reason for this could be that the increasing daily functions of mobile phones mean that people use them more frequently and for longer periods. Mobile phones are widely used around the world. However, compared with traditional mobile phones, smartphones with their numerous functions and multiple modes of use are superior [38,39]. Today, we are all immersed in a digital world of new technologies that are also making their way into the education process. The root causes of procrastination are most likely related to this deeper change in civilization and educational systems.

Students today are born into the digital world, and they learn how to operate it at an early age. At school, they enter an environment completely different from the family setting: there are many children of the same age who they need to communicate with in college. Using a mobile phone may also be the easiest way for them to seek more social interaction. Contemporary college students tend to study in a paperless and electronic manner [40] and as a result, their screen time and rates of mobile phone addition are also increasing [19]. This view could explain the findings of this study— that there exists a significant correlation between mobile phone addiction and irrational procrastination, thus tracing a serious increase in irrational procrastination. At the same time, the appearance of irrational procrastination behaviors is an explicit manifestation of some mental health conditions [2]. According to previous studies, mobile phone addiction is likely related to anxiety, stress, and sleep quality [41,42], and such studies have also shown an association between screen time and mental health [43]. Therefore, we posit that mobile phone addiction affects mental activities and mental health, thus causing irrational procrastination; this could be a potential mechanism for the correlation between mobile phone addiction and irrational procrastination.

In the present study, the level of physical activity among the participants was significantly correlated with irrational procrastination. The IPS scores of the participants with sufficient physical activity were significantly lower than those of the participants with insufficient physical activity. Similar to Zhong's study, we found that physical activity is correlated with procrastination among college students [9]. In addition, we found that the intensities of physical activity that the students engaged in were significantly correlated with the effect of physical activity on irrational procrastination. Figure 1 illustrates that higher intensities of physical activity had a greater effect on irrational procrastination. Previous studies have shown that higher intensities of physical activity result in a greater improvement in suppressing passive emotion; this suppression of passive emotions could be a potential reason for physical activity preventing procrastination [41,44]. Numerous studies have documented the correlation between passive emotion—such as anxiety [45], stress [5], and depression [11,46]—and procrastination.

In addition, more time spent engaging in VPA and MPA significantly contributed to lower degrees of irrational procrastination among male students. In contrast, among female students, more time spent engaging in MPA was seen to significantly contribute to a lower degree of irrational procrastination, but the fitting effect of VPA on irrational procrastination was not significant. Studies have found that female students prefer moderate physical activity, such as dance or gymnastics, as their daily exercise [47]. This greater inclination toward moderate physical activity could be a potential reason for MPA being the best exercise intensity to minimize irrational procrastination.

However, after controlling for mobile phone addiction, we found that physical activity levels were significantly correlated with serious irrational procrastination among male but not female students. We also found previous evidence stating that students of different genders have different physical activity habits, which could be one of the main reasons for our finding [47,48]. In particular, even though we found no link between physical activity levels and serious irrational procrastination in female students after controlling for mobile phone addiction, the participants who were both insufficiently active and mobile phone addicts also reported more serious irrational procrastination than those with only a mobile phone addiction. It is possible that these participants took longer to engage in physical activity and instead used their mobile phones, which increased their irrational procrastination behavior. In other words, our study suggests that females who engage in insufficient physical activity could indirectly increase their odds of experiencing serious irrational procrastination.

Furthermore, this study determined that a relationship exists between insufficient physical activity, mobile phone addiction, and irrational procrastination, even after adjusting for the effects of age, BMI, tobacco use, alcohol use, and sedentary time. The presence of both insufficient physical activity and mobile phone addiction significantly increased the odds of irrational procrastination when compared with the presence of behaviors from other categories. Meanwhile, mobile phone addiction had a significant positive correlation with serious irrational procrastination even after controlling for the adjustment factor and physical activity time.

5. Conclusions

In this study, irrational procrastination was found to display a significant correlation between both physical activity and mobile phone addiction. Therefore, from a public health and behavior perspective, if college students want to reduce irrational procrastination to improve efficiency, increasing physical activity and reducing mobile phone addiction are vital steps.

The main strength of the present study is that it considers the effects of different intensities of physical activity on irrational procrastination; physical activity and mobile phone addiction are included as mutually confounding factors and are seen as a whole; we also discuss their impact on irrational procrastination separately by studying a large sample of Chinese college students.

One limitation of our present study is the cross-sectional research design: the possibility of reversal causation cannot be ruled out. Another limitation involves the self-reported measures; although they were assessed using standardized questionnaires, this method of information collection is susceptible to measurement errors, memory bias, and the social expectation effect. Therefore, future research should use more objective methods of data collection and measurement, such as accelerometers and mobile monitoring software. Another limitation is that the classification of covariates such as smoking and drinking was not clear enough and should be considered more comprehensively in future research.

Author Contributions: The authors' contributions are as follows: X.F. was the principal investigator; he designed the study and oversaw the implementation of the project. M.S. drafted the manuscript and completed the data analyses. X.Z. participated in the revision of the manuscript and improved its quality. S.L. and Y.S. participated in data collection and the discussion of statistical methods. All authors have read and agreed to the published version of the manuscript.

Funding: This study is a social science research project which was supported in part by the National Key R&D Program of China (NO.2020YFC2007004).

Institutional Review Board Statement: The study was conducted according to the guidelines of the Declaration of Helsinki and approved by the Ethics Committee of Shanghai Jiao Tong University (NO.H2020043I) on 14 October 2020.

Informed Consent Statement: Informed consent was obtained from all the subjects involved in the study.

Data Availability Statement: The data in the study are not publicly available in order to protect privacy of the participants.

Acknowledgments: The authors are very grateful to Shun Song, Can Wang, Wenxian Zhou, and other fieldworkers for their help and contribution in data collection. We thank Tong Zhao and Mei Ye for their guidance on statistical methods and data consolidation. We also thank all the students who participated in this study.

Conflicts of Interest: The authors declare no conflict of interest.

References

1. Steel, P. The nature of procrastination: A meta-analytic and theoretical review of quintessential self-regulatory failure. *Psychol. Bull.* **2007**, *133*, 65–94. [CrossRef] [PubMed]
2. Owens, S.; Bowman, C.; Dill, C. Overcoming Procrastination: The Effect of Implementation Intentions1. *J. Appl. Soc. Psychol.* **2008**, *38*, 366–384. [CrossRef]
3. Küchler, A.-M.; Albus, P.; Ebert, D.; Baumeister, H. Effectiveness of an internet-based intervention for procrastination in college students (StudiCare Procrastination): Study protocol of a randomized controlled trial. *Internet Interv.* **2019**, *17*, 100245. [CrossRef] [PubMed]
4. Lukas, C.; Berking, M. Reducing procrastination using a smartphone-based treatment program: A randomized controlled pilot study. *Internet Interv.* **2017**, *12*. [CrossRef]
5. Sirois, F. "I'll look after my health, later": A replication and extension of the procrastination-health model with community-dwelling adults. *Personal. Individ. Differ.* **2007**, *43*, 15–26. [CrossRef]
6. Malouff, J. The Efficacy of Interventions Aimed at Reducing Procrastination: A Meta-Analysis of Randomized Controlled Trials. *J. Couns. Dev.* **2019**, *97*, 117–127. [CrossRef]
7. Day, V.; Mensink, D.; O'Sullivan, M. Patterns of Academic Procrastination. *J. Coll. Read. Learn.* **2000**, *30*, 120–134. [CrossRef]
8. Geng, J.; Han, L.; Gao, F.; Jou, M.; Huang, C.-C. Internet addiction and procrastination among Chinese young adults: A moderated mediation model. *Comput. Hum. Behav.* **2018**, *84*. [CrossRef]
9. Zhong, J. Preliminary Study on the Relationship between Procrastination and Physical Activity of College Students. *J. Beijing Sport Univ.* **2013**, *20*, 21–22.
10. Chu, Q.; Xiao, R.; Lin, Q. Research on the Procrastination Behavior and Characteristics of University Students. *China J. Health Psychol.* **2010**, *18*, 970–972.
11. Solomon, L.; Rothblum, E. Academic procrastination: Frequency and cognitive-behavioral correlates. *J. Couns. Psychol.* **1984**, *31*, 503–509. [CrossRef]
12. Wan, H.; Downey, L.; Stough, C. Understanding non-work presenteeism: Relationships between emotional intelligence, boredom, procrastination and job stress. *Personal. Individ. Differ.* **2014**, *65*. [CrossRef]
13. Flett, G.L.; Blankstein, K.R.; Martin, T.R. Procrastination, Negative Self-Evaluation, and Stress in Depression and Anxiety. In *Procrastination and Task Avoidance: Theory, Research, and Treatment*; Ferrari, J.R., Johnson, J.L., McCown, W.G., Eds.; The Springer Series in Social Clinical Psychology; Springer: Boston, MA, USA, 1995; pp. 137–167. ISBN 978-1-4899-0227-6.
14. Burka, J.; Yuen, L. *Procrastination: Why You Do It; What to Do about It, NOW*; Burka, J., Yuen, L., Eds.; Da Capo Press: Cambridge, MA, USA, 2008; ISBN 978-0-7382-1170-1.

15. Bull, F.C.; Al-Ansari, S.S.; Biddle, S.; Borodulin, K.; Buman, M.P.; Cardon, G.; Carty, C.; Chaput, J.-P.; Chastin, S.; Chou, R.; et al. World Health Organization 2020 guidelines on physical activity and sedentary behaviour. *Br. J. Sports Med.* **2020**, *54*, 1451–1462. [CrossRef] [PubMed]

16. WHO. Prevalence of Insufficient Physical Activity among Adults-Data by Country. Available online: https://apps.who.int/gho/data/node.main.A893?lang=en# (accessed on 10 January 2021).

17. Jiang, Z.; Zhao, X. Self-control and problematic mobile phone use in Chinese college students: The mediating role of mobile phone use patterns. *BMC Psychiatry* **2016**, *16*, 416. [CrossRef] [PubMed]

18. Junaid, S.; Jangda, Z. Successful Deployment of a Laboratory Information Management System LIMS. In *Striding towards Modern; Paperless Labs*: Dhahran, Saudi Arabia, 2020.

19. Thomée, S.; Härenstam, A.; Hagberg, M. Mobile phone use and stress, sleep disturbances, and symptoms of depression among young adults—A prospective cohort study. *BMC Public Health* **2011**, *11*, 66. [CrossRef]

20. Lee, S.; Kang, H.; Shin, G. Head flexion angle while using a smartphone. *Ergonomics* **2015**, *58*, 220–226. [CrossRef] [PubMed]

21. Wang, P.; Liu, S.; Zhao, M.; Yang, X.; Zhang, G.; Chu, X.; Wang, X.; Zeng, P.; Lei, L. How is problematic smartphone use related to adolescent depression? A moderated mediation analysis. *Child. Youth Serv. Rev.* **2019**, *104*, 104384. [CrossRef]

22. Qaisar, S.; Akhter, N.; Masood, A.; Rashid, S. Problematic Mobile Phone Use, Academic Procrastination and Academic Performance of College Students. *J. Educ. Res.* **2017**, *20*, 201–214.

23. Zhao, Y.; Chen, F.; Yuan, C.; Luo, R.; Ma, X.; Zhang, C. Parental favoritism and mobile phone addiction in Chinese adolescents: The role of sibling relationship and gender difference. *Child. Youth Serv. Rev.* **2021**, *120*, 105766. [CrossRef]

24. Craig, C.L.; Marshall, A.L.; Sjöström, M.; Bauman, A.E.; Booth, M.L.; Ainsworth, B.E.; Pratt, M.; Ekelund, U.; Yngve, A.; Sallis, J.F.; et al. International Physical Activity Questionnaire: 12-Country Reliability and Validity. *Med. Sci. Sports Exerc.* **2003**, *35*, 1381–1395. [CrossRef]

25. Macfarlane, D.J.; Lee, C.C.Y.; Ho, E.Y.K.; Chan, K.L.; Chan, D.T.S. Reliability and validity of the Chinese version of IPAQ (short, last 7 days). *J. Sci. Med. Sport* **2007**, *10*, 45–51. [CrossRef] [PubMed]

26. Qu, N.; Li, K. Study on the reliability and validity of international physical activity questionnaire (Chinese Vision, IPAQ). *Zhonghua Liu Xing Bing Xue Za Zhi Zhonghua Liuxingbingxue Zazhi* **2004**, *25*, 265–268. [PubMed]

27. Fan, M.; Lyu, J.; He, P. Chinese guidelines for data processing and analysis concerning the International Physical Activity Questionnaire. *Zhonghua Liu Xing Bing Xue Za Zhi Zhonghua Liuxingbingxue Zazhi* **2014**, *35*, 961–964. [PubMed]

28. Leung, L. Linking psychological attributes to addiction and improper use of the mobile phone among adolescents in Hong Kong. *J. Child. Media* **2008**, *2*, 93–113. [CrossRef]

29. Mobile Phone Addiction and College Students' Procrastination: Analysis of a Moderated Mediation Model—Psychological Development and Education. Available online: http://en.cnki.com.cn/Article_en/CJFDTotal-XLFZ201805010.htm (accessed on 29 April 2021).

30. Eerde, W.; Klingsieck, K. Overcoming Procrastination? A Meta-Analysis of Intervention Studies. *Educ. Res. Rev.* **2018**, *25*. [CrossRef]

31. Kroese, F.; Ridder, D.; Evers, C.; Adriaanse, M. Bedtime Procrastination: Introducing a New Area of Procrastination. *Front. Psychol.* **2014**, *5*, 611. [CrossRef] [PubMed]

32. Ni, S.; Xu, J.; Ye, L. Revision of Irrational Procrastination Scale in Chinese College Students and Its Relations with Self-efficacy and Health Behavior. *Chin. J. Clin. Psychol.* **2012**, *20*, 603–605, 655.

33. Ni, S.; Zhang, P.; Zhao, G. Mediating role of stress in the relation between undergraduates' negative procrastination and physical health. *Chin. J. Sch. Health* **2012**, *33*, 682–683.

34. George, D.; Mallery, P. *SPSS for Windows Step by Step: A Simple Study Guide and Reference, 17.0 Update*, 10th ed.; Allyn & Bacon, Inc.: Boston, MA, USA, 2009; ISBN 978-0-205-75561-5.

35. Syabilla, Y.; Suryanda, A.; Sigit, D. A correlation between self concept and procrastination based on gender in neuroscience perspective. *Biosfer* **1970**, *11*, 114–120. [CrossRef]

36. Long, J.; Liu, T.-Q.; Liao, Y.; Qi, C.; He, H.; Shubao, C.; Billieux, J. Prevalence and correlates of problematic smartphone use in a large random sample of Chinese undergraduates. *BMC Psychiatry* **2016**, *16*, 408. [CrossRef]

37. De Sola, J.; Fonseca, F.; Rubio, G. Cell-Phone Addiction: A Review. *Front. Psychiatry* **2016**, *7*. [CrossRef]

38. Grant, J.E.; Lust, K.; Chamberlain, S.R. Problematic Smart Phone Use is Associated with Greater Alcohol Consumption, Mental Health Issues, Poorer Academic Performance, and Impulsivity. *J. Behav. Addict.* **2019**, *8*, 335–342. [CrossRef] [PubMed]

39. Lv, S.; Li, C.; Xiong, Y.; Zhou, C.; Li, X.; Ye, M. Smartphone use disorder and future time perspective of college students: The mediating role of depression and moderating role of mindfulness. *Child Adolesc. Psychiatry Ment. Health* **2020**, *14*. [CrossRef]

40. Keengwe, J.; Schnellert, G.; Jonas, D. Mobile phones in education: Challenges and opportunities for learning. *Educ. Inf. Technol.* **2012**, *19*, 441–450. [CrossRef]

41. Zhai, X.; Wu, N.; Koriyama, S.; Wang, C.; Shi, M.; Huang, T.; Wang, K.; Sawada, S.S.; Fan, X. Mediating Effect of Perceived Stress on the Association between Physical Activity and Sleep Quality among Chinese College Students. *Int. J. Environ. Res. Public. Health* **2021**, *18*, 289. [CrossRef]

42. Zhai, X.; Ye, M.; Wang, C.; Gu, Q.; Huang, T.; Wang, K.; Chen, Z.; Fan, X. Associations among physical activity and smartphone use with perceived stress and sleep quality of Chinese college students. *Ment. Health Phys. Act.* **2020**, *18*, 100323. [CrossRef]

43. Goldbacher, E.; Matthews, K. Are psychological characteristics related to risk of the metabolic syndrome? A review of the literature. *Ann. Behav. Med. Publ. Soc. Behav. Med.* **2007**, *34*, 240–252. [CrossRef] [PubMed]
44. Arent, S.; Landers, D.; Etnier, J. The Effects of Exercise on Mood in Older Adults: A Meta-Analytic Review. *J. Aging Phys. Act.* **2000**, *8*, 407–430. [CrossRef]
45. Flett, G.; Stainton, M.; Hewitt, P.; Sherry, S.; Lay, C. Procrastination Automatic Thoughts as a Personality Construct: An Analysis of the Procrastinatory Cognitions Inventory. *J. Ration. Emot. Cogn. Behav. Ther.* **2012**, *30*, 233–236. [CrossRef]
46. Martin, T.; Flett, G.; Hewitt, P.; Krames, L.; Szanto, G. Personality Correlates of Depression and Health Symptoms: A Test of a Self-Regulation Model. *J. Res. Personal.* **1996**, *30*, 264–277. [CrossRef]
47. Plaza, M.; Boiché, J.; Brunel, L.; Ruchaud, F. Sport = Male . . . But Not All Sports: Investigating the Gender Stereotypes of Sport Activities at the Explicit and Implicit Levels. *Sex Roles* **2017**, *76*. [CrossRef]
48. Portman, R.M.; Bradbury, J.; Lewis, K. Social physique anxiety and physical activity behaviour of male and female exercisers. *Eur. J. Sport Sci.* **2018**, *18*, 257–265. [CrossRef] [PubMed]

International Journal of
*Environmental Research
and Public Health*

MDPI

Review

Physical Activity during COVID-19 Lockdown in Italy: A Systematic Review

Luciana Zaccagni [1,2], Stefania Toselli [3,*] and Davide Barbieri [1]

[1] Department of Neuroscience and Rehabilitation, University of Ferrara, 44121 Ferrara, Italy; luciana.zaccagni@unife.it (L.Z.); davide.barbieri@unife.it (D.B.)
[2] Center of Sport and Exercise Sciences, University of Ferrara, 44123 Ferrara, Italy
[3] Department of Biomedical and Neuromotor Science, University of Bologna, 40126 Bologna, Italy
* Correspondence: stefania.toselli@unibo.it

Abstract: The recent COVID-19 pandemic has imposed a general lockdown in Italy, one of the most affected countries at the beginning of the outbreak, between 9 March and 3 May 2020. As a consequence, Italian citizens were confined at home for almost two months, an unprecedented situation, which could have negative effects on both psychological and physical health. The aim of this study was to review the published papers concerning the effects of the lockdown on physical activity and the consequences on general health. As expected, most studies highlighted a significant reduction in the amount of performed physical activity compared to before lockdown, in both the general population and in individuals with chronic conditions. This fact had negative consequences on both general health, in terms of increased body mass, and on specific chronic conditions, especially obesity and neurological diseases.

Keywords: pandemic; coronavirus; physical exercise; general health

check for
updates

Citation: Zaccagni, L.; Toselli, S.; Barbieri, D. Physical Activity during COVID-19 Lockdown in Italy: A Systematic Review. *Int. J. Environ. Res. Public Health* **2021**, *18*, 6416. https://doi.org/10.3390/ ijerph18126416

Academic Editor: Ann M. Swartz

Received: 30 April 2021
Accepted: 11 June 2021
Published: 13 June 2021

Publisher's Note: MDPI stays neutral with regard to jurisdictional claims in published maps and institutional affiliations.

1. Introduction

A new coronavirus (SARS-CoV-2), causing severe acute respiratory syndrome, was discovered at the end of 2019. The first cases were reported in the province of Wuhan, China. On the 30 January 2020, the Director-General of the World Health Organization (WHO) declared the SARS-CoV-2 outbreak a public health emergency of international concern, the WHO's highest level of alarm. The associated syndrome was named coronavirus disease 2019, abbreviated as COVID-19, and it is currently a major global health issue. At the time of the submission there have been 147,539,302 confirmed cases of COVID-19, including 3,116,444 deaths.

Given the exponential growth in cases and deceased during February and early March 2020, and the risk of collapse of the national health system (in particular of intensive care units), the Italian government declared a general lockdown on 9 March 2020. This resolution remained in place until 3 May 2020, and implied that Italian citizens all over the country had to remain confined at home for almost two months, and could go out only for primary necessities, such as buying food or medicines, or seeking medical care. National policies imposed social distancing, the closure of schools and universities, and the suspension of any social event. All activities practiced in gyms, sports centers, and swimming pools were suspended. Additionally, jogging or walking in parks and cycling were prohibited. Essential activities (agriculture, biomedical manufactory, information and communication technology, medical care, energy production, and similar) were maintained, but when possible, work (for example administrative tasks and meetings) had to be performed from home, using internet connections and web conferencing tools. The imperative was "stay at home" to curb the spread of the virus.

As a consequence, a drastic reduction in the amount of performed individual physical activity (PA) was expected, which could have negative consequences on the population's

213

general health. It is well known that physical inactivity causes over 5 million deaths worldwide and represents damage to the economy of the public health systems. In particular, it could have a negative effect on glycemic control, which is especially dangerous for subjects with diabetes [1,2]. In obese patients, a further increase in body mass would worsen their condition. In cardiac patients, daily moderate PA is essential to reduce cardiovascular risk [3,4]. In patients with neurological disorders, exercising is of fundamental importance to control their conditions, in terms of both motor control [5,6] and cognitive impairment [7]. Furthermore, being locked at home implied a reduction in the exposure to open air and sunlight, which could undermine the immune system and its capability to react to the new viral infection. In particular, PA has immune benefits, especially in older adults [8] who are more at risk during the current pandemic.

The aim of this paper was to perform a systematic review of the studies on PA carried out at home by the Italian population during the lockdown period, in order to evaluate the changes in lifestyle, whether sedentary or active, and the eventual consequences on general health. The underlying hypothesis is that home confinement has diminished the amount of PA, with negative consequences on individual health in general.

2. Materials and Methods

A systematic review of the studies investigating the PA practiced during the lockdown by Italian citizens was conducted in accordance with the *Preferred Reporting Items for Systematic Reviews and Meta-Analyses* (PRISMA) guidelines [9].

An electronic literature search was performed on the PubMed and Scopus online databases, using the following search string: ("physical activity" OR "physical exercise") AND (COVID-19 OR lockdown) AND Italy. As inclusion criteria, we considered only articles in the English language and only studies performed on Italian citizens; concerning the publication type, we considered only research papers, excluding reviews, letters to editor, comments, editorials, and recommendations. The search was conducted till 31 December 2020.

A total of 136 records were collected. Two reviewers (D.B. and L.Z.) selected the relevant studies separately on the basis of the titles and abstracts. Then, they independently reviewed the full text of the selected studies to decide on their final suitability according to the inclusion and exclusion criteria. In case of disagreement, the decision was made collegially with the contribution of a third investigator (S.T.).

After the exclusion of 92 articles, we included 23 full-text papers to critically evaluate the role of PA during the lockdown caused by COVID-19 in Italy. The following information was collected from each of the included articles: focus, study design, sample description (size, age, sex (% males)), method of PA-related data collection (type of PA assessment tool and type of survey), amount of PA (before and during lockdown, when reported), and main findings.

The assessment of the methodological quality of the selected studies was carried out using the Newcastle–Ottawa scale (NOS) [10], adapted for cross-sectional studies [11], independently by two reviewers. A third reviewer was available to resolve any disagreements. Each study was assessed with the following items: selection (representativeness of the sample, sample size, non-respondents, and ascertainment of the exposure), comparability, and outcome (assessment of the outcome and statistics). Scores range from 0 to 10, with higher scores indicating better quality research.

The whole process is described in Figure 1 by means of the standard PRISMA flowchart.

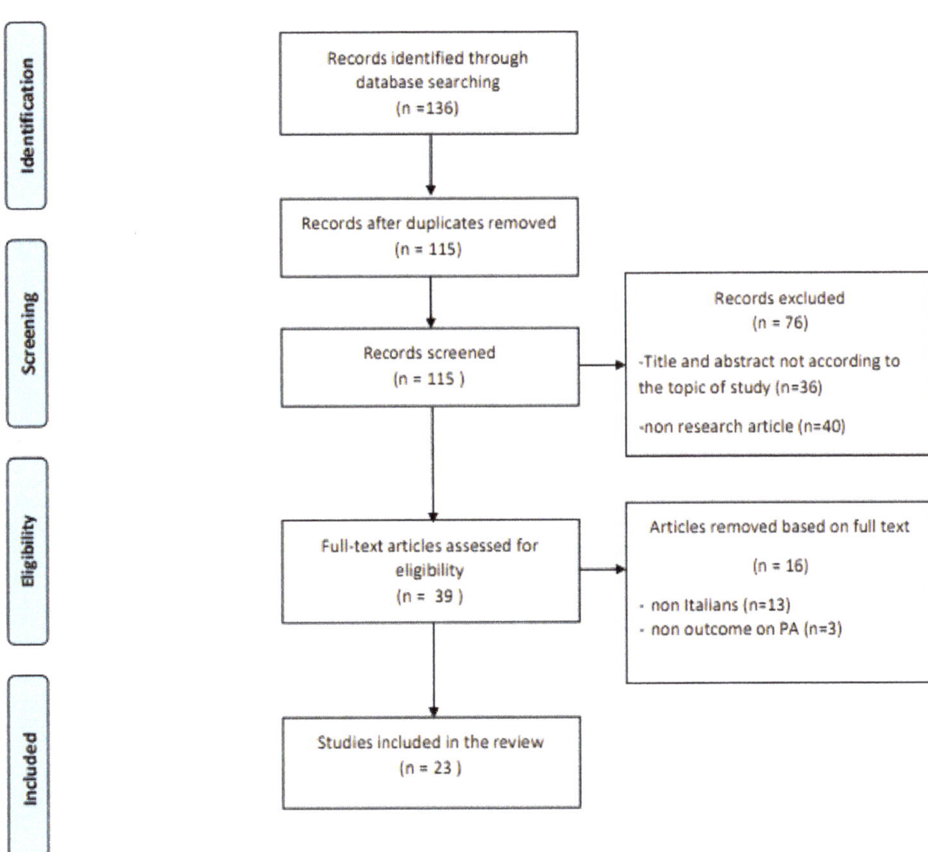

Figure 1. Flow diagram of the literature search strategy and review process, following PRISMA rules [9].

3. Results

Twenty-three published research papers that met inclusion criteria were selected and included in the review. The collected information of the included studies is summarized in Table 1.

Table 1. *Cont.*

Samples with Pathologies	Focus	Study Design	Sample, Pathology, Size, Age, % Males	PA Assessment Tool Survey Type	Amount of PA (Mean ± SD)	Main Findings
Gallè et al. [28]	Sedentary behaviors and PA	Observational, cross-sectional	N = 1430 undergraduate students, 22.9 ± 4.5 y, 34.5% males	IPAQ-SF Online survey	Before L: 520 ± 820 min/w During L: 270 ± 340 min/w	Significantly increased sedentary lifestyle, decreased PA
Gallè et al. [29]	Health-related behaviors PA	Observational, cross-sectional	N = 2125 undergraduate students, 22.5 ± 0.08 y, 37.2% males	Questionnaire online survey	NR	Significant reduction in PA
Giustino et al. [30]	Level of PA	Observational, cross-sectional	N = 802, 32.27 ± 12.81 y, 49% males	IPAQ-SF online survey	Before L: 3006 MET-min/w During L: 1483.8 MET-min/w	Significant reduction of PA, especially in males and in overweight
Luciano et al. [31]	Behaviors during lockdown (PA, sedentariness, sleep)	Observational, cross-sectional	N = 1471 medicine students 23 ± 2 y, 30% males	IPAQ-SF online survey	Before L: 1588 MET-min/w During L: 960 MET-min/w	Decreased PA, and increased sitting and sleep time
Maugeri et al. [32]	PA on psychological	Observational, cross-sectional	N = 2524 43.6% males	IPAQ Online survey	Before L: 2429 MET-min/w During L: 1577 MET-min/w	PA level decreased with negative impact on psychological health
Raiola et al. [33]	Changes in PA	Observational, cross-sectional	N = 268 Mean age = 26 y	Q not validated Online survey	NR	No change in PA
Tornaghi et al. [34]	PA levels	Observational, cross-sectional	N = 1568 students Aged 15–18	IPAQ Online survey	Before L: 1676.37 ± 20.6 MET-min/w After L: 1774.50 ± 33.93 MET-min/w	Inactive or moderately active students unchanged their PA level; highly active ones increased PA level

T1D: Type 1 diabetes; PA: physical activity; IPAQ: International Physical Activity Questionnaire; IPAQ-SF: International Physical Activity Questionnaire Short-Form; NR: not reported; L: lockdown.

All papers were observational studies. Specifically, the study by Predieri et al. [16] was longitudinal, while all others were cross-sectional.

Samples were different according to size (from a minimum of 24 participants in [20] to a maximum of 7847 in [27], age (from adolescents [16,17,34] to old adults [20], sex percentage (males vary from 16% in [24] to 62% in [13], and type (subjects with and without pathologies, students).

Ten studies (43.5%) were carried out on individuals with pathologies: six samples with Type 1 diabetes, one with neuromuscular disease, one with implantable cardioverter-defibrillator, one with Parkinson disease, and one with obesity. Studies with diabetes were concerned with glycemic control during lockdown. In one study [17], Type 1 diabetes patients continued to exercise regularly as before during the lockdown period without significant variations, even if they were confined at home, and were able to maintain good glycemic control. Notwithstanding the reduction of performed PA, three [14–16] of such studies found an improvement in glycemic control. Two studies [12,13] found a reduction in both glycemic control and PA.

Obese subjects worsened their condition [19]: body mass increased significantly and was associated with a reduction on PA. In addition, patients with a cardiovascular disease [20] reduced the amount of performed PA, which could increase their health-related risk.

Concerning neuromuscular diseases, one study [18] found a worsening of the patients' condition, associated to a reduction in PA. One study [21] did not find significant changes before and during lockdown in the number of Parkinson's disease patients who practiced PA. Nonetheless, those who declared a worsening of their condition reported a lower amount of daily PA compared to those who did not experience a worsening, highlighting the potential protective effect of PA.

Four studies were carried out on students: one on high school [34] and three on university students [28,29,31]: all of them reported a reduction in PA and an increase in sedentary behavior.

All studies but one used self-reported data collected during some form of interview (telemedicine visits or telephone contact) or by means of online questionnaires. Only one study [20] used smart devices to automatically and directly collect the data. Another study [23] used smart devices to measure the number of daily performed steps, but measurements were self-reported by study participants. Concerning the questionnaire used to assess the level of PA, only eight studies (34.8%) administered the validated IPAQ or IPAQ-SF; Assaloni et al. [12] administered the Godin Leisure Time Exercise Questionnaire and the remainders did not use validated questionnaires.

In Di Corrado et al. [25], the amount of individuals who began PA during lockdown was significantly greater than those who stopped. Di Renzo et al. [26] reported an increase in the percentage of individuals who trained five or more times a week. The studies which quantified the amount of performed PA found a relevant reduction during lockdown, compared to pre-lockdown.

The results of the assessment of the quality of studies are detailed in Table 2; the mean NOS score of the included studies was 5.04 (SD = 1.36; range 2–8).

Table 2. NOS scores for all included studies (range: 0–10, with higher scores indicating better quality research).

REFERENCE	Representativeness of Sample	Sample Size	Non-Respondents	Ascertainment of the Exposure	Comparability	Assessment of the Outcome	Statistics	NOS Score
Assaloni et al. [12]	1	0	0	2	0	1	1	6
Barchetta et al. [13]	1	0	0	1	0	1	1	4
Barrea et al. [22]	1	1	0	1	0	1	1	5
Buoite Stella et al. [23]	1	0	0	2	0	1	1	5
Cancello et al. [24]	1	0	0	1	0	1	1	4
Capaldo et al. [14]	1	0	0	0	0	1	0	2
Caruso et al. [15]	1	0	0	0	0	1	1	3
Di Corrado et al. [25]	1	0	0	0	0	1	1	3
Di Renzo et al. [26]	1	1	1	1	1	1	1	7
Di Stefano et al. [18]	1	0	0	2	1	1	1	6
Ferrante et al. [27]	1	1	1	1	2	1	1	8
Gallè et al. [28]	1	1	0	2	0	1	1	6
Gallè et al. [29]	1	1	0	2	0	1	1	6
Giustino et al. [30]	1	1	0	1	0	1	1	5
Luciano et al. [31]	1	1	0	2	0	1	1	6
Maugeri et al. [32]	1	0	0	2	0	1	1	5
Pellegrini et al. [19]	1	0	0	1	0	1	1	4
Predieri et al. [16]	1	0	0	1	0	1	1	4
Raiola et al. [33]	1	1	0	1	0	1	1	5
Sassone et al. [20]	1	0	0	2	0	2	1	6
Schirinzi et al. [21]	1	0	0	2	0	1	1	5
Tornaghi et al. [34]	1	0	0	2	1	1	1	6
Tornese et al. [17]	1	0	0	1	1	1	1	5

4. Discussion

There is a lack of consistency in the findings of the selected studies, in terms of the amount of performed PA, even if a majority declared a significant reduction during lockdown compared to before, as expected. Overall, the two-month lockdown imposed as a consequence of the COVID-19 pandemic had a detrimental effect on general health in Italians, especially those with chronic conditions such as obesity and neurological diseases. Coping with such diseases has been a challenge, especially because of issues in supplies and lack of access to health facilities and healthcare providers. Overall, a reduction of PA was detected as a consequence of COVID-19 lockdown, with a worsening of health status.

An increase of PA was only reported by Di Corrado et al. [25] and Di Renzo [26]. The results of Di Renzo et al. [26] suggested that highly active people maintained or increased their PA level; Di Corrado et al. [25] reported an increase of PA beginners during lockdown. This may be that highly physically active people have reached an acknowledgement of healthy lifestyle which allows them to maintain it, including in critical conditions such as lockdown or confinement.

Studies on diabetes and glycemic control have shown contradictory results. Notwithstanding the diminished amount of performed PA reported by most patients, glycemic control improved, possibly because of a more regular lifestyle and daily timetable. It is also possible that the use of telemedicine (video and phone calls), which allows patients to conduct virtual visits, has been a useful support to manage this chronic disease [16]. Still, one study reported a reduction in glycemic control, and therefore the association between meal schedule and PA in the treatment of diabetes should be investigated further.

The role of diet during lockdown was examined, especially in studies on obese and diabetic subjects; in particular, Pellegrini et al. [19], in their study on obese people, highlighted that even if all the patients received personalized nutritional advice, they reported many unhealthy dietary habits, such eating more, not paying attention to the healthiness of the consumed food, consuming more sweets, more snacks, more frozen/canned foods, and less fruit and vegetables than before. Moreover, Tornese et al. [17], in a study on diabetic adolescents, found a negative change in eating behavior.

Concerning neuromuscular diseases, the subjects who practiced PA have not shown signs of worsening. Schirinzi et al. [21] pointed out that higher educational level and a mild motor impairment are significant predictors of daily PA in Parkinson's disease patients, since they seem to have a greater awareness of the importance of regular exercise and a more resilient response to COVID-19-related changes. Particular attention should be paid to involve more advanced, cognitively impaired, or uneducated patients who could be excluded from telecommunications.

The main limitation of most of the studies was that data were self-reported by the participants by means of a questionnaire, a web survey, or a phone interview. Only in one study [20] were all data collected automatically by means of electronic sensors. In another one [22], parts of the data were collected before lockdown by means of direct anthropometric measurements, while during lockdown, data were collected by means of a phone interview. The studies that reported the amount of PA are limited and heterogeneous, and this represents an obstacle in the interpretation of the results.

Schirinzi et al. [21], in their study regarding the use of available technology-based tools to assist physical exercise and its implications in such a critical period, reported that levels of MET did not differ between users and the nonusers, probably due to the old age of the sample. However, in our opinion, an increased adoption of remote monitoring systems should be envisaged to keep the amount of PA performed by both patients and healthy subjects under direct control, avoiding the collection of self-reported—i.e., non-observed—data. Data on PA can also be provided by popular fitness monitoring cell phone applications. There are at least two main advantages in their adoption: (i) a more objective evaluation of both PA and biomedical parameters, and (ii) the possibility for physicians to avoid close contact with suspected or infected patients, thus diminishing their exposure to the risk of contagion and improving the timeliness and continuity of their monitoring activities. This should be the case, especially, for subjects with a chronic condition.

In case of another lockdown, which may further diminish the general health of the population, simple workouts may be proposed in the form of instructional videos and/or online tutorials to people confined in their homes. In addition, personal training may be supplied by means of the remote supervision of fitness professionals, using videoconferencing tools. As one of the studies [26] which showed an increase in PA during lockdown suggested, the main choice for such workouts may be body-weight training (calisthenics), which can be easily performed at home without any equipment but a simple yoga mat. Yoga itself may be a good option, given the positive effects which it may have on inflammatory conditions and on the immune system.

Another interesting aspect that emerges from the 23 selected studies is linked to the psychology and motivation that drive PA: higher anxiety scores during the COVID-19 lockdown negatively influenced commitment to exercise. This may lead to practical implications when considering the confinement period that people are held at home for long periods of time, since psychological conditions (anxiety and depression in particular) must also be taken into consideration. Unfortunately, motivation to exercise seems to diminish in individuals who are confined at home, and home workouts do not have the same benefits as outdoor PA for the immune system [35]. Exposure to sunlight has proved to have important beneficial effects on the immune system response, including—of course—reaction to viral infections [36]. Subjects without access to a garden, courtyard, or terrace may be at a further disadvantage when confined at home.

This is especially important for older individuals, even in the case of a forthcoming vaccine. Infections have a greater incidence in the elderly [37], and their immune systems respond less effectively to vaccines, while physical exercise improves the effectiveness of the immune system response not only to infections but also to vaccinations [38]. Recent studies have found a positive effect of PA on patients with chronic conditions, including autoimmune diseases [39]. It must be highlighted that COVID-19 may lead to an intense and fatal cytokine response, akin to an autoimmune reaction.

As it is usually the case, any policy or measure which is enforced in order to diminish a risk (including a health-related risk, such as that of a viral infection) has, as a side effect, the potential of increasing another risk. In case of subsequent lockdowns, health-related risks, for both healthy individuals and patients with chronic diseases, may increase because of lack of PA and exposure to sunlight. In a seemingly paradoxical way, even infection-related risks may specifically increase. In fact, those measures which are put in place to diminish the spread of an infection in a population in the short run may hinder the capability of the population to react to the same infection—or even to other kinds of infections—in the long run.

The limitations of the selected studies can be summarized as follows. First, there was no consistency in the tools and methods of assessing PA, making the comparison of results difficult. Second, all studies but one [20] used self-reported questionnaires to assess the performed PA, and asked individuals about their pre-lockdown behavior retrospectively. These facts diminish both the accuracy and objectivity of the assessment. Finally, the selected samples were not representative of the population at a national level.

5. Conclusions

Physical activity is a prerequisite for the prevention and treatment of most medical conditions, especially metabolic, cardiovascular, and neurodegenerative diseases. Furthermore, it is an established fact that outdoor exercise, during sunlight hours, has a positive impact on the immune system, which is of great importance in case of a viral infection. It appears instead that during the outbreak of the pandemic, the whole Italian national health service focused solely and exclusively on the risk posed by the SARS-CoV-2 infection. While this is understandable in the short run, given the terrific burden that it put on hospitals and intensive care units, the negative effects that prolonged home confinement have on the global national health should not be neglected.

A general lockdown, imposed repeatedly and maintained in the long period—without giving alternative and viable options for the practice of physical activity, particularly outdoor—may lead to a worsening of the population general health, in both healthy subjects and those with a chronic condition, creating further medical emergencies and an increased burden on the national health service. People suffering from chronic diseases need special attention during a pandemic, and there should be some plan for them during lockdown to reduce the impact on their health. We suggest, therefore, that at least individual outdoor exercise should be allowed and promoted, especially during daylight hours, while maintaining physical distancing in case another lockdown will be enforced for the containment of current and future pandemics.

Further research is needed in order to assess the amount of performed PA in a more standardized and quantifiable way so that study results can be compared. In addition, it should be necessary to have more information about the use of technological tools and personalized and supervised PA, in order to value their effectiveness.

Author Contributions: Conceptualization, L.Z., D.B., and S.T.; methodology, L.Z., D.B., and S.T.; validation, L.Z., D.B., and S.T.; investigation, L.Z., D.B., and S.T.; data curation, L.Z., D.B., and S.T.; writing—original draft preparation, L.Z., D.B., and S.T.; writing—review and editing, L.Z., D.B., and S.T.; supervision, L.Z. All authors have read and agreed to the published version of the manuscript.

Funding: This research received no external funding.

Institutional Review Board Statement: Not applicable.

Informed Consent Statement: Not applicable.

Data Availability Statement: Data is contained within the article.

Conflicts of Interest: The authors declare no conflict of interest.

References

1. Bohn, B.; Herbst, A.; Pfeifer, M.; Krakow, D.; Zimny, S.; Kopp, F.; Melmer, A.; Steinacker, J.M.; Holl, R.W. DPV Initiative. Impact of Physical Activity on Glycemic Control and Prevalence of Cardiovascular Risk Factors in Adults With Type 1 Diabetes: A Cross-sectional Multicenter Study of 18,028 Patients. *Diabetes Care.* **2015**, *38*, 1536–1543. [CrossRef] [PubMed]
2. Riddell, M.C.; Gallen, I.W.; Smart, C.E.; Taplin, C.E.; Adolfsson, P.; Lumb, A.N.; Kowalski, A.; Rabasa-Lhoret, R.; McCrimmon, R.J.; Hume, C.; et al. Exercise management in type 1 diabetes: A consensus statement. *Lancet Diabetes Endocrinol.* **2017**, *5*, 377–390. [CrossRef]
3. Gielen, S.; Laughlin, M.H.; O'Conner, C.; Duncker, D.J. Exercise training in patients with heart disease: Review of beneficial effects and clinical recommendations. *Prog. Cardiovasc. Dis.* **2015**, *57*, 347–355. [CrossRef] [PubMed]
4. Wang, L.; Ai, D.; Zhang, N. Exercise Benefits Coronary Heart Disease. *Adv. Exp. Med. Biol.* **2017**, *1000*, 3–7. [PubMed]
5. Anziska, Y.; Inan, S. Exercise in neuromuscular disease. *Semin. Neurol.* **2014**, *34*, 542–556. [CrossRef] [PubMed]
6. Wu, P.L.; Lee, M.; Huang, T.T. Effectiveness of physical activity on patients with depression and Parkinson's disease: A systematic review. *PLoS ONE* **2017**, *27*, e0181515. [CrossRef]
7. Lauzé, M.; Daneault, J.F.; Duval, C. The Effects of Physical Activity in Parkinson's Disease: A Review. *J. Parkinsons Dis.* **2016**, *19*, 685–698. [CrossRef]
8. Duggal, N.A.; Niemiro, G.; Harridge, S.; Simpson, R.J.; Lord, J.M. Can physical activity ameliorate immunosenescence and thereby reduce age-related multi-morbidity? *Nat. Rev. Immunol.* **2019**, *19*, 563–572. [CrossRef]
9. Moher, D.; Liberati, A.; Tetzlaff, J.; Altman, D.G. The PRISMA Group. Preferred reporting items for systematic reviews and meta-analyses: The PRISMA statement. *PLoS Med.* **2009**, *6*, e1000097. [CrossRef]
10. Wells, G.A.; Shea, B.; O'Connell, D. *The Newcastle-Ottawa Scale (NOS) for Assessing the Quality of Non Randomised Studies in Meta-Analyses*; Ottawa Hospital Research Institute: Ottawa, ON, Canada, 2009.
11. Modesti, P.A.; Reboldi, G.; Cappuccio, F.P.; Agyemang, C.; Remuzzi, G.; Rapi, S.; Perruolo, E.; Parati, G.; ESH Working Group on CV Risk in Low Resource Settings. Panethnic Differences in Blood Pressure in Europe: A Systematic Review and Meta-Analysis. *PLoS ONE* **2016**, *11*, e0147601. [CrossRef]
12. Assaloni, R.; Pellino, V.C.; Puci, M.V.; Ferraro, O.E.; Lovecchio, N.; Girelli, A.; Vandoni, M. Coronavirus disease (Covid-19): How does the exercise practice in active people with type 1 diabetes change? A preliminary survey. *Diabetes Res. Clin. Pract.* **2020**, *166*, 108297. [CrossRef] [PubMed]

13. Barchetta, I.; Cimini, F.A.; Bertoccini, L.; Ceccarelli, V.; Spaccarotella, M.; Baroni, M.G.; Cavallo, M.G. Effects of work status changes and perceived stress on glycaemic control in individuals with type 1 diabetes during COVID-19 lockdown in Italy. *Diabetes Res. Clin. Pract.* **2020**, *170*, 108513. [CrossRef] [PubMed]

14. Capaldo, B.; Annuzzi, G.; Creanza, A.; Giglio, C.; De Angelis, R.; Lupoli, R.; Masulli, M.; Riccardi, G.; Rivellese, A.A.; Bozzetto, L. Blood Glucose Control During Lockdown for COVID-19: CGM Metrics in Italian Adults With Type 1 Diabetes. *Diabetes Care* **2020**, *43*, e88–e89. [CrossRef]

15. Caruso, I.; Di Molfetta, S.; Guarini, F.; Giordano, F.; Cignarelli, A.; Natalicchio, A.; Perrini, S.; Leonardini, A.; Giorgino, F.; Laviola, L. Reduction of hypoglycaemia, lifestyle modifications and psychological distress during lockdown following SARS-CoV-2 outbreak in type 1 diabetes. *Diabetes Metab. Res. Rev.* **2020**, e3404. [CrossRef] [PubMed]

16. Predieri, B.; Leo, F.; Candia, F.; Lucaccioni, L.; Madeo, S.F.; Pugliese, M.; Vivaccia, V.; Bruzzi, P.; Iughetti, L. Glycemic Control Improvement in Italian Children and Adolescents With Type 1 Diabetes Followed Through Telemedicine During Lockdown Due to the COVID-19 Pandemic. *Front. Endocrinol.* **2020**, *11*, 595735. [CrossRef] [PubMed]

17. Tornese, G.; Ceconi, V.; Monasta, L.; Carletti, C.; Faleschini, E.; Barbi, E. Glycemic Control in Type 1 Diabetes Mellitus During COVID-19 Quarantine and the Role of In-Home Physical Activity. *Diabetes Technol Ther.* **2020**, *22*, 462–467. [CrossRef]

18. Di Stefano, V.; Battaglia, G.; Giustino, V.; Gagliardo, A.; D'Aleo, M.; Giannini, O.; Palma, A.; Brighina, F. Significant reduction of physical activity in patients with neuromuscular disease during COVID-19 pandemic: The long-term consequences of quarantine. *J. Neurol.* **2021**, *268*, 20–26. [CrossRef]

19. Pellegrini, M.; Ponzo, V.; Rosato, R.; Scumaci, E.; Goitre, I.; Benso, A.; Belcastro, S.; Crespi, C.; De Michieli, F.; Ghigo, E.; et al. Changes in weight and nutritional habits in adults with obesity during the "lockdown" period caused by the COVID-19 virus emergency. *Nutrients* **2020**, *12*, 2016. [CrossRef]

20. Sassone, B.; Mandini, S.; Grazzi, G.; Mazzoni, G.; Myers, J.; Pasanisi, G. Impact of COVID-19 Pandemic on Physical Activity in Patients With Implantable Cardioverter-Defibrillators. *J. Cardiopulm. Rehabil. Prev.* **2020**, *40*, 285–286. [CrossRef]

21. Schirinzi, T.; Di Lazzaro, G.; Salimei, C.; Cerroni, R.; Liguori, C.; Scalise, S.; Alwardat, M.; Mercuri, N.B.; Pierantozzi, M.; Stefani, A.; et al. Physical activity changes and correlate effects in patients with Parkinson's disease during COVID-19 lockdown. *Mov. Disord. Clin. Pract.* **2020**, *7*, 797–802. [CrossRef]

22. Barrea, L.; Pugliese, G.; Framondi, L.; Di Matteo, R.; Laudisio, D.; Savastano, S.; Colao, A.; Muscogiuri, G. Does Sars-Cov-2 threaten our dreams? Effect of quarantine on sleep quality and body mass index. *J. Transl. Med.* **2020**, *18*, 318. [CrossRef]

23. Buoite Stella, A.; AjČeviĆ, M.; Furlanis, G.; Cillotto, T.; Menichelli, A.; Accardo, A.; Manganotti, P. Smart technology for physical activity and health assessment during COVID-19 lockdown. *J. Sports Med. Phys. Fitness.* **2021**, *61*, 452–460. [CrossRef] [PubMed]

24. Cancello, R.; Soranna, D.; Zambra, G.; Zambon, A.; Invitti, C. Determinants of the Lifestyle Changes during COVID-19 Pandemic in the Residents of Northern Italy. *Int J. Environ. Res. Public Health* **2020**, *17*, 6287. [CrossRef] [PubMed]

25. Di Corrado, D.; Magnano, P.; Muzii, B.; Coco, M.; Guarnera, M.; De Lucia, S.; Maldonato, N.M. Effects of social distancing on psychological state and physical activity routines during the COVID-19 pandemic. *Sport Sci. Health* **2020**, 1–6. [CrossRef]

26. Di Renzo, L.; Gualtieri, P.; Pivari, F.; Soldati, L.; Attinà, A.; Cinelli, G.; Leggeri, C.; Caparello, G.; Barrea, L.; Scerbo, F.; et al. Eating habits and lifestyle changes during COVID-19 lockdown: An Italian survey. *J. Transl Med.* **2020**, *18*, 229. [CrossRef] [PubMed]

27. Ferrante, G.; Camussi, E.; Piccinelli, C.; Senore, C.; Armaroli, P.; Ortale, A.; Garena, F.; Giordano, L. Did social isolation during the SARS-CoV-2 epidemic have an impact on the lifestyles of citizens?. L'isolamento sociale durante l'epidemia da SARS-CoV-2 ha avuto un impatto sugli stili di vita dei cittadini? *Epidemiol. Prev.* **2020**, *44* (Suppl. S2), 353–362. [PubMed]

28. Gallè, F.; Sabella, E.A.; Da Molin, G.; De Giglio, O.; Caggiano, G.; Di Onofrio, V.; Ferracuti, S.; Montagna, M.T.; Liguori, G.; Orsi, G.B.; et al. Understanding Knowledge and Behaviors Related to CoViD-19 Epidemic in Italian Undergraduate Students: The EPICO Study. *Int J. Environ. Res. Public Health* **2020**, *17*, 3481. [CrossRef]

29. Gallè, F.; Sabella, E.A.; Ferracuti, S.; De Giglio, O.; Caggiano, G.; Protano, C.; Valeriani, F.; Parisi, E.A.; Valerio, G.; Liguori, G.; et al. Sedentary Behaviors and Physical Activity of Italian Undergraduate Students during Lockdown at the Time of CoViD-19 Pandemic. *Int J. Environ. Res. Public Health* **2020**, *17*, 6171. [CrossRef]

30. Giustino, V.; Parroco, A.M.; Gennaro, A.; Musumeci, G.; Palma, A.; Battaglia, G. Physical activity levels and related energy expenditure during COVID-19 quarantine among the sicilian active population: A cross-sectional online survey study. *Sustainability* **2020**, *12*, 4356. [CrossRef]

31. Luciano, F.; Cenacchi, V.; Vegro, V.; Pavei, G. COVID-19 lockdown: Physical activity, sedentary behaviour and sleep in Italian medicine students. *Eur. J. Sport Sci.* **2020**, 1–10. [CrossRef]

32. Maugeri, G.; Castrogiovanni, P.; Battaglia, G.; Pippi, R.; D'Agata, V.; Palma, A.; Di Rosa, M.; Musumeci, G. The impact of physical activity on psychological health during Covid-19 pandemic in Italy. *Heliyon* **2020**, *6*, e04315. [CrossRef]

33. Raiola, G.; Aliberti, S.; Esposito, G.; Altavilla, G.; D'Isanto, T.; D'Elia, F. How has the practice of physical activity changed during the covid-19 quarantine? a preliminary survey. *Teor. Metod. Fiz. Vihov.* **2020**, *20*, 242–247. [CrossRef]

34. Tornaghi, M.; Lovecchio, N.; Vandoni, M.; Chirico, A.; Codella, R. Physical activity levels across COVID-19 outbreak in youngsters of Northwestern Lombardy. *J. Sports Med. Phys. Fitness* **2020**. [CrossRef]

35. Chirico, A.; Lucidi, F.; Galli, F.; Giancamilli, F.; Vitale, J.; Borghi, S.; La Torre, A.; Codella, R. COVID-19 Outbreak and Physical Activity in the Italian Population: A Cross-Sectional Analysis of the Underlying Psychosocial Mechanisms. *Front. Psychol.* **2020**, *11*, 2100. [CrossRef]

36. Hart, P.H.; Gorman, S.; Finlay-Jones, J.J. Modulation of the immune system by UV radiation: More than just the effects of vitamin D? *Nat. Rev. Immunol.* **2011**, *19*, 584–596. [CrossRef] [PubMed]
37. Dewan, S.K.; Zheng, S.B.; Xia, S.J.; Bill, K. Senescent remodeling of the immune system and its contribution to the predisposition of the elderly to infections. *Chin. Med. J.* **2012**, *125*, 3325–3331. [PubMed]
38. Simpson, R.J.; Lowder, T.W.; Spielmann, G.; Bigley, A.B.; LaVoy, E.C.; Kunz, H. Exercise and the aging immune system. *Ageing Res. Rev.* **2012**, *11*, 404–420. [CrossRef]
39. Sharif, K.; Watad, A.; Bragazzi, N.L.; Lichtbroun, M.; Amital, H.; Shoenfeld, Y. Physical activity and autoimmune diseases: Get moving and manage the disease. *Autoimmun. Rev.* **2018**, *17*, 53–72. [CrossRef]

International Journal of
*Environmental Research
and Public Health*

Article

Continuous Compared to Accumulated Walking-Training on Physical Function and Health-Related Quality of Life in Sedentary Older Persons

Pablo Monteagudo [1,2,*], Ainoa Roldán [1,3], Ana Cordellat [1,3], Mari Carmen Gómez-Cabrera [4] and Cristina Blasco-Lafarga [1,3,*]

1   Sport Performance and Physical Fitness Research Group (UIRFIDE), University of Valencia, 46010 Valencia, Spain; Ainoa.Roldan@uv.es (A.R.); Ana.Cordellat@uv.es (A.C.)
2   Department of Education and Specific Didactics, Jaume I University, 12071 Castellon, Spain
3   Physical Education and Sports Department, University of Valencia, 46010 Valencia, Spain
4   Freshage Research Group, Department of Physiology, faculty of Medicine, University of Valencia, CIBERFES, Fundación Investigación Hospital Clínico Universitario/INCLIVA, 46010 Valencia, Spain; Carmen.Gomez@uv.es
*   Correspondence: pmonteag@uji.es (P.M.); M.Cristina.Blasco@uv.es (C.B.-L.); Tel.: +34-696-030-938 (P.M.); +34-658-591-854 (C.B.-L.)

Received: 22 July 2020; Accepted: 18 August 2020; Published: 20 August 2020

Abstract: The present study aimed to analyze the impact of overground walking interval training (WIT) in a group of sedentary older adults, comparing two different dose-distributions. In this quasi-experimental and longitudinal study, we recruited twenty-three sedentary older adults (71.00 ± 4.10 years) who were assigned to two groups of WIT. The continuous group (CWIT) trained for 60 min/session in the morning, while the accumulated group (AWIT) performed the same duration and intensity of exercise, but it was distributed twice a day (30 min in the morning and 30 more in the afternoon). After 15 weeks of an equal external-load training (3 days/week), Bonferroni post-hoc comparisons revealed significant ($p < 0.050$) and similar large improvements in both groups in cardiorespiratory fitness and lower limb strength; even larger gains in preferred walking speed and instrumental daily life activity, which was slightly superior for CWIT; and improvements in agility, which were moderate for CWIT and large for AWIT. However, none of the training protocols had an impact on the executive function in the individuals, and only the AWIT group improved health-related quality of life. Although both training protocols induced a general significant improvement in physical function in older adults, our results showed that the accumulative strategy should be recommended when health-related quality of life is the main target, and the continuous strategy should be recommended when weakness may be a threat in the short or medium term.

Keywords: physical activity; public health; aging; dose-response; cardiorespiratory fitness; agility test; executive function; strength; older adults

---

1. Introduction

Increasing evidence suggests that prolonged sedentary bouts [1,2] and/or few breaks in sedentary time [3,4] may damage metabolic health, independently of total sedentary time and moderate–vigorous physical activity (PA) [5]. High volumes of accumulated sedentary time contribute to increase the risk of cardiovascular disease, certain cancers, and premature mortality [6]. It is also associated with chronic, sterile, low-grade inflammation underlying the pathogenesis of many age-related diseases, described as inflammaging in older adults [7]. Since PA diminishes with age [8], sedentary time has become a world leading cause of death and disability according to the World Health Organization [9]. Similarly, research evidence confirms the negative impact of sedentary behaviors on physical fitness, especially in older people [10,11]. This is associated with the development of functional limitations, independently of PA levels [12]. Sedentary behaviors represent thus a serious public health problem in these older individuals.

Conversely, regular PA positively influences almost every human's physiologic system or psychological aspect [13–15]. Exercise training is one of the most effective strategies for decreasing the likelihood of sedentary lifestyles and age-related diseases, thereby promoting independence in daily life activities and enhancing the quality of life in the growing older adult populations of many countries [16]. In this scenario, walking or brisk walking is the main example of moderate-intensity activity recommended by public health guidelines [17], because it is most frequently chosen by seniors [18], it requires minimal equipment, and it offers opportunities for friendship and social support [19]. Walking interventions, apart from being effective in increasing PA [20], are related to the prevention of cognitive decline [21] and to the improvement of health-related quality of life in older people [22]. Moreover, this kind of exercise may be a convenient activity to circumvent barriers to training, since it can be performed in the individual's proximity zone, and it can be performed at a variety of intensities or modalities, either alone or in a group [23].

Classic proposals in walking programs to improve physical fitness in older persons have been established on the following basis: moderate intensity (5–6 on a 10 point scale where all-out effort is valued with a 10), minimum of 30 min, five days a week [24]. However, experimental findings have demonstrated that moderate-to-vigorous PA, accumulated in short bouts (>10 min) and totaling at least 30 min in duration, may be as effective as longer bouts in improving some disease risk factors, like plasma lipid profile, fasting plasma insulin levels, or body composition [25,26]. As a consequence, many PA guidelines have evolved to incorporate this recommendation [17,27]. In addition, some authors have proposed that PA guidelines in older adults should take into account the physical–behavioral binomial [28], applying the training principles based on people's activity or sedentary patterns. In this way, exercise interventions in senior individuals should maintain active morning bouts and reduce sedentary behaviors in the afternoon and evening hours [29]. Accumulative exercise (distributing exercise training in the morning and in the afternoon) could be a good strategy to break and reduce sedentary time, especially in the less active time-slots. Furthermore, to our knowledge, no-studies have investigated whether accumulative proposals convey some advantage compared to similar doses of continuous exercise regarding physical function in sedentary older people. In addition, it remains unknown whether accumulated training enhances, more than does continuous training, the benefits in functional outcomes or health-related quality of life.

The aim of the present study was to analyze the impact of an overground walking program on physical function and health-related quality of life in a group of sedentary older adults comparing two different dose-distributions (accumulative versus continuous). Given the analyzed changes in body composition [25], we hypothesize that both strategies, tailored and periodized regarding their capacities, will be effective with differences in effects size.

2. Materials and Methods

2.1. Participants

Older adults from the Health Care Centre of Buñol (Manises Hospital area) were recruited to participate in the present study, which was approved by the ethics committee of the University of Valencia (H1484058781638). Inclusion criteria were as follows: ≥60 years old and fit to participate in a regular exercise program according to the medical referral; and sedentary (no participation in a regular exercise program or intentional activities beyond normal daily habits within the previous 4 months), and reporting a gait speed higher than 0.6 m/s. Exclusion criteria were presence of any disorder that would prevent the participant from being able to complete a training program, missing 4 or more consecutive training sessions, and adherence lower than 75% to the training sessions. A total of thirty-five older adults were screened, but only twenty-seven individuals met the inclusion criteria and signed the written informed consent. Participants were homogeneously stratified into 2 groups in terms of age, gender, body mass index (BMI), and gait speed in 6 m (this last categorized according to the *"Practical Guide for Prescribing a Multi-Component Physical Training Program to prevent weakness and falls in People over 70"*) [30]. Since three men and one woman did not complete the investigation for reasons not related to the study, 23 participants ($N = 23$; 71.0 ± 4.1 years; 75.6 ± 13.1 kg; 10 female) were included in the statistical analyses.

2.2. Research Design

The participants were enrolled to participate in this quasi-experimental and longitudinal study in January 2017. During February 2017, a baseline multidisciplinary team assessment of each participant was performed, comprising the gathering of relevant demographic and biological information, functional ability, health-related quality of life, executive function, and instrumental activities of daily living. Participants were assigned to two groups of supervised and tailored walking interval training (WIT) for 15 weeks. The continuous walking interval training (CWIT) group trained for 60 min/session, always in the morning, while the accumulated walking interval training (AWIT) group performed exactly the same duration and intensity of exercise, but distributed twice a day (30 min in the morning and 30 more in the afternoon) with at least 5-h separating each exercise bout. WIT programs started in March 2017. After the 15 weeks had ended (June 2017), all the participants were re-assessed by repeating the initial protocol.

2.3. Walking Interval Program (WIT)

All participants in the WIT program trained 3 times a week for 15 weeks (divided in 7 + 7 with a week of rest between weeks 7 and 8). Sets, repetitions, intensity, duration of work, and active recovery intervals are reported in Table 1. As previously described by our research group [25], intervals and intensities were increased and scheduled considering the rating of perceived effort (RPE 1–10) and adjusted by heart rate (HR) monitoring. Participants were instructed to walk close to the programmed RPE, reinforcing the first sessions with some RPE familiarization tasks. In order to control their HR (target: $\leq 80\%$ HR_{max}), they were also provided with a Beurer PM-15 HR monitor, without chest strap (Beurer, Ulm, Germany), and one individualized card with the HR estimated for every RPE zone. All sessions began with a brief warm-up period and ended with a cool down including breathing, stability, and joints mobility exercises.

Table 1. Walking Interval Training.

Week	Session	Session Description (Sets × Repetitions (Work + Recovery))	Total Session Duration (min)	Week	Session	Session Description (Sets × Repetitions (Work + Recovery))	Total Session Duration (min)
Week 1	1	* 2 × 5 (2 min 4 RPE + 2 min 2 RPE)	40	Week 9	20	1 × 7 (2 min 6 RPE + 4 min 4 RPE)	42
Week 1	2	1 × 10 (2 min 4 RPE + 2 min 2 RPE)	40	Week 9	21	1 × 10 (2 min 6 RPE + 2 min 4 RPE)	40
	3	1 × 10 (2 min 4 RPE + 1.5 min 2 RPE)	35		22	1 × 5 (2 min 7 RPE + 3 min 4 RPE)	45
	4	1 × 5 (4 min 4 RPE + 4 min 2 RPE)	40	Week 10	23	1 × 5 (2 min 7 RPE + 3 min 4 RPE)	40
Week 2	5	1 × 5 (4 min 4 RPE + 4 min 3 RPE)	35		24	1 × 5 (3 min 6 RPE + 3 min 4 RPE)	37.5
	6	1 × 7 (4 min 4 RPE + 4 min 2 RPE)	42		25	1 × 6 (2 min 7 RPE + 1.5 min 4 RPE)	42
Week 3	7	* 2 × 5 (2 min 5 RPE + 2 min 3 RPE)	40	Week 11	26	1 × 7 (3 min 7 RPE + 3 min 7 RPE)	42
Week 3	8	1 × 10 (2 min 5 RPE + 2 min 3 RPE)	40		27	1 × 10 (2 min 7 RPE + 1 min 5 RPE)	40
	9	1 × 5 (4 min 5 RPE + 4 min 3 RPE)	40		28	1 × 6 (4 min 6 RPE + 2 min 4 RPE)	42
Week 4	10	1 × 6 (4 min 5 RPE + 3 min 3 RPE)	42	Week 12	29	1 × 7 (2 min 7 RPE + 1 min 5 RPE)	42
	11	1 × 8 (4 min 5 RPE + 2 min 3 RPE)	48		30	1 × 8 (2.5 min 7 RPE + 3 min 4 RPE)	44
Week 5	12	1 × 10 (2 min 6 RPE + 2 min 4 RPE)	40	Week 13	31	1 × 5 (4 min 6 RPE + 1.5 min 5 RPE)	45
Week 5	13	1 × 10 (2 min 6 RPE + 2 min 4 RPE)	40	Week 13	32	1 × 5 (3 min 6 RPE + 3 min 4 RPE)	40
	14	1 × 5 (4 min 6 RPE + 4 min 4 RPE)	40		33	1 × 6 (2 min 7 RPE + 1.5 min 4 RPE)	42
Week 6	15	1 × 6 (4 min 6 RPE + 3 min 4 RPE)	42	Week 14	34	1 × 5 (3 min 7 RPE + 3 min 7 RPE)	40
Week 6	16	1 × 8 (4 min 6 RPE + 2 min 4 RPE)	48		35	1 × 6 (3 min 7 RPE + 1 min 5 RPE)	42
	17	1 × 9 (2 min 7 RPE + 2 min 4 RPE)	36		36	1 × 7 (4 min 7 RPE + 3 min 5 RPE)	42
Week 7	18	1 × 10 (2 min 7 RPE + 2 min 4 RPE)	40	Week 15	37	1 × 7 (3 min 7 RPE + 2.5 min 5 RPE)	38.5
Week 7	19	1 × 10 (2 min 7 RPE + 2 min 4 RPE)	40		38	1 × 8 (2 min 7 RPE + 2 min 5 RPE)	40

* When two blocks were performed, 1 to 3 min of break were considered in order to drink water and rest. RPE: the rating of perceived effort.

2.4. Outcomes

We assessed different functional ability and psychosocial parameters before and after the intervention by the following tests and questionnaires.

2.4.1. Grip Strength (GS)

Grip strength (GS) was evaluated by the Takei 5401 adaptable dynamometer (Takei Scientific Instruments CO., LTD, Tokyo, Japan). Following previous protocol [31], the contraction was maintained for 5 s, with the arm stretched along the body. Two measurements were taken on each side, with 1-min rest between them, and the best value was considered for the final analysis.

2.4.2. Six Minute Walk Test (6MWT)

Cardiorespiratory fitness was evaluated with the six minute walk test (6MWT), according to the standard protocol [32], in a walking course of 30 m. Participants walked as fast as possible for 6 min, without running, being encouraged in each lap. They were warned of the time at 3 and 5 min.

2.4.3. Five Times Sit-To-Stand Test (FTSST)

The five times sit-to-stand test (FTSST) was used to assess lower limb strength [33,34]. To perform the test, the subject was instructed to cross both arms across his/her chest and stand up completely, then sit and stand up a total of five times as quickly and safely as possible. Video recordings of the test were analyzed with the sports analysis video player software Kinovea (https://www.kinovea.org). Timing began when the subject's buttocks took off from the seat and stopped when they returned to the seat after the fifth repetition. The test was performed once. If the subject did not perform part of the test correctly, he/she was stopped immediately by the investigator, and the test was restarted. The subject was closely guarded by the investigator to ensure the correct performance of the test and to prevent any injurious events.

2.4.4. Preferred Walking Speed (PWS)

Preferred walking speed (PWS), also known as "most comfortable" or "self-paced" walking speed, was determined over ground on a 4.5 m walkway, using a system of two electric photocells by means of the Chronojump Software (Velleman PEM10D photocell, Cronojump Bosco System, response time 5–100 ms). Participants completed the distance (without acceleration but with a 2 m deceleration zone) walking at a comfortable and usual pace, and the mean of three attempts was taken as the PWS. This PWS was evaluated in the screening phase and after the intervention.

2.4.5. Timed Up and Go Test (TUG)

We recorded the time that the participants took to rise from a chair, walk three meters, turn around, and walk back to the seated position. Every participant repeated the timed up and go test (TUG) three times and, when necessary, a rest period of up to 1 min was allowed in between tests. The stopwatch was started on the command "Go", and the time was stopped when the test subject's buttocks touched the chair seat again. The fastest of 3 timed trials were used for the reported testing. The fastest test was selected, and the participants did not receive verbal encouragement during the protocol [35,36].

2.4.6. Executive Function

Executive function was assessed through the Stroop Color and Word Test [37]. This test consists of three parts that provide information on reading ability and psychomotor speed executive function, and that allow interference to be found in order to control the possible contaminating effect of the first 2 parts. Thus, the interference (IN) was used as a representative value of the executive function, according to the formula proposed by other authors [38].

2.4.7. Instrumental Activities of Daily Living (IADL)

The VIDA questionnaire was used to assess instrumental activities of daily living (IADL). This questionnaire assesses the autonomous realization of 10 activities, using a Likert scale with 3–4 responses. The total summative score can range between 10 and 38 points. The VIDA questionnaire correlates well with the results of other tests assessing functioning like TUG and the Lawton and Brody scale [39,40].

2.4.8. Health-Related Quality of life

The EQ-5D-5L was used to assess the health related quality of life [41]. This questionnaire has two parts: the EQindex, a descriptive profile that can be converted into an index-summary which defines health in terms of 5 dimensions (mobility, self-care, daily activities, pain/discomfort, and anxiety/depression); and the EQVAS, where respondents rate their overall health using a vertical visual analog scale from 0 to 100.

2.4.9. Other Variables

The following parameters were also collected during the study: age, sex, weight, height, blood pressure, oxygen saturation, and HR. Arterial oxygen saturation (SpO_2) and HR were determined with a pulse-oximeter attached to the fourth finger of the left hand (WristOx2-3150; Nonin, Plymouth, MN, USA), in a sitting position. Blood pressure was measured on the left arm with an Omron M3 Intellisense (HEM-7051-E) (Omron Healthcare, Kyoto, Japan) tensiometer. Standing height (m) was registered by means of a stadiometer (SECA 222, Hamburg, Germany). Subjects were measured without shoes with arms at their sides, looking straight, with knees together, and heels together. Shoulder blades, buttocks, and heels touched the measuring board. Measurement was taken at maximum inspiration with the head positioned in the Frankfort horizontal plane. After this, body weight (kg) was also registered by bioimpedance (TANITA, model BC-545N, Tokyo, Japan). Participants were weighed in light clothing controlling food intake in the previous hours to reproduce the evaluation conditions. BMI was calculated by dividing weight (kg) by height squared (m^2).

2.5. Statistical Analysis

The analysis of the data was performed with the SPSS statistics package version 23 (IBM SPSS Statistics for Windows, Chicago, IL, USA). After testing for normality (Shapiro–Wilks), Student's *t* test or the Mann–Whitney U test (SpO_2) were first applied for baseline group comparisons. A repeated measures ANOVA was then conducted to analyze changes in health-related quality of life and functional measures, considering the main effect of the intervention (pre-post overall comparison) and the interaction between type*dose-distribution (CWIT vs. AWIT). Within-subjects effects tests at the first level, followed by Bonferroni post-hoc tests, were performed with statistical significance set at the level of $p \leq 0.05$. Later on, in order to homogenize and analyze these changes, the effect size (ES) was calculated by means of the Cohen's d, where the effect was considered small (d = 0.20–0.40), medium (d = 0.50–0.70), or large (d = 0.80–2.0) according to Cohen [42]. Descriptive statistics were expressed as mean ± standard deviation (SD).

Changes in functional and health-related quality of life variables were further expressed as percentage of change (calculated by means of the formula (post-score − pre-score)/pre-score × 100). Student's t test or the Mann–Whitney U test were applied looking for group comparisons within deltas. Individual variables were checked for homogeneity of variance using Levene's test. When differences were found in any functional variable, Spearman correlations were performed considering BMI to give light to these changes.

3. Results

Table 2 includes the baseline characteristics of the participants. The CWIT and AWIT groups were homogenous at the baseline features, with no statistically significant differences observed in terms of age, gender, weight, height, BMI, systolic blood pressure (SBP), diastolic blood pressure (DBP), SpO_2, and HR. Functional and health-related quality of life outcomes were neither different at baseline ($p > 0.05$). Participants completed the intervention with an adherence rate of 83.75%.

Table 2. Physical characteristics at baseline.

	Total, $N = 23$	CWIT, $N = 11$	AWIT, $N = 12$
Age, years	71.0 ± 4.1	71.7 ± 3.3	70.3 ± 4.7
Weight, kg	75.9 ± 13.2	71.5 ± 11.1	79.9 ± 14.2
Height, m	1.6 ± 0.1	1.6 ± 0.1	1.6 ± 0.1
BMI, kg/m^2	29.1 ± 4.0	27.8 ± 3.1	30.3 ± 4.5
SBP, mmHg	152.4 ± 16.0	151.4 ± 14.4	153.4 ± 17.9
DBP, mmHg	82.5 ± 10.5	83.4 ± 9.5	81.7 ± 11.7
SpO_2, %	94.4 ± 4.5	95.3 ± 2.6	93.6 ± 5.7
HR, bpm	73.4 ± 10.7	75.2 ± 12.4	71.7 ± 9.0
Gender			
Females, % (n)	39.1 (9)	45.5 (5)	33.3 (4)
Males, % (n)	60.9 (14)	54.5 (6)	66.7 (8)

BMI: body mass index; SBP: systolic blood pressure; DBP: diastolic blood pressure; SpO_2: arterial oxygen saturation; HR: heart rate. CWIT: Continuous Walking Interval Training; AWIT: Accumulated Walking Interval Training.

The within-subjects effects test showed a significant main effect of "intervention" ($p \leq 0.05$), but there was not effect of the interaction "intervention × dose-distribution" strategy for any variable. Only PWS, as revealed by a trend to significance ($p < 0.100$) (Table 3).

Table 3. Tests of within-subjects' effects.

	Variable	Type III Sum of Squares	df	Mean Square	F	p	Partial Eta Squared
Intervention	BMI	4.906	1	4.906	7.875	0.011 *	0.273
	GS	42.13	1	42.13	6.72	0.017 *	0.243
	FTSST	80.29	1	80.29	21.07	0.001 **	0.501
	6MWT	65,140.41	1	65,140.41	78.48	0.001 **	0.789
	PWS	0.68	1	0.68	90.58	0.001 **	0.812
	TUG	7.22	1	7.22	20.17	0.001 **	0.490
	IN	1.69	1	1.69	0.04	0.836	0.002
	IADL	122.66	1	122.66	66.35	0.001 **	0.760
	EQindex	0.04	1	0.04	4.33	0.050 *	0.171
	EQVAS	145.52	1	145.52	1.44	0.243	0.064
Intervention × Dose-distribution	BMI	0.045	1	0.045	0.073	0.790	0.003
	GS	0.63	1	0.63	0.10	0.755	0.005
	FTSST	2.85	1	2.85	0.75	0.397	0.034
	6MWT	15.62	1	15.62	0.02	0.892	0.001
	PWS	0.03	1	0.03	3.63	0.070 †	0.147
	TUG	0.47	1	0.47	1.32	0.263	0.059
	IN	1.99	1	1.99	0.05	0.822	0.002
	IADL	0.39	1	0.39	0.21	0.648	0.010
	EQindex	0.01	1	0.01	0.78	0.387	0.036
	EQVAS	41.17	1	41.17	0.41	0.530	0.019

BMI: body mass index; GS: grip strength; FTSST: five times sit to stand test; 6MWT: 6 min walk test; PWS: preferred walking speed; TUG: timed up and go; IN: interference; IADL: instrumental activities of daily living; EQindex: descriptive index of Euroqol; EQVAS: visual analogue scale of Euroqol. ** $p \leq 0.001$; * $p \leq 0.050$; † $p \leq 0.100$.

Regarding the main effect of the intervention, in the whole sample (Table 3), the repeated-measures ANOVA showed significant improvements in lower limb strength (FTSST: 11.77 ± 3.36 vs. 9.10 ± 1.34 s), cardiorespiratory fitness (6MWT: 522.26 ± 64.60 vs. 597.54 ± 80.28 m), PWS (1.16 ± 0.19 vs. 1.40 ± 0.15 m/s), agility (TUG: 7.40 ± 1.28 vs. 6.60 ± 0.89 s), autonomy (IADL: 32.83 ± 2.64 vs. 36.09 ± 1.76), health-related quality of life (EQindex: 0.85 ± 0.14 vs. 0.91 ± 0.09), and BMI (29.11 ± 3.99

vs. 28.45 ± 3.86). Significant decrements were found in GS (34.65 ± 9.55 vs. 32.72 ± 10.21 kg). Non-significant changes were found for executive function or EQVAS.

Further Bonferroni analyses of pre-post differences (Table 4) showed significant improvements for both strategies on FTSST, 6MWT, PWS, TUG and IADL with large and moderate ES. The 6MWT and FTSST showed similar large ES (d ≈ 1.0), while PWS and IADL showed large ES with a slight and superior difference for the CWIT group. Conversely, the AWIT group revealed a large ES for the TUG (d = 1.00), while the CWIT group presented a moderate ES (d = 0.49).

Table 4. Measures for continuous (*n* = 11) and accumulated groups (*n* = 12).

	Pre-CWIT	Post-CWIT	ES	Pre-AWIT	Post-AWIT	ES
BMI, kg/m^2	27.8 ± 3.1	27.3 ± 3.1 †	0.2	30.3 ± 4.5	29.5 ± 4.3 *	0.2
GS, kg	33.1 ± 9.1	31.4 ± 10.1	0.2	36.1 ± 10.1	33.93 ± 10.6 *	0.2
FTSST, s	10.9 ± 2.2	8.7 ± 1.5 *	1.1	12.6 ± 4.1	9.5 ± 1.2 **	1.0
6MWT, m	529.9 ± 68.5	606.4 ± 80.6 **	1.0	515.2 ± 63.0	589.4 ± 82.6 **	1.0
PWS, m/s	1.1 ± 0.2	1.4 ± 0.2 **	1.7	1.2 ± 0.2	1.4 ± 0.1 **	1.2
TUG, s	7.1 ± 1.4	6.5 ± 1.0 *	0.5	7.7 ± 1.2	6.7 ± 0.8 **	1.0
IN	−4.7 ± 8.7	−3.9 ± 10.3	0.1	−8.3 ± 5.7	−8.3 ± 6.5	0.0
IADL	32.5 ± 2.6	36.0 ± 1.6 **	1.6	33.1 ± 2.7	36.2 ± 1.9 **	1.3
EQindex	0.9 ± 0.1	0.9 ± 0.1	0.3	0.8 ± 0.2	0.9 ± 0.1 *	0.7
EQVAS	74.1 ± 17.1	79.6 ± 18.1	0.3	77.5 ± 14.8	79.2 ± 16.0	0.1

BMI: body mass index; GS: grip strength; FTSST: five times sit to stand test; 6 MWT: 6 min walk test; PWS: preferred walking speed; TUG: timed up and go; IN: interference; IADL: instrumental activities of daily living; EQindex: descriptive index of Euroqol; EQVAS: visual analogue scale of Euroqol; ES: effect size. ** $p \leq 0.001$; * $p \leq 0.050$; † $p \leq 0.100$.

The EQindex showed a significant increase only for the AWIT group with a moderate ES (d = 0.66). No statistically significant variations were detected in any group for IN or EQVAS. Finally, GS showed a significant decrease in the AWIT group but with a small ES (d = 0.21).

The percentage of change in the main outcomes measured in our study are reported in Figure 1 (physical function parameters) and Figure 2 (executive function, IADL and health-related quality of life). We did not find statistically significant differences between CWIT and AWIT in any of the measurements performed. Only PWS showed a trend to significance (*p* = 0.097), with a higher change for the CWIT (27%) with regard the AWIT group (18%). Moreover, no-significant correlations (*p* > 0.05) were found between the percentage of change in functional outcomes and the BMI (deltas).

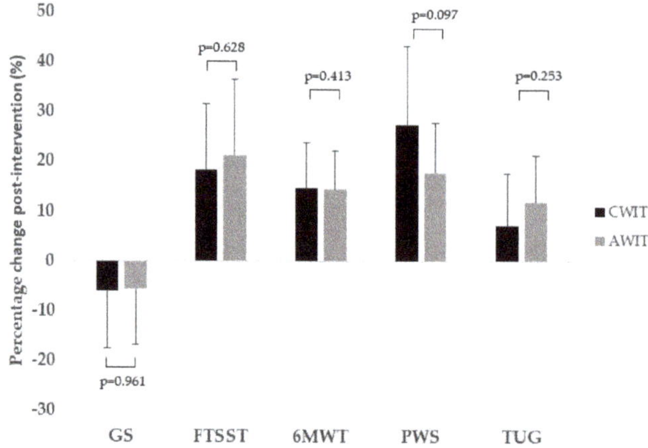

Figure 1. Percentage of change in physical function measurements following CWIT (solid bar) and AWIT (gray bar). GS indicates grip strength; FTSST, lower limb strength (time in seconds); 6MWT, 6 min walk test (m); PWS, preferred walking speed (time in seconds); TUG, timed up and go (time in seconds); CWIT, continuous walking interval training; AWIT, accumulated walking interval training.

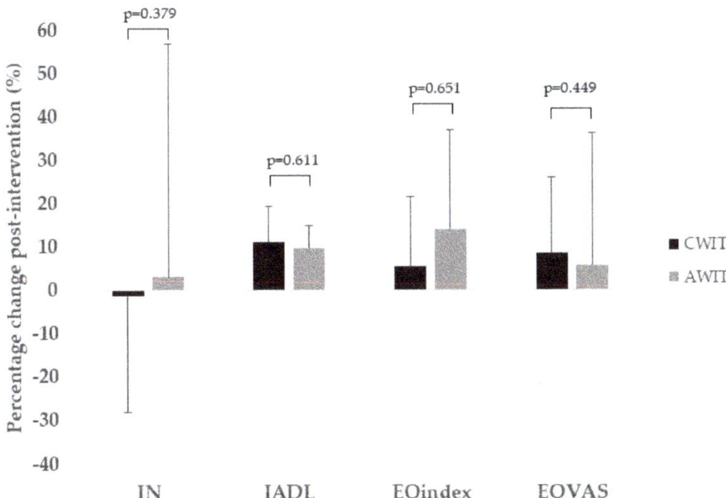

Figure 2. Percentage of change in executive function, instrumental activities of daily living, and health-related quality of life measurements following CWIT (solid bar) and AWIT (gray bar). IN, inhibition; IADL, instrumental activities of daily living; CWIT, continuous walking interval training; AWIT, accumulated walking interval training.

Finally, although we did not find significant differences for executive function, we observed that the CWIT group showed a negative percentage of change, while the AWIT group showed a positive one.

4. Discussion

The purpose of this study was to investigate the effects of continuous versus accumulated bouts of walking interval training on physical function and health-related quality of life in older persons. Our first hypothesis was that both strategies would be effective in the improvement of physical function and health related quality of life. In accordance with this hypothesis, our results showed that both interventions evoke similar benefits on physical function and daily life activities, so the accumulation of morning/afternoon exercise bouts does not exceed the beneficial effects of continuous walking training in sedentary older adults. However, PWS, IADL, and TUG reported different ES regarding both strategies. Additionally, we found no-improvements in GS, and only the accumulated strategy was able to improve the health related quality of life after 15 weeks of walking training.

The effect of starting physical exercise on physical function seems evident in sedentary older adults when it is properly programmed, tailored, and supervised by professionals [43,44], even at very advanced ages [45]. Despite the medium duration of the intervention (15 weeks) [46], we found similar improvements in cardiorespiratory fitness and lower limb strength, independently of the protocol. Enhancing these variables over 15–20%, with a large ES (whatever the group and outcome) is very important for older adults, since lower limb strength is related to falls prevention [47], and cardiorespiratory fitness is an independent risk factor in terms of mortality from any cause [48].

Previous studies have evaluated the influence of continuous versus accumulated exercise during the day, especially in cardiorespiratory fitness, with similar results for both strategies [49–51]. Nevertheless, these studies were conducted with younger subjects (<65 years) and/or with specific pathologies (obese, diabetic, etc.) [49–51]. In addition, the duration of the exercise bouts may vary among different studies between 10 and 30 min. For instance, Karstoft et al. [52] found that both the accumulated walking bouts group (3 × 10 min/day) and continuous walking group (30 min/day) significantly increased the aerobic condition after eight weeks. In the same way, other authors found improvements in both strategies in obese sedentary adults (continuous group achieved increases of 8% and accumulated group increased by 6%) in cardiorespiratory fitness [53]. Whatever the strategy, the improvements were lower than the ones found in our study (about 15%). However, some studies have shown slight differences, with the continuous strategy showing superior results in the first six months training [54], which were equated with the accumulated strategy at 18 months. These data suggest that to induce short-term changes, continuous strategies might be more effective.

The gradual increase in exercise-intensity and the modulation of the lap-duration through the whole WIT, with walking intensities close to the anaerobic threshold in the last weeks of the intervention (RPE of 7), can explain some of these differences. In the same way, as the walking speed increases during exercise, the length of the step and the amplitude of the movement also increase, which is an extra stimulus for the neuromuscular system [55]. Hence, the intensity of exercise in older adults is an important matter [56–58]. Previous studies have already theorized that both accumulated and continuous strategies could induce different adaptations due to differences in the effective intensity of exercise (internal load). Concentrating the load may lead to more pronounced physiological alterations [50], so the continuous dose would induce higher demands for exercises with similar volume and external intensity, or even RPE, improving cardiovascular responses.

Regarding the changes in agility, it was significantly improved for both doses of WIT without significance differences between groups, even though the ES was larger in the accumulated strategy when compared to the continuous one. The most time spent in a sedentary behavior may explain this result [59]. Agility usually comprises accelerations, decelerations, stop-and-go patterns, changes of direction (cutting maneuvers), and eccentric loads [60]. Therefore, it could be hypothesized that older adults who break sedentary behaviors more frequently could have slight benefits in this variable not only related to exercise training, but due to displacement and higher motor-time associated with the assistance to the training place.

Conversely, PWS and IADL showed a greater ES for the continuous strategy. Self-regulating walking speed over a longer consecutive training time (60 min vs. 30 min) may have resulted in participants in the continuous group being able to automate and apply that walking speed over a shorter distance (such as 6 m). This could explain the different ES in PWS (CWIT = 1.66 vs. AWIT = 1.24), which does not vary in aerobic capacity between strategies (CWIT = 1.02 vs. AWIT = 1.01). Regarding IADL, improvements in this outcome may be associated with greater physical function due to the increase in gait speed [61]. Therefore, our results support this association. In any case, improvements in these variables are very important in older adults. For example Perera et al. [62] highlighted that a higher walking speed is associated with a lower incidence of disability or mobility problems in three years; and the ability to perform complex tasks that involve IADL (such as the responsibility for preparation and taking medication, the control of domestic economy, the use of transport or telephone, etc.) are basic in older adult lives and tend to decline quickly [63]. Unfortunately, the sample size was not enough to consider any possible influence of gender in this different response to exercise training. However, according to Scaglioni et al. [64], there are no gender differences in the older adults' gait speed, despite the fact that several domains of this gait are sex-related, and hence sex differences should be considered when preparing exercise programs in older adults.

As mentioned above, no dose-distribution has managed to produce positive effects on GS. In fact, some authors have already pointed out that in programs aimed at aerobic conditioning, improvements in strength depend largely on the muscle group involved in this type of exercise [44]. Specifically, our data reflect a slight and significant decrease in GS in the accumulated group (low ES). Loss of GS has been shown to vary according to BMI trajectory [65], which was significant for AWIT, and just a trend for CWIT, but we have found no-correlation between the percentage of change in GS and BMI in our study. Further studies will elucidate if 60 min are always enough stimuli to ensure the reaching of the minimum neuromuscular (or metabolic) demands to maintain upper limb strength, since accumulating this duration does not prevent loses in this capacity. Consistent with these results,

Rooks et al. [66] found that an intervention of continuous walks (45 min) at an intensity chosen by the participants maintained the levels of GS compared to the controls.

Although positive effects of aerobic training programs have been demonstrated to elicit selective impacts in areas of executive functions (such as multi-tasking, planning, and inhibition) [11–13], surprisingly, in this study we found no-significant improvements in executive function (interference) after the intervention. Recently, it has been observed how cognitive performance (inhibition function) could be maintained or even improved after a period of training cessation [67], experimenting with a slower and more delayed evolution regarding physical function [68,69]. Given that detraining has not been evaluated in this study, we cannot confirm this effect, but in any case, some authors consider that maintaining executive function at these ages can also be considered a beneficial effect after this type of intervention [70]. In addition, we point out the different percentage of change for each exercise group: negative for continuous group, and positive for the accumulated (but without significant differences). The high dispersion in this variable can provide an idea of the heterogeneity of older adults and suggest that WIT could have benefited some participants, so exercise-training individualization matters.

Our results regarding health-related quality of life show a significant improvement for the descriptive index of the Euroqol, but in the posterior analysis of Bonferroni comparisons, we found that this difference is due to the AWIT group. Several factors can influence the quality of life in older populations. Functional capacity, an active lifestyle, and good social relationships are some factors that explained subjective quality of life [71]. Some authors have identified that factors such as interaction with family and friends, enjoying nature, and being helpful to others are also important to the quality of life of older adults [72]. Therefore, it is possible that the accumulated groups, having to carry out a greater number of sessions, generate more opportunities for social encounters and mutual help before and after the exercise session (such as sharing private transport to go to the training sessions). Moreover, the walking program took place in a natural environment, with the psychosocial advantages that this entails [72,73].

The main limitation of our study is that we could not compare PA levels or changes in sedentary behavior after the intervention. Quantifying PA levels through accelerometry could have given more information about the transfer of both strategies towards more active lifestyles. In addition, the small size of the sample implicates some uncertainty when extrapolating the results and becomes another limitation of our study, jointly with the slightly different sexual composition between groups in the final sample (in the second group the males were twice as many as the females). Future research is needed to confirm that accumulated and continuous exercise produce similar but not equal improvements in older adults following walking interval training, and whether other types of exercise (like multicomponent programs) evoke differences in functional or quality of life outcomes.

5. Conclusions

As far as we know, this is the first study to compare the benefits of accumulating or concentrating the dose of a WIT on physical function and health-related quality of life in sedentary older persons. Public health policies aimed to prescribe walking exercise programs must account for the specific implications of exercise dose-distributions also in this population, knowing that exercise prescription is a complex process and requires manipulating individual training loads.

According to our results, when comparing the effect of splitting a continuous bout of interval-walking into shorter bouts (of equivalent total duration dispersed throughout the day), the accumulated program provides, at least, similar benefits on physical function compared to the continuous program. These findings provide further evidence that bout length is not a determinant of the health functional effects associated with exercise. Moreover, while continuous exercise can provide a slight difference for PWS, fractionalizing a single exercise into two series throughout the day can do so for autonomy, agility, and health-related quality of life. It is noteworthy that accumulated strategies may also have additional consequences, helping to change sedentary behavior in the short-term.

Author Contributions: Conceptualization, P.M. and C.B.-L.; Methodology, P.M. and C.B.-L.; Formal Analysis, C.B.-L. and P.M.; Funding Acquisition, P.M. and C.B.-L.; Investigation, P.M., A.R., A.C., and C.B.-L.; Data Curation, P.M., A.C., and A.R.; Writing—Original Draft Preparation, P.M., C.B.-L., A.C., A.R., and M.C.G.-C.; Writing—Review and Editing, P.M., C.B.-L., A.C., A.R., and M.C.G.-C.; Project Administration, C.B.-L. All authors have read and agreed to the published version of the manuscript.

Funding: This work was partially supported by a predoctoral grant of Conselleria d'Investigació de la Generalitat Valenciana (ACIF/2016/423); and by the European Social Fund (BEFPI-2018).

Acknowledgments: The authors want to thank the Buñol local politicians and the University of Valencia for their support, as well as the older adults association "Entrenamiento con Mayores".

Conflicts of Interest: The authors declare no conflict of interest. The funders had no role in the design of the study; in the collection, analyses, or interpretation of data; in the writing of the manuscript; or in the decision to publish the results.

References

1. Bergouignan, A.; Rudwill, F.; Simon, C.; Blanc, S. Physical inactivity as the culprit of metabolic inflexibility: Evidence from bed-rest studies. *J. Appl. Physiol.* **2011**, *111*, 1201–1210. [CrossRef] [PubMed]

2. Dunstan, D.W.; Kingwell, B.A.; Larsen, R.; Healy, G.N.; Cerin, E.; Hamilton, M.T.; Shaw, J.E.; Bertovic, D.A.; Zimmet, P.Z.; Salmon, J. Breaking up prolonged sitting reduces postprandial glucose and insulin responses. *Diabetes Care* **2012**, *35*, 976–983. [CrossRef] [PubMed]

3. Cooper, A.; Sebire, S.; Montgomery, A.; Peters, T.; Sharp, D.; Jackson, N.; Fitzsimons, K.; Dayan, C.M.; Andrews, R. Sedentary time, breaks in sedentary time and metabolic variables in people with newly diagnosed type 2 diabetes. *Diabetologia* **2012**, *55*, 589–599. [CrossRef] [PubMed]

4. Healy, G.N.; Matthews, C.E.; Dunstan, D.W.; Winkler, E.A.; Owen, N. Sedentary time and cardio-metabolic biomarkers in US adults: NHANES 2003–2006. *Eur. Heart J.* **2011**, *32*, 590–597. [CrossRef] [PubMed]

5. Yerrakalva, D.; Cooper, A.; Westgate, K.; Khaw, K.-T.; Wareham, N.; Brage, S.; Griffin, S.; Wijndaele, K. The descriptive epidemiology of the diurnal profile of bouts and breaks in sedentary time in older English adults. *Int. J. Epidemiol.* **2017**, *46*, 1871–1881. [CrossRef]

6. Dempsey, P.C.; Larsen, R.N.; Dunstan, D.W.; Owen, N.; Kingwell, B.A. Sitting less and moving more: Implications for hypertension. *Hypertension* **2018**, *72*, 1037–1046. [CrossRef]

7. Franceschi, C.; Garagnani, P.; Parini, P.; Giuliani, C.; Santoro, A. Inflammaging: A new immune–metabolic viewpoint for age-related diseases. *Nat. Rev. Endocrinol.* **2018**, *14*, 576–590. [CrossRef]

8. Sun, F.; Norman, I.J.; While, A.E. Physical activity in older people: A systematic review. *BMC Public Health* **2013**, *13*, 449. [CrossRef]

9. Mathers, C.; Stevens, G.; Mascarenhas, M. *Global Health Risks: Mortality and Burden of Disease Attributable to Selected Major Risks*; World Health Organization: Geneva, Switzerland, 2009.

10. McDermott, M.M.; Liu, K.; Ferrucci, L.; Tian, L.; Guralnik, J.M.; Liao, Y.; Criqui, M.H. Greater sedentary hours and slower walking speed outside the home predict faster declines in functioning and adverse calf muscle changes in peripheral arterial disease. *J. Am. Coll. Cardiol.* **2011**, *57*, 2356–2364. [CrossRef]

11. Santos, D.A.; Silva, A.M.; Baptista, F.; Santos, R.; Vale, S.; Mota, J.; Sardinha, L.B. Sedentary behavior and physical activity are independently related to functional fitness in older adults. *Exp. Gerontol.* **2012**, *47*, 908–912. [CrossRef]

12. Sardinha, L.B.; Santos, D.A.; Silva, A.M.; Baptista, F.; Owen, N. Breaking-up sedentary time is associated with physical function in older adults. *J. Gerontol. A Biol. Sci. Med. Sci.* **2015**, *70*, 119–124. [CrossRef] [PubMed]

13. Barr-Anderson, D.J.; AuYoung, M.; Whitt-Glover, M.C.; Glenn, B.A.; Yancey, A.K. Integration of short bouts of physical activity into organizational routine: A systematic review of the literature. *Am. J. Prev. Med.* **2011**, *40*, 76–93. [CrossRef] [PubMed]

14. Galli, F.; Chirico, A.; Mallia, L.; Girelli, L.; De Laurentiis, M.; Lucidi, F.; Giordano, A.; Botti, G. Active lifestyles in older adults: An integrated predictive model of physical activity and exercise. *Oncotarget* **2018**, *9*, 25402. [CrossRef] [PubMed]

15. Windle, G. Exercise, physical activity and mental well-being in later life. *Rev. Clin. Gerontol.* **2014**, *24*, 319–325. [CrossRef]

16. Blair, S.N.; Kohl, H.W.; Paffenbarger, R.S.; Clark, D.G.; Cooper, K.H.; Gibbons, L.W. Physical fitness and all-cause mortality: A prospective study of healthy men and women. *JAMA* **1989**, *262*, 2395–2401. [CrossRef]

17. Health, D.O. *Stay Active: A Report on Physical Activity for Health from the Four Home Countries' Chief Medical Officers 2011*; Department of Health, London: London, UK, 2011.
18. Hughes, J.P.; McDowell, M.A.; Brody, D.J. Leisure-time physical activity among US adults 60 or more years of age: Results from NHANES 1999–2004. *J. Phys. Activ. Health* **2008**, *5*, 347–358. [CrossRef]
19. Floegel, T.A.; Giacobbi Jr, P.R.; Dzierzewski, J.M.; Aiken-Morgan, A.T.; Roberts, B.; McCrae, C.S.; Marsiske, M.; Buman, M.P. Intervention markers of physical activity maintenance in older adults. *Am. J. Health Behav.* **2015**, *39*, 487–499. [CrossRef]
20. Rosenberg, D.E.; Kerr, J.; Sallis, J.F.; Norman, G.J.; Calfas, K.; Patrick, K. Promoting walking among older adults living in retirement communities. *J. Aging Phys. Activ.* **2012**, *20*, 379–394. [CrossRef]
21. Maki, Y.; Ura, C.; Yamaguchi, T.; Murai, T.; Isahai, M.; Kaiho, A.; Yamagami, T.; Tanaka, S.; Miyamae, F.; Sugiyama, M. Effects of intervention using a community-based walking program for prevention of mental decline: A randomized controlled trial. *J. Am. Geriatr. Soc.* **2012**, *60*, 505–510. [CrossRef]
22. Awick, E.A.; Wójcicki, T.R.; Olson, E.A.; Fanning, J.; Chung, H.D.; Zuniga, K.; Mackenzie, M.; Kramer, A.F.; McAuley, E. Differential exercise effects on quality of life and health-related quality of life in older adults: A randomized controlled trial. *Qual. Life Res.* **2015**, *24*, 455–462. [CrossRef]
23. Blain, H.; Jaussent, A.; Picot, M.-C.; Maimoun, L.; Coste, O.; Masud, T.; Bousquet, J.; Bernard, P. Effect of a 6-month brisk walking program on walking endurance in sedentary and physically deconditioned women aged 60 or older: A randomized trial. *J. Nutr. Health Aging* **2017**, *21*, 1183–1189. [CrossRef] [PubMed]
24. Nelson, M.E.; Rejeski, W.J.; Blair, S.N.; Duncan, P.W.; Judge, J.O.; King, A.C.; Macera, C.A.; Castaneda-Sceppa, C. Physical activity and public health in older adults: Recommendation from the American College of Sports Medicine and the American Heart Association. *Med. Sci. Sports Exerc.* **2007**, *39*, 1435–1445. [CrossRef] [PubMed]
25. Blasco-Lafarga, C.; Monteagudo, P.; Roldán, A.; Cordellat, A.; Pesce, C. Strategies to change body composition in older adults: Do type of exercise and dose distribution matter? *J. Sport Med. Phys. Fit.* **2020**, *60*, 552.
26. Haskell, W.L.; Lee, I.-M.; Pate, R.R.; Powell, K.E.; Blair, S.N.; Franklin, B.A.; Macera, C.A.; Heath, G.W.; Thompson, P.D.; Bauman, A. Physical activity and public health: Updated recommendation for adults from the American College of Sports Medicine and the American Heart Association. *Med. Sci. Sports Exerc.* **2007**, *39*, 1423–1434. [CrossRef]
27. Committee, P.A.G.A. *Physical Activity Guidelines Advisory Committee Report, 2008*; Department of Health and Human Services: Washington, DC, USA, 2008; pp. A1–H14.
28. Manini, T.M.; Carr, L.J.; King, A.C.; Marshall, S.; Robinson, T.N.; Rejeski, W.J. Interventions to reduce sedentary behavior. *Med. Sci. Sports Exerc.* **2015**, *47*, 1306. [CrossRef]
29. Sartini, C.; Wannamethee, S.G.; Iliffe, S.; Morris, R.W.; Ash, S.; Lennon, L.; Whincup, P.H.; Jefferis, B.J. Diurnal patterns of objectively measured physical activity and sedentary behaviour in older men. *BMC Public Health* **2015**, *15*, 609. [CrossRef]
30. Izquierdo, M.; Casas, A.; De Navarra, C. *Guía Práctica Para la Prescripción de un Programa de Entrenamiento Físico Multicomponente Para la Prevención de la Fragilidad y Caídas en Mayores de 70 Años*; VIVIFRAIL: Navarra, Spain, 2017.
31. Vianna, L.C.; Oliveira, R.B.; Araújo, C.G.S. Age-related decline in handgrip strength differs according to gender. *J. Strength Cond. Res.* **2007**, *21*, 1310–1314.
32. Rikli, R.E.; Jones, C.J. Development and validation of a functional fitness test for community-residing older adults. *J. Aging Phys. Act.* **1999**, *7*, 129–161. [CrossRef]
33. Bohannon, R.W. Sit-to-stand test for measuring performance of lower extremity muscles. *Percept. Mot. Skills* **1995**, *80*, 163–166. [CrossRef]
34. Trommelen, R.D.; Buttone, L.F.; Dicharry, D.Z.; Jacobs, R.M.; Karpinski, A. The use of five repetition sit to stand test (FRSTST) to assess fall risk in the assisted living population. *Phys. Occup. Ther. Geriatr.* **2015**, *33*, 152–162. [CrossRef]
35. Bloch, M.L.; Jønsson, L.R.; Kristensen, M.T. Introducing a third timed up & go test trial improves performances of hospitalized and community-dwelling older individuals. *J. Geriatr. Phys. Ther.* **2017**, *40*, 121. [PubMed]
36. Podsiadlo, D.; Richardson, S. The timed "Up & Go": A test of basic functional mobility for frail elderly persons. *J. Am. Geriatr. Soc.* **1991**, *39*, 142–148. [PubMed]
37. Golden, C.J. *Stroop: Test deCcolores y Palabras*; TEA: Milano, Italy, 1999.

38. Rivera, D.; Perrin, P.; Stevens, L.; Garza, M.; Weil, C.; Saracho, C.; Rodriguez, W.; Rodriguez-Agudelo, Y.; Rabago, B.; Weiler, G. Stroop color-word interference test: Normative data for the Latin American Spanish speaking adult population. *NeuroRehabilitation* **2015**, *37*, 591–624. [CrossRef] [PubMed]

39. Martín-Lesende, I.; Quintana-Cantero, S.; Urzay-Atucha, V.; Ganzarain-Oyarbide, E.; Aguirre-Minana, T.; Pedrero-Jocano, J. Fiabilidad del cuestionario VIDA, para valoración de Actividades Instrumentales de la Vida Diaria (AIVD) en personas mayores. *Aten. Primaria* **2012**, *44*, 309–317. [CrossRef]

40. Martín-Lesende, I.; Vrotsou, K.; Vergara, I.; Bueno, A.; Diez, A.; Nuñez, J. Design and validation of the vida questionnaire, for assessing instrumental activities of daily living in elderly people. *J. Gerontol. Geriatr. Res.* **2015**, *4*, 2.

41. Herdman, M.; Gudex, C.; Lloyd, A.; Janssen, M.; Kind, P.; Parkin, D.; Bonsel, G.; Badia, X. Development and preliminary testing of the new five-level version of EQ-5D (EQ-5D-5L). *Qual. Life Res.* **2011**, *20*, 1727–1736. [CrossRef]

42. Cohen, J. *Statistical Power Analysis for the Behavioral Sciences*; Stat. Lawrence Erlbaum Associates: New York, NY, USA, 1988; p. 567.

43. Bouaziz, W.; Lang, P.; Schmitt, E.; Kaltenbach, G.; Geny, B.; Vogel, T. Health benefits of multicomponent training programmes in seniors: A systematic review. *Int. J. Clin. Pract.* **2016**, *70*, 520–536. [CrossRef]

44. Bouaziz, W.; Vogel, T.; Schmitt, E.; Kaltenbach, G.; Geny, B.; Lang, P.O. Health benefits of aerobic training programs in adults aged 70 and over: A systematic review. *Arch. Gerontol. Geriat.* **2017**, *69*, 110–127. [CrossRef]

45. Stessman, J.; Hammerman-Rozenberg, R.; Cohen, A.; Ein-Mor, E.; Jacobs, J.M. Physical activity, function, and longevity among the very old. *Arch. Intern. Med.* **2009**, *169*, 1476–1483. [CrossRef]

46. Mian, O.; Baltzopoulos, V.; Minetti, A.; Narici, M. The impact of physical training on locomotor function in older people. *Sports Med.* **2007**, *37*, 683–701. [CrossRef]

47. Zhang, F.; Ferrucci, L.; Culham, E.; Metter, E.J.; Guralnik, J.; Deshpande, N. Performance on five times sit-to-stand task as a predictor of subsequent falls and disability in older persons. *J. Aging Health* **2013**, *25*, 478–492. [CrossRef] [PubMed]

48. Harber, M.; Kaminsky, L.; Arena, R.; Blair, S.; Franklin, B.; Myers, J.; Ross, R. Impact of cardiorespiratory fitness on all-cause and disease-specific mortality: Aadvances since 2009. *Prog. Cardiovasc. Dis.* **2017**, *60*, 11–20. [CrossRef] [PubMed]

49. Asikainen, T.; Miilunpalo, S.; Oja, P.; Rinne, M.; Pasanen, M.; Vuori, I. Walking trials in postmenopausal women: Effect of one vs. two daily bouts on aerobic fitness. *Scand. J. Med. Sci. Sports* **2002**, *12*, 99–105. [CrossRef] [PubMed]

50. Campbell, L.; Wallman, K.; Green, D. The effects of intermittent exercise on physiological outcomes in an obese population: Continuous versus interval walking. *J. Sports Sci.* **2010**, *9*, 24.

51. Serwe, K.M.; Swartz, A.M.; Hart, T.L.; Strath, S.J. Effectiveness of long and short bout walking on increasing physical activity in women. *J. Women's Health* **2011**, *20*, 247–253. [CrossRef]

52. Karstoft, K.; Winding, K.; Knudsen, S.; Nielsen, J.; Thomsen, C.; Pedersen, B.; Solomon, T. The effects of free-living interval-walking training on glycemic control, body composition, and physical fitness in type 2 diabetic patients: A randomized, controlled trial. *Diabetes Care* **2013**, *36*, 228–236. [CrossRef]

53. Donnelly, J.; Jacobsen, D.; Heelan, K.; Seip, R.; Smith, S. The effects of 18 months of intermittent vs. continuous exercise on aerobic capacity, body weight and composition, and metabolic fitness in previously sedentary, moderately obese females. *Int. J. Obes.* **2000**, *24*, 566–572. [CrossRef]

54. Jakicic, J.M.; Winters, C.; Lang, W.; Wing, R.R. Effects of intermittent exercise and use of home exercise equipment on adherence, weight loss, and fitness in overweight women: A randomized trial. *JAMA* **1999**, *282*, 1554–1560. [CrossRef]

55. López-Chicharro, J.; Vicente-Campos, D.; Cancino, J. *Fisiología del Entrenamiento Aeróbico. Una visión integrada*; Editorial Médica Panamericana: Madrid, Spain, 2013.

56. García-Pinillos, F.; Laredo-Aguilera, J.; Muñoz-Jiménez, M.; Latorre-Román, P. Effects of 12-Week Concurrent High-Intensity Interval Strength and Endurance Training Program on Physical Performance in Healthy Older People. *J. Strength Cond. Res.* **2019**, *33*, 1445–1452. [CrossRef]

57. Hatori, M.; Hasegawa, A.; Adachi, H.; Shínozaki, A.; Hayashi, R.; Okano, H.; Mizunuma, H.; Murata, K. The effects of walking at the anaerobic threshold level on vertebral bone loss in postmenopausal women. *Calcif. Tissue Int.* **1993**, *52*, 411–414. [CrossRef]

58. Mangione, K.; McCully, K.; Gloviak, A.; Lefebvre, I.; Hofmann, M.; Craik, R. The effects of high-intensity and low-intensity cycle ergometry in older adults with knee osteoarthritis. *J. Gerontol. A Biol. Sci. Med. Sci.* **1999**, *54*, 184–190. [CrossRef] [PubMed]

59. Hamilton, M.T.; Hamilton, D.G.; Zderic, T.W. Role of low energy expenditure and sitting in obesity, metabolic syndrome, type 2 diabetes, and cardiovascular disease. *Diabetes* **2007**, *56*, 2655–2667. [CrossRef] [PubMed]

60. Morat, M.; Faude, O.; Hanssen, H.; Ludyga, S.; Zacher, J.; Eibl, A.; Albracht, K.; Donath, L. Agility Training to Integratively Promote Neuromuscular, Cognitive, Cardiovascular and Psychosocial Function in Healthy Older Adults: A Study Protocol of a One-Year Randomized-Controlled Trial. *Int. J. Environ. Res. Public Health* **2020**, *17*, 1853. [CrossRef] [PubMed]

61. Chou, C.-H.; Hwang, C.-L.; Wu, Y.-T. Effect of exercise on physical function, daily living activities, and quality of life in the frail older adults: A meta-analysis. *Arch. Phys. Med. Rehab.* **2012**, *93*, 237–244. [CrossRef] [PubMed]

62. Perera, S.; Patel, K.V.; Rosano, C.; Rubin, S.M.; Satterfield, S.; Harris, T.; Ensrud, K.; Orwoll, E.; Lee, C.G.; Chandler, J.M. Gait speed predicts incident disability: A pooled analysis. *J. Gerontol. A Biol. Sci. Med. Sci.* **2016**, *71*, 63–71. [CrossRef]

63. Lesende, I.M. Escalas de valoración funcional y cognitivas. In *Grupo de Trabajo de la semFYC de Atención al Mayor. Atención a las Personas Mayores Desde la Atención Primaria*; sem FYC Ediciones: Barcelona, Spain, 2004.

64. Scaglioni-Solano, P.; Aragón-Vargas, L.F. Gait characteristics and sensory abilities of older adults are modulated by gender. *Gait Posture* **2015**, *42*, 54–59. [CrossRef]

65. Reinders, I.; Murphy, R.; Martin, K.; Brouwer, I.; Visser, M.; White, D.; Newman, A.; Houston, D.; Kanaya, A.; Nagin, D. Body mass index trajectories in relation to change in lean mass and physical function: The health, aging and body composition study. *J. Am. Geriatr. Soc.* **2015**, *63*, 1615–1621. [CrossRef]

66. Rooks, D.; Kiel, D.; Parsons, C.; Hayes, W. Self-paced resistance training and walking exercise in community-dwelling older adults: Effects on neuromotor performance. *J. Gerontol. A Biol. Sci. Med. Sci.* **1997**, *52*, M161–M168. [CrossRef]

67. Blasco-Lafarga, C.; Cordellat, A.; Forte, A.; Roldán, A.; Monteagudo, P. Short and Long-term Trainability in Older Adults: Training and Detraining Following Two Years of Multicomponent Cognitive–physical Exercise Training. *Int. J. Environ. Res. Public Health* **2020**, *17*, 5984. [CrossRef]

68. Eggenberger, P.; Schumacher, V.; Angst, M.; Theill, N.; de Bruin, E. Does multicomponent physical exercise with simultaneous cognitive training boost cognitive performance in older adults? A 6-month randomized controlled trial with a 1-year follow-up. *Clin. Interv. Aging* **2015**, *10*, 1335.

69. Rodrigues, L.; Bherer, L.; Bosquet, L.; Vrinceanu, T.; Nadeau, S.; Lehr, L.; Bobeuf, F.; Kergoat, M.; Vu, T.; Berryman, N. Effects of an 8-week training cessation period on cognition and functional capacity in older adults. *Exp. Gerontol.* **2020**, *134*, 110890. [CrossRef] [PubMed]

70. Kelly, M.; Loughrey, D.; Lawlor, B.; Robertson, I.; Walsh, C.; Brennan, S. The impact of exercise on the cognitive functioning of healthy older adults: A systematic review and meta-analysis. *Ageing Res. Rev.* **2014**, *16*, 12–31. [CrossRef] [PubMed]

71. Nejati, V.; Shirinbayan, P.; Akbari Kamrani, A.; Foroughan, M.; Taheri, P.; Sheikhvatan, M. Quality of life in elderly people in Kashan, Iran. *Middle East J. Age Aging.* **2008**, *5*, 21–25.

72. Guse, L.W.; Masesar, M.A. Quality of life and successful aging in long-term care: Perceptions of residents. *Issues Ment. Health Nurs.* **1999**, *20*, 527–539.

73. Blasco-Lafarga, C. Ejercicio físico en el medio natural: La terapia que se anda. In *Paisaje, Turismo & Salud*; Iranzo-García, E., Ed.; Tirant lo Blanch: Valencia, Spain, 2017; pp. 48–65.

International Journal of
*Environmental Research
and Public Health*

Article

Foresight for the Fitness Sector: Results from a European Delphi Study and Its Relevance in the Time of COVID-19

Louis Moustakas [1,*], Anna Szumilewicz [2], Xian Mayo [3], Elisabeth Thienemann [4] and Andrew Grant [5]

[1]   Institute for European Sport Development and Leisure Studies, German Sport University, 50933 Cologne, Germany
[2]   Faculty of Physical Culture, Gdansk University of Physical Education and Sport, 80-336 Gdansk, Poland; anna.szumilewicz@awf.gda.pl
[3]   Observatory of Healthy and Active Living of Spain Active Foundation, Centre for Sport Studies, King Juan Carlos University, 28942 Madrid, Spain; xian.mayo@urjc.es
[4]   Independent Scholar, 1000 Brussels, Belgium; elisabeth.thienemann@gmail.com
[5]   Independent Consultant, Sunderland SR5 1SN, UK; andygrantconsultancy@gmail.com
*   Correspondence: l.moustakas@dshs-koeln.de

Received: 11 November 2020; Accepted: 27 November 2020; Published: 1 December 2020

Abstract: The fitness sector is an essential player in the promotion of physical activity and healthy behaviour in Europe. However, the sector is confronted with numerous socio-demographic trends that will shape its ability to be financially successful and contribute to public health. The sector must understand current drivers of change and the skills its workforce needs to navigate them. As such, using the results of a 2019 Delphi Survey of over 50 fitness experts from 26 countries, we aim to define the drivers of change facing the sector and identify the skills needed by the fitness workforce to navigate these changes. We find that several technological, social, health and economic trends affect the sector. As a result, so-called soft skills such as communication or customer service, along with digital technology skills, are becoming increasingly important. There is also growing recognition that fitness professionals need to be trained to work with a number of special populations. Furthermore, we argue that many of the trends identified here—such as the increasing use of technology or the focus on individual customer needs—have been accelerated by the COVID-19 pandemic. We conclude by arguing that well-developed, pan-European qualifications are needed to address these common issues.

Keywords: fitness; exercise; foresight; delphi study; Europe

---

1. Introduction

Physical activity helps prevent and treat a plethora of non-communicable diseases while also improving mental health, quality of life, and well-being. Despite these well-known benefits, over the last decade in Europe, there has been a decrease in overall physical activity and a concomitant increase in obesity. Between 25 and 35% of the European population is physically inactive [1], and similar percentages are obese [2]. In its latest Global Action Plan, the World Health Organisation emphasised the importance of building active societies as a strategic objective to strengthen the promotion of physical activity. In this regard, supporting pre- and in-service training to increase knowledge and skills related to the fitness sector professionals, along with other health sector stakeholders, was proposed as a policy instrument to help reduce worldwide physical inactivity by 2030 [1]. Similarly, the Regional Office for Europe of the World Health Organization suggests that states should support membership in fitness centres as a way to promote physical activity amongst younger and more vulnerable demographic groups [2].

These facts reinforce the importance of providing high-quality, responsive fitness offers in a variety of communities and settings. To some extent, the fitness industry has attempted to address these needs and support the current public health agenda. For instance, the industry has increased the number of facilities available and has also promoted the development of recognised qualifications. As a result, over the last decade, fitness users have increased by 72%, and almost 10% of all European adults are users of fitness services [3]. Authors of previous reports have tried to analyse the ongoing professionalisation of the industry, pointing out areas where the workforce excel, what the emerging trends are, and where the industry should turn its focus [4]. In this sense, and despite this business growth, fitness employers face numerous challenges that impede service delivery. There are significant skill shortages and mismatches within the fitness workforce. Based on Cedefop's skills online vacancy analysis tool, the European fitness sector alone is seeking to fill 115,000 vacant positions [5]. Employers also struggle to find staff with the right skill set to meet the industry's evolving needs. One in five employers report difficulties recruiting the trainers they want to work in their clubs [6]. Furthermore, fitness employers must consider how their current and future workforce will need to evolve to support changing public health needs. In short, though the industry has been active in trying to sustain its growth and support public health, there remain significant present and future challenges.

It is against this backdrop that EuropeActive (EA)—the European body for the fitness industry—launched the Sector Skills Alliance (SSA) for Active Leisure [7] according to the Council of the European Union declaration on the European Alliance for Apprenticeships [8]. From the SSA, the Blueprint for Skills Cooperation and Employment in Active Leisure project emerged [9]. These initiatives aim at developing a sector skills strategy for the fitness sector in collaboration with several European organisations, representing employers and employees, national authorities, training providers, higher education, and research institutions, as well as other relevant sector stakeholders.

Building on findings from the Blueprint project [10], the following paper aims to answer two main questions: (1) what are the main drivers for change and development in the fitness sector by 2030 and how will this impact fitness services? and (2) what skills will fitness professionals need by 2030 in response to these drivers? Identifying these drivers and skill requirements not only allows the fitness industry to ensure its continued sustainability, but it also provides valuable insights for the sector to continue supporting the public health challenges described above. Based on the results of a European Delphi study of experts from across the fitness sector, our study aims to identify the drivers most likely to impact the fitness sector and map out the specific skills needed within the fitness workforce to address them. Furthermore, as this study was conducted in 2019, we will argue that the trends and skills identified remain highly relevant despite the massive global changes created by the COVID-19 pandemic.

2. Materials and Methods

2.1. Design

The Delphi method has traditionally been used for several purposes, including aggregating ideas, generating consensus and forecasting [11,12]. The current study is aligned with this latter purpose. As outlined above, our goal is to identify the main drivers for change and development in the fitness sector as well as the related skill requirements for fitness professionals. Here, we define fitness professionals as instructors, trainers, and other exercise specialists, qualified to deliver diverse, structured exercise programmes that help people of all ages and abilities to improve their fitness, and physical and mental health [4].

In the brainstorming phase, a focus group discussion allowed experts to elaborate on the influences affecting the fitness sector. Based on the qualitative results generated, we implemented two survey rounds for a broader panel of experts. Within this survey, experts were asked to establish preliminary priorities among the survey items. Specifically, respondents were asked to evaluate, on a four-point scale, if a particular trend or challenge was either significant (4 = Significant Impact to 1 = No Impact) or likely (4 = Extremely likely to 1 = Unlikely). After the initial survey round, a second survey was sent whereby each expert received a questionnaire with the priorities summarised from the previous round. This second survey allowed experts to confirm or revise their opinions. Typically, the literature on the Delphi method suggests that consensus is achieved somewhere between 70% and 80% [11–13]. Given this, here, a consensus was determined when at least 75% of respondents ranked an item as either extremely likely/likely or as having significant/moderate impact. Finally, to further validate findings, survey results were presented and discussed in a smaller face-to-face expert focus group.

Finally, it should be noted that our data was collected shortly before the COVID-19 pandemic. Nevertheless, we analyse them later in the context of the needs and changes in the fitness sector caused by the situation. Additionally, in November 2020, our outcomes were discussed by a broad group of experts during the 3rd Meeting of the Sector Skills Alliance in Active Leisure [14] and The 11th EuropeActive International Standards Meeting [15].

2.2. Sampling

Purposive and snowball sampling were used to identify participants. As the Delphi method relies more on group consensus building than on statistical power, we aimed to achieve at least the recommended minimum sample size of 20 [13]. Experts were selected based on their experience and engagement in the European fitness sectors, be it in terms of advocacy, management, implementation, or research. As such, experts could represent national fitness associations, fitness providers, governmental bodies, European organisations, or academic institutions. Thus, the experts possessed both a wealth of experience as well as the diverse set of perspectives needed to build industry-wide consensus. Ultimately, this range of experts enhanced the credibility and applicability of the results, increasing the overall reliability of the findings. Similarly, the inclusion of experts who have knowledge of the sector and the multiple rounds of data collection support the validity of the results [12].

We targeted experts for the brainstorming amongst participants of the Sector Skills Alliance (SSA) for Active Leisure meeting held on the margins of the 9th EA International (Fitness) Standards Meeting (ISM) in Warsaw, Poland, in 2018. We further identified experts based on references provided by staff at EuropeActive and partners of the Blueprint project, leading to a list of 152 experts. Prior to taking part in the study, we explained the purpose of the study and that participation in the process was voluntary. Furthermore, all responses have been anonymised for publication.

2.3. Data Collection

2.3.1. Brainstorming

The brainstorming phase took place amongst members of SSA during the 2018 ISM in Warsaw, Poland, where a focus group discussion was held. The brainstorming looked to identify drivers of change in the fitness sector and how these drivers might impact the provision of services as well as employment in fitness. This discussion took place on 14 November 2018 with 17 participants and was jointly conducted by the fourth and fifth authors.

2.3.2. Online Survey

The survey was sent out in two rounds, and each round was sent to the 152 identified experts. The first round was conducted between 19 February 2019 and 3 March 2019, leading to 54 replies (response rate of 35.5%). After completion of the first round, we reviewed the data and reformulated open-ended responses into statements for verification in the second round. The second round was held between 11 and 24 March 2019, generating 50 replies (response rate of 30.4%). Overall, the sampled experts included a broad range of relevant sub-fields and represented over 25 European countries. Most prominently, vocational training providers, government representatives, and employers represented more than half of responses. Furthermore, 72% of respondents reported having more than ten years of experience in the industry. The breakdown of the background and location of experts are presented in Tables 1 and 2, respectively.

Table 1. Background of Experts.

	Round 1 (*n* = 54)		Round 2 (*n* = 50)	
Description	**Number**	**Per Cent**	**Number**	**Per Cent**
Vocational Training Provider	15	27.78%	7	14.00%
Government/Government Agency/NGO/EU institution (EU/national/regional/local)	12	22.22%	11	22.00%
Operator (employer)	10	18.52%	10	20.00%
Other (e.g., consultant, federation)	9	16.67%	10	20.00%
Researcher/higher education institute/university	5	9.26%	8	16.00%
Other fitness professional	2	3.70%	2	4.00%
Personal Trainer or higher-skilled fitness professional	1	1.85%	1	2.00%
Public Employment Service (National/EURES)	0	0.00%	1	2.00%

Table 2. Location of Experts.

Country	Round 1 (*n* = 54)	Round 2 (*n* = 50)
Belgium	5	5
Bulgaria	1	0
Croatia	1	0
Cyprus	1	0
Czech Republic	0	1
Denmark	1	0
Finland	5	5
France	1	2
Germany	4	4
Greece	2	2
Hungary	1	1
Iceland	1	1
Ireland	1	3
Italy	4	2
Malta	1	1
Moldova	1	0
Other	0	2
Poland	4	1
Portugal	1	2
Russia	2	0
Slovenia	0	1
Spain	5	6
Sweden	1	3
The Netherlands	7	4
Turkey	1	1
United Kingdom	3	3

2.3.3. Validation

In April 2019, we compiled data from the two survey rounds and sent a summary to 21 members of the Sector Skills Alliance for online review and feedback. Additionally, we further validated our findings through a supplementary focus group discussion on 27 June 2019 with 16 members of the Sector Skills Alliance. This discussion was jointly conducted by the first author and an external consultant.

2.4. Data Analysis

Qualitative (i.e., open-ended responses) generated throughout the Delphi survey process were analysed following a process of thematic analysis. This process helped us identify the main trends and issues that emerged throughout the open-ended responses and focus group discussions. We used the process of thematic analysis described by Braun and Clarke [16] to organise the data. This included becoming familiar with the data, searching for themes, reviewing potential themes, defining final themes, and writing the data. Quantitative, Likert scale data obtained from the Delphi surveys were entered into a tabular format, and descriptive statistics were generated.

3. Results

3.1. Main Drivers for Change and Development in the European Fitness Market Over the Next Decade

As can be observed in Figure 1, 69% of the experts found that digitalisation and technologies will have a significant impact on the fitness sector over the next decade. Likewise, 65% of experts felt that health and demographics will have a significant impact on the industry by 2030. Other potentially significant drivers such as society and communities (48% of the experts) and economy and innovation (28% of the experts) were rated by the experts less highly in comparison.

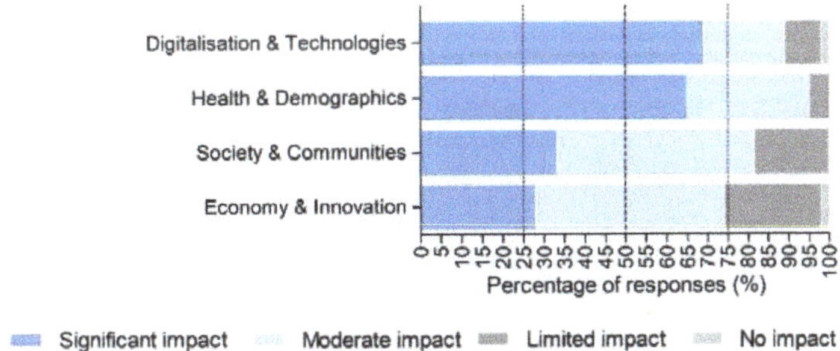

Figure 1. Level of impact of the main drivers for change and development over the next decade in percentages (%) by the experts (*n* = 50). In dark blue, significant impact. In light blue, moderate impact. In black grey, limited impact. In light grey, no impact.

3.2. Characteristics of Fitness Places Over the Next Decade

Figure 2 shows the potential characteristics of fitness places and how likely they will be by 2030. In this regard, more than half of the experts (52%) perceived it as extremely likely that fitness places will provide an experience to the users. More specifically, this experience should be inherently social and support the creation of fitness communities. Following closely behind, 45% of the experts reported it as extremely likely that fitness centres will work more on customer retention than at present. Lastly, 41% of the experts reported that fitness places will adopt a special concept such as micro-gyms, boutiques, or high-tech gyms.

3.3. Skills of Fitness Professionals Over the Next Decade

Figure 3 depicts how the experts rated the likeliness of several variables related to the future skills and requirements of fitness professionals. In this regard, the responses indicated that higher levels of agreement exist over two main issues. First, there was a broad agreement over the need for fitness professionals to work on a variety of health issues and with a range of populations. Items including skilled to work with special populations (e.g., older people, people with non-communicable diseases), skilled to work in health prevention, and being more of a lifestyle coach were rated as extremely likely by 59%, 50%, and 42% of the experts, respectively. Second, various soft skills were seen as increasingly required. Items including being highly skilled, having better communication skills, maintaining an attitude of continued learning and career development, and providing more accurate information, were rated as extremely likely by 52%, 50%, 50%, and 46% of the experts, respectively. Elsewhere, variables related to technology did not generate the same levels of consensus. Items such as being booked and rated online and skills connected with self-marketing were rated as extremely likely by 41%, 41%, and 26% of experts, respectively.

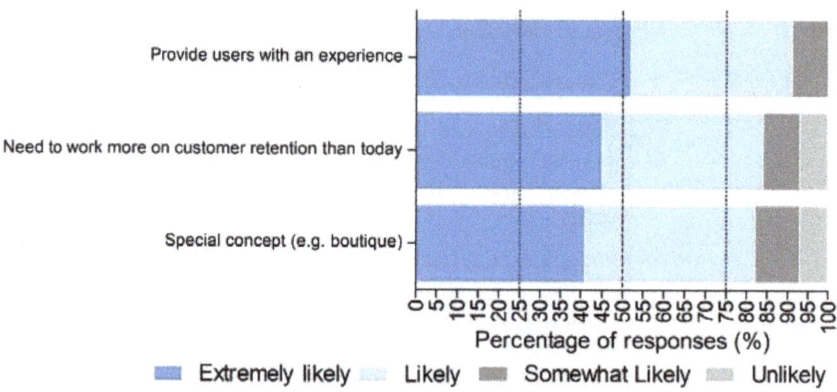

Figure 2. Levels of likeliness rated in percentages (%) by the experts (*n* = 50) about how fitness places will be by 2030. In dark blue, extremely likely. In light blue, likely. In black grey, somewhat likely. In light grey, unlikely.

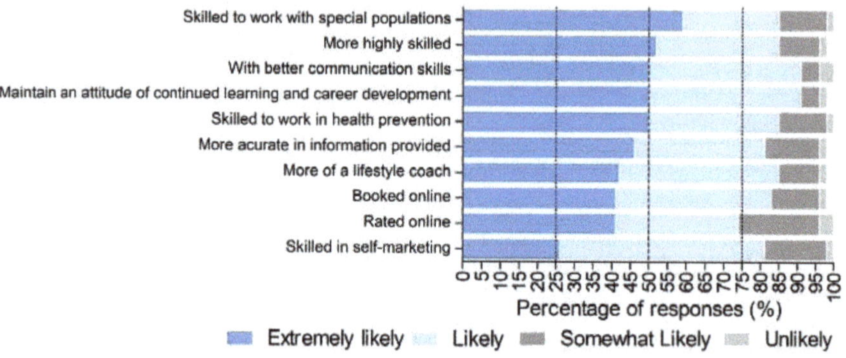

Figure 3. Levels of likeliness rated in percentages (%) by the experts (*n* = 50) about the required skills/characteristics of fitness professionals by 2030. In dark blue, extremely likely. In light blue, likely. In black grey, somewhat likely. In light grey, unlikely.

4. Discussion

Using the results of a 2019 Delphi Survey, our study aimed to define the drivers of change facing the fitness sector, understand how these drivers will influence fitness places, and outline the professional skills needed to navigate these changes. Thus, our work fits in with other research looking at trends in the industry. However, much of this other work focuses on a single group of respondents such as fitness trainers [17] or employers [6]. In that sense, our present study is unique as it generates findings and consensus across a broad range of experts, hence ensuring a wider variety of perspectives and increased validity.

First and foremost, our results highlight the need for the fitness sector and fitness professionals to be better aligned with the health sector to contribute to the emerging public health agenda. For instance, the importance of working with special populations (e.g., older people, people with non-communicable diseases), working in health prevention and being more of a lifestyle coach rated very highly in our study. This aligns strongly with other studies, both pre- and post-pandemic, that also found that skills relevant to disease prevention and health promotion are critical for fitness professionals [6,17,18]. Furthermore, the desire to improve the health of the clients and reduce non-communicable diseases are some of the primary motivators behind the work and further education of fitness professionals [19].

The continued development of links between the fitness industry and the health sector is not only relevant in light of the challenges posed by obesity or physical inactivity, but emerging research suggests that the fitness sector can play a critical role in mitigating the effects of COVID-19. Physical activity can play a significant role in reducing COVID-related mortality and will have a crucial role in recovering health lost during the pandemic. Physical fitness enhances the immune system [20] and helps reduce obesity, which is a predictor of mortality amongst COVID patients [21]. In addition, physical activity can help to recover from the effects of severe lockdowns and restrictions on physical and mental health [22]. In any event, the survey and focus group results highlight the need for the fitness sector to work more closely with the health sector to support primary prevention of non-communicable diseases and supporting behaviours change for healthy lifestyles more generally. Though in the past, there have been medical referral schemes connected to the fitness sector, this referral was specialized more in clinical exercise interventions helping to manage secondary and tertiary prevention (i.e., preventing a disease from getting worse or managing an existing disease) [23]. We must also note that medical professionals must likewise become better equipped to interact with the fitness sector. Based on recent studies, general practitioners show very low levels of knowledge about physical activity guidelines and exercise promotion in primary health care is very rare [24–26]. According to Füzéki et al. [24] "strengthening the topics of physical activity and health and physical activity counselling in medical curriculum is strongly recommended" as it would prepare health care providers for effective collaboration with fitness professionals to support increased physical activity levels.

Another group of variables increasingly needed by fitness professionals are related with the soft skills (i.e., personal transversal competencies such as social aptitudes, communication, teamwork, and other personality traits) that improve the provision of services and products. In our study, being highly skilled, having better communication skills, maintaining an attitude of continued learning and career development, and providing more accurate information were rated as extremely likely requirements for fitness professionals by around 46–52% of the respondents. Yet, though these skills are seen as more and more essential, research has suggested that fitness professionals lack many of the required communication, social, and counselling skills [19,27]. This mirrors concerns from employers in other industries [28], and most fitness professionals need additional training when starting a job. This reflects the changing nature of the fitness profession, whereby how to communicate and engage with clients is now a cornerstone of the provision of services. Indeed, most fitness professionals possess entry-level (e.g., EQF level 3 or 4) instructor or personal trainer qualifications [29]. Educational standards for these levels contain learning outcomes related to communication, client motivation, and counselling [30]. The low level of soft skills among fitness professionals, observed by other authors, may result from the way their education and training is carried out and how the required competencies are verified. Indeed, it is much more challenging to assess the achievement of soft skills compared to technical skills. This conclusion has been incorporated into work on the development of learning outcomes assessment strategy for fitness professionals under the Blueprint project [9]. Yet the need for interpersonal skills and continued learning are embedded in the ever-changing notions of the fitness profession. Fitness trainers now routinely take on a multitude of roles, including those of teacher, trainer, counsellor, coach, supervisor, supporter, nutritionist, life management advisor, weight controller, or even personal life consultant [31–33]. In addition, fitness professionals will need to be given the skills and knowledge to deliver services in a variety of new settings. At present, fitness centres are predominantly large-scale facilities solely dedicated to the provision of exercise services. Moving forward, numerous new fitness concepts are emerging, including integrated wellness offerings, outdoor gyms, smaller boutique gyms, and even gyms within retail locations [17,34,35]. Technology also directly impacts the scope and nature of the skills required. As noted as early as 1960 by Kerr [36], technology directly influences skill requirements. Today, institutions must therefore find a balance between physical workplace demands and the skill sets essential to meeting current technological needs.

While technology was reported in this study and preceding reports as a critical driver for change in the fitness sector, in our analysis, technology-related skills were rated less highly compared to other skills for fitness professionals (i.e., being booked and rated online or skills related with marketing). This is in line with a previous report where technology-related trends were rated less highly compared to other issues [17]. However, almost 80% of owners, operators, and gym managers report that online training and the use of technology are emerging trends in the European region [6]. In other words, though technology is perceived as a critical driver, it is not entirely clear to fitness experts exactly how that will translate to fitness places or the skills of fitness professionals.

Nonetheless, based on developments in the broader health sector and the impact of the COVID-19 pandemic, we can anticipate some of the directions technology might take within the fitness sector. Experts from the World Health Organization noted that digital health, or the use of digital technologies for health, "has become a salient field of practice for employing routine and innovative forms of information and communications technology to address health needs" [37]. Related to this is the concept of mobile health (mHealth), defined as "the use of mobile wireless technologies for health" [37]. In the current landscape, we can already see manifestations of "digital fitness" or "mFitness". For instance, individuals can continuously monitor their physical activity with the use of wearable technology [38]. Elsewhere, specialised applications provide quick feedback on the implementation of the prescribed exercise programme and may support the maintenance of client motivation [39].

The COVID-19 pandemic has further focused the industry's attention towards providing digital or mobile services. The closure of fitness clubs and the compulsory limitation of direct social contact resulted in many people experiencing online home fitness for the first time. Consequently, during lockdown, many exercise professionals and gyms began to provide their services exclusively online. The technological drivers of change identified within our survey remain relevant but have indeed been greatly accelerated. Further, this technological shift is likely to continue after the pandemic as the benefits and risks of online training become ever better understood [40–42].

In response to the growing need for digital fitness offerings, EuropeActive has been developing a supplementary qualification concerning the online provision of fitness services (9). It includes the professional competencies necessary for fitness professionals to be able to carry out their professional tasks safely and effectively while using online tools [43]. Here, online fitness services should be understood as something much broader than just conducting live exercise sessions via online technology. It includes all the stages related to guiding and coaching clients to positive lifestyle behavioural change and integrating healthy behaviours into daily routines [44].

Similar qualifications for various fitness professionals, aligned with the European Qualification Framework (EQF) and its associated recommendations, have been developed over the last many years to address many of the special, or clinical, populations identified above. At present, specialised qualifications related to active ageing, diabetes, and pre- and post-natal exercise are formally in place [45]. Ultimately, these current and future qualifications support the delivery of improved, increasingly relevant professional skills.

However, there remain substantial challenges in the delivery and recognition of these qualifications. Not all qualifications are delivered in all countries. Further, often, qualifications that are aligned to the EQF and recognised on the National Qualification Framework (NQF) of one country are not formally recognised in other EU member states. In turn, these factors conspire to limit the development and mobility of fitness professionals—which then further impedes the sector's ability to address the drivers of change identified here. As a result, this provokes a shortage of qualified professionals in the sector and limits individual career development. In response, we argue that more European coordination, development, and mutual recognition of qualifications are needed to help the fitness sector become a holistic, well-developed sector able to support the current public health agenda. To that end, both the European Commission and individual countries have a responsibility to support the development of relevant qualifications and ensure their recognition across the continent. As our study makes clear, there is consensus about the drivers of change and skills required across a broad thematic and geographic range of experts. Though we do not wish to dismiss country-specific needs or realities, our results make clear that many pan-European challenges would benefit from pan-European solutions.

Finally, we recognize that there are important financial implications involved in the development and delivery of qualifications, as well as in the remuneration of an increasingly educated and professionalized workforce. It is certainly beyond the purview of our paper to engage in a comprehensive cost-benefit analysis. However, it is clear to us that current difficulties in attracting and retaining qualified staff as well as the importance of client retention would more than likely justify these investments. Staff turnover is a high cost for businesses, and recent research in the fitness sector suggests that more educated staff can support client retention [46,47]. If the fitness industry does not upgrade and update its qualifications to meet current challenges and ensures a satisfied, well-paid workforce, the challenges, and threats, posed by lacklustre staff and client retention will remain. In addition, investment in high-quality exercise specialists and the promotion of physical activity can potentially lower the social costs associated with treating diseases caused by physical inactivity. This issue is still being analysed by other authors [23].

As the first Delphi study on foresight for the fitness sector, our work has its limitations. For instance, some factors, based on the brainstorming phase, have been combined, like "innovation and economy" or "health and demographics". In hindsight, analysing them separately would have provided more information on the main drivers of development in the fitness sector. There is also a need to integrate fitness clients into such studies, as these end-users are often absent in research on the sector. Similarly, many forecasting studies are predicated on pan-European perspectives. However, there remains room for nationally-focused, comparative research that looks to identify commonalities and differences across individual European states. Ultimately, such multi-directional analysis would allow for a better understanding of the sector's needs and more effective preparation against upcoming changes.

5. Conclusions

Building on the results of a 2019 Delphi Survey conducted in the context of EuropeActive's Blueprint project, our study aimed to define the drivers of change facing the fitness sector, understand how these drivers will influence fitness places, and outline the professional skills needed to navigate these changes. Ultimately, we find that technology, healthcare needs, and customer retention are critical drivers of change in the fitness industry. In response to this, fitness professionals must improve both their professional skills, especially as they relate to service provision for special populations, as well as their soft skills.

Though the global landscape has changed significantly since this study was conducted due to the COVID-19 pandemic, we argue that the results here are as relevant as ever. The ability to engage with technology and to have an understanding of specific health-related issues have only come into sharper focus due to the pandemic and its related social changes.

Moving forward, it will be imperative for the fitness sector to continue developing pan-European solutions and qualifications to address these continent-wide trends and challenges. Likewise, government agencies must redouble their efforts to recognise such qualifications both within their countries and from other European countries.

Author Contributions: L.M.: Validation, Formal Analysis, Investigation, Writing—Original Draft, Writing—Review and Editing, Visualization; X.M.: Validation, Formal Analysis, Writing—Original Draft, Writing—Review and Editing, Visualization; A.S.: Validation, Formal Analysis, Investigation, Writing—Original Draft, Writing—Review and Editing;

E.T.: Conceptualisation, Methodology, Funding Acquisition, Formal Analysis, Investigation, Writing—Review and Editing, Project Administration; A.G.: Investigation, Writing—Review and Editing. All authors have read and agreed to the published version of the manuscript.

Funding: The European Commission funded this research and associated APC through an Erasmus+ Project Grant (590345-EPP-1-2017-1-BE-SPO-SCP). The funder did not influence the design or implementation of this study. The European Commission support for the production of this publication does not constitute an endorsement of the contents which reflects the views only of the authors, and the Commission cannot be held responsible for any use which may be made of the information contained therein.

Acknowledgments: Special thanks go out to the EuropeActive Team, including Cliff Collins, Julian Berriman, Kevin Haddad, and Antigone Vatylioti, for their work on the Blueprint Project and in supporting this publication. Our gratitude also extends to the partners of the Blueprint project, including Herman Smulders, Jean-Yves Lapeyrere, Steve McQuaid, Michael McGeehin, Tony Wright, and Sergio Lara-Bercial. We would further like to acknowledge the support of the European Network of Sport Education (ENSE) in this publication. Finally, thanks go to Arda Alan Isik for his competent assistance in sourcing relevant literature. We also gratefully acknowledge the cooperation of all the fitness sector experts who volunteered for the study.

Conflicts of Interest: The authors declare no conflict of interest.

References

1. World Health Organisation. Global Action Plan on Physical Activity 2018–2030. At a Glance. 2018. Available online: https://apps.who.int/iris/bitstream/handle/10665/272721/WHO-NMH-PND-18.5-eng.pdf (accessed on 18 November 2020).
2. World Health Organisation. *Physical Activity Strategy for the WHO European Region 2016–2025*; World Health Organisation: Geneva, Switzerland, 2015.
3. Deloitte. *European Health & Fitness Market Report 2019*; Deloitte: London, UK, 2019.
4. Jimenez, A.; Berriman, J.; Collins, C.; Thienemann, E.; Szumilewicz, A.; Smulders, H. *The Relevance of the Active Leisure Sector and International Qualification Framework to the EQF (SIQAF): Final Report*; EuropeActive: Brussels, Belgium, 2018.
5. Cedefop. Skills-OVATE: Skills Online Vacancy Analysis Tool for Europe. Available online: https://www.cedefop.europa.eu/en/data-visualisations/skills-online-vacancies (accessed on 20 October 2020).
6. EuropeActive. *European Employers Skills Survey 2019*; EuropeActive: Brussels, Belgium, 2019.
7. Active Leisure Alliance. Home. Available online: https://www.active-leisure-alliance.eu (accessed on 4 November 2020).
8. Council of the European Union. *European Alliance for Apprenticeships. Council Declaration. 14986/13*; Council of the European Union: Brussels, Belgium, 2013.
9. EuropeActive. Blueprint for Skills Cooperation and Employment in Active Leisure. Available online: https://www.europeactive-euaffairs.eu/projects/blueprint (accessed on 18 November 2020).
10. Thienemann, E.; Lapeyrère, J.-Y. *Skills Foresight (Research) in the Active Leisure Sector*; EuropeActive: Brussels, Belgium, 2020.
11. Hsu, C.-C.; Sandford, B.A. The Delphi Technique: Making Sense of Consensus. *Pract. Assess. Res. Eval.* **2007**, *12*, 10. [CrossRef]
12. Hasson, F.; Keeney, S.; McKenna, H. Research guidelines for the Delphi survey technique. *J. Adv. Nurs.* **2000**, *32*, 1008–1015. [CrossRef]
13. Okoli, C.; Pawlowski, S.D. The Delphi method as a research tool: An example, design considerations and applications. *Inf. Manag.* **2004**, *42*, 15–29. [CrossRef]
14. EuropeActive. Sector Skills Alliance for Active Leisure Meeting to Review Its Strategy amid Covid-19 Pandemic. Available online: https://www.europeactive.eu/news/sector-skills-alliance-active-leisure-meeting-review-its-strategy-amid-covid-19-pandemic (accessed on 25 November 2020).
15. EuropeActive. 2020 ISM: Facing the New Normal Together. Available online: https://www.europeactive.eu/news/2020-ism-facing-new-normal-together (accessed on 25 November 2020).
16. Braun, V.; Clarke, V. Thematic analysis. In *APA Handbook of Research Methods in Psychology, Vol 2: Research Designs: Quantitative, Qualitative, Neuropsychological, and Biological*; Cooper, H., Camic, P.M., Long, D.L., Panter, A.T., Rindskopf, D., Sher, K.J., Eds.; American Psychological Association: Washington, DC, USA, 2012; pp. 57–71, ISBN 1-4338-1005-0.
17. Batrakoulis, A. European Survey of Fitness Trends for 2020. *ACSM's Health Fit. J.* **2019**, *23*, 28–35. [CrossRef]

18. EuropeActive. Sectoral Manifesto for EuropeActive's Horizon 2025. Available online: www.europeactive.eu/sectoral-manifesto-horizon-202 (accessed on 4 November 2020).

19. UK Active. *Going the Distance: Exercise Professionals in the Wider Public Health Workforce*; UK Active: London, UK, 2018.

20. Simpson, R.J.; Kunz, H.; Agha, N.; Graff, R. Exercise and the Regulation of Immune Functions. *Prog. Mol. Biol. Transl. Sci.* **2015**, *135*, 355–380. [CrossRef] [PubMed]

21. Popkin, B.M.; Du, S.; Green, W.D.; Beck, M.A.; Algaith, T.; Herbst, C.H.; Alsukait, R.F.; Alluhidan, M.; Alazemi, N.; Shekar, M. Individuals with obesity and COVID-19: A global perspective on the epidemiology and biological relationships. *Obes. Rev.* **2020**, *21*, e13128. [CrossRef] [PubMed]

22. Smith, A.; Biddle, S.; Bird, S. *The BASES Expert Statement on Physical Activity and Exercise during Covid-19 "Lockdowns" and "Restrictions"*; British Association of Sport and Exercise Sciences: Leeds, UK, 2020.

23. Campbell, F.; Holmes, M.; Everson-Hock, E.; Davis, S.; Buckley Woods, H.; Anokye, N.; Tappenden, P.; Kaltenthaler, E. A systematic review and economic evaluation of exercise referral schemes in primary care: A short report. *Health Technol. Assess.* **2015**, *19*, 1–110. [CrossRef]

24. Füzéki, E.; Weber, T.; Groneberg, D.A.; Banzer, W. Physical Activity Counseling in Primary Care in Germany-An Integrative Review. *Int. J. Environ. Res. Public Health* **2020**, *17*, 5625. [CrossRef]

25. Chatterjee, R.; Chapman, T.; Brannan, M.G.; Varney, J. GPs' knowledge, use, and confidence in national physical activity and health guidelines and tools: A questionnaire-based survey of general practice in England. *Br. J. Gen. Pract.* **2017**, *67*, e668–e675. [CrossRef] [PubMed]

26. Alahmed, Z.; Lobelo, F. Correlates of physical activity counseling provided by physicians: A cross-sectional study in Eastern Province, Saudi Arabia. *PLoS ONE* **2019**, *14*, e0220396. [CrossRef] [PubMed]

27. UK Active. *Raising the Bar 2018*; UK Active: London, UK, 2018.

28. Minton-Eversole, T. Concerns Grow Over Workforce Retirements and Skills Gap. Available online: www.shrm.org/resourcesandtools/hr-topics/talent-acquis (accessed on 1 November 2020).

29. European Register of Exercise Professionals. Member Directory. Available online: https://www.ereps.eu/member-directory (accessed on 21 November 2020).

30. EuropeActive. EuropeActive Standards. Available online: https://www.europeactive-standards.eu/es-standards (accessed on 21 November 2020).

31. Chiu, W.-Y.; Lee, Y.-D.; Lin, T.-Y. Performance Evaluation Criteria for Personal Trainers: An Analytical Hierarchy Process Approach. *Soc. Behav. Pers.* **2010**, *38*, 895–905. [CrossRef]

32. Maguire, J.S. *Fit for Consumption: Sociology and the Business of Fitness*; Routledge: London, UK; New York, NY, USA, 2008; ISBN 0415421810.

33. McKean, M.R.; Slater, G.; Oprescu, F.; Burkett, B.J. Do the Nutrition Qualifications and Professional Practices of Registered Exercise Professionals Align? *Int. J. Sport Nutr. Exerc. Metab.* **2015**, *25*, 154–162. [CrossRef] [PubMed]

34. IHRSA. 2019 Fitness Industry Trends Shed Light on 2020 & Beyond. Available online: https://www.ihrsa.org/improve-your-club/industry-news/2019-fitness-industry-trends-shed-light-on-2020-beyond/ (accessed on 25 November 2020).

35. Jansson, A.; Lubans, D.; Smith, J.; Duncan, M.; Haslam, R.; Plotnikoff, R. A systematic review of outdoor gym use: Current evidence and future directions. *J. Sci. Med. Sport* **2019**, *22*, S103. [CrossRef]

36. Kerr, C. *Industrialism and Industrial Man: The Problems of Labor and Management*; Harvard University Press: Cambridge, MA, USA, 1960.

37. World Health Organization. *WHO Guideline Recommendations on Digital Interventions for Health System Strengthening*; World Health Organization: Geneva, Switzerland, 2019; ISBN 9789241550505.

38. Norman, G.J.; Heltemes, K.J.; Heck, D.; Osmick, M.J. Employee Use of a Wireless Physical Activity Tracker Within Two Incentive Designs at One Company. *Popul. Health Manag.* **2016**, *19*, 88–94. [CrossRef] [PubMed]

39. Silva, A.G.; Simões, P.; Queirós, A.; Rocha, N.P.; Rodrigues, M. Effectiveness of Mobile Applications Running on Smartphones to Promote Physical Activity: A Systematic Review with Meta-Analysis. *Int. J. Environ. Res. Public Health* **2020**, *17*, 2251. [CrossRef] [PubMed]

40. Eickhoff-Shemek, J.M.; White, C.J. Internet Personal Training and/or Coaching: What are the Legal Issues? *ACSM's Health Fit. J.* **2005**, *3*, 5–31.

41. Innes, A.Q.; Thomson, G.; Cotter, M.; King, J.A.; Vollaard, N.B.J.; Kelly, B.M. Evaluating differences in the clinical impact of a free online weight loss programme, a resource-intensive commercial weight loss

programme and an active control condition: A parallel randomised controlled trial. *BMC Public Health* **2019**, *19*, 1732. [CrossRef] [PubMed]

42. Abbott, A.A. Online Impersonal Training Risk versus Benefit. *ACSM's Health Fit. J.* **2016**, *20*, 34–38. [CrossRef]

43. EuropeActive. Consultation—Online Provision of Fitness Services. Available online: www.europeactive-standards.eu/online-fitness-consultatio (accessed on 4 November 2020).

44. Council Recommendation of 22 May 2017 on the European Qualifications Framework for lifelong learning and repealing the recommendation of the European Parliament and of the Council of 23 April 2008 on the establishment of the European Qualifications Framework for lifelong learning. Available online: https://eur-lex.europa.eu/legal-content/EN/TXT/?uri=CELEX%3A32017H0615%2801%29 (accessed on 4 November 2020).

45. European Register of Exercise Professionals. Provider Directory. Available online: https://www.ereps.eu/acc-provider-directory (accessed on 5 November 2020).

46. Preston, J. An Investigation of the Relationship between Education, Credentials and Knowledge of Personal Trainers and Client Retention. Master's Thesis, Arizona State University, Tempe, AZ, USA, 2018.

47. Aguilar Royo, P.T. *Actas I Congreso Internacional Marca, Territorio y Deporte*, 1st ed.; Tirant Humanidades: Valencia, Spain, 2019; ISBN 978-84-17973-05-6.

Publisher's Note: MDPI stays neutral with regard to jurisdictional claims in published maps and institutional affiliations.

International Journal of
*Environmental Research
and Public Health*

Article

Providing Sports Venues on Mainland China: Implications for Promoting Leisure-Time Physical Activity and National Fitness Policies

Kai Wang [1,2,*] **and Xuhui Wang** [1]

1   College of Landscape Architecture and Arts, Northwest A&F University, Yangling 712100, China;
     fywxuh@nwafu.edu.cn
2   Centre for Urban Research, RMIT University, Melbourne, VIC 3001, Australia
*   Correspondence: kai.wang@nwsuaf.edu.cn; Tel.: +86-29-8708-0269

Received: 26 June 2020; Accepted: 12 July 2020; Published: 16 July 2020

Abstract: Leisure-time physical activity (LTPA) has been well documented as having substantial health benefits. The 2014 Chinese Fitness Survey Report stated that a lack of physical activity (PA) spaces is the most important non-human factor, leading to 10% of leisure-time physical inactivity in people aged 20 and above. We investigated the provision of sports venues in China and discussed the development of sports venues and national fitness policies in the context of promoting LTPA and public health. We analyzed information from China's most recent sport venue census, the Sixth National Sports Venues Census, conducted in 2013. The number of sports venues increased between 2000 and 2013, with an inflection point around the year 2008. At the end of 2013, there were 12.45 venues for every 10,000 residents, and the per capita area was 1.46 m^2. However, numbers were still small compared with the United States and Japan. The percentages of full-time access, part-time access and membership venues were 51.5%, 14.3% and 34.2% respectively. Only half of sports venues were fully open to the public, meaning that the realized number and area per capita could be even lower. A lack of sports venues forces people who want to engage in PA to occupy other urban spaces that are not planned and designed for PA. Urban parks had 119,750 fitness station facilities (3.32% of the total), and 2366 urban fitness trails (19.24%), with a combined length of 6450 km (32.91%). On average, urban and rural areas had 13.17 and 10.80 venues per 10,000 persons, and 1.83 m^2 and 0.97 m^2 per capita. The urban-rural gap in sports venues exactly embodies some aspects of the "urban-rural dual structure" in China's society. Measures to promote PA should focus on new and existing sports venues. In the policy making process, Chinese governments need to pay attention to the potential impact of related, external factors such as the gap between the urban and the rural and the potential advantage of indoor venues against summer heat and air pollution.

Keywords: National Fitness Programs; per capita area; school sports facilities; urban parks; rural sports venues

---

1. Introduction

Globally, more than 1.4 billion adults over the age of 18 did not reach the recommended levels of physical activity (PA) in 2016. According to the first study of worldwide trends in insufficient PA, this increases their risk of contracting the world's major non-communicable diseases (NCDs), such as hypertension, cardiovascular disease, type 2 diabetes, dementia and some cancers [1]. As one of the most populous countries and a typical upper middle-income country, mainland China (hereafter "China") faces a prevalence of NCDs and physical inactivity. Based on the Global Cancer Statistics 2018, China leads in cancer incidence and mortality rates in the world, and around one in five cancer patients come from China [2]. In 2017, over 114 million adults with diabetes lived in China (one in every 10 adults had diabetes), which was more than one-fourth of diabetes patients worldwide [3]. NCDs account for 88% of mortality in China, causing more than 70% of the total disease burden [4]. Zhang and Chaaban [5] suggest that physical inactivity contributes to an increased risk of five major NCDs (i.e., coronary heart disease, stroke, hypertension, cancer and type 2 diabetes), which in China constitute between 12% and 19%, and which alone account for more than 15% of the medical and non-medical yearly costs of the principal NCDs.

PA can be divided into leisure-time, occupational, transport-related and domestic activity categories. Occupational and domestic PA comprises the vast majority of PA among Chinese adults, while the amount of leisure-time and transport-related PA is very small in comparison [6]. A sharp decline in PA levels was observed by the China Health and Nutrition Survey using repeat measures and largely driven by a reduction in occupational and domestic PA. Transport-related PA, such as walking and cycling to and from work, has also sharply declined due to the skyrocketing car and e-bike ownership in China. The proportion of bicycle trips in Beijing dropped from 62.7% in 1986 to 11.3% in 2014 [7]. Bicycle lanes and sidewalks have been routinely compressed in order to tackle congestion problems caused by high car use. The increase in traffic injuries is a serious threat to the safety of pedestrians and cyclists. Wang et al. [8] investigated road traffic mortality in China from 2006 to 2016 and found pedestrians to be the most vulnerable road users. An in increase in sedentary jobs and an increased reliance on motorized transport have made leisure-time physical activity (LTPA) more important in fulfilling recommended PA levels [9].

LTPA has been well documented as having substantial health benefits, including improving physical wellness, lowering stress and depression, and leading to faster healing from medical conditions (also called recreational therapy) [10,11]. Although LTPA increased from 2.2 MET hours/week in 1991 to 4.8 MET hours/week in 2009 for Chinese adults, it was at 11.9 MET hours/week in 2009 for American adults and 14.8 MET hours/week in 2005 for British adults [12]. Additionally, this gap is projected to further widen in the future. The 2014 Chinese Fitness Survey Report (CFSR) states that the lack of PA spaces is the most important non-human factor, leading to 10% of leisure-time physical inactivity for people aged 20 and above [13]. The report also shows that about 34% of people aged 20 years and older who participate in LTPA use public sports venues, and another 12% use workplace or community sports venues that are only open to specific people. Sports venues play a vital role in promoting LTPA, especially public venues that are of no or low cost [14].

Chinese society has attached great importance to the development of sports venues. One key target of the Healthy China 2030 Plan and the National Fitness Programs is to enhance the growth of sports facilities. Since the founding of the People's Republic of China in 1949, much effort has been made to build sports venues [15]. However, the insufficient resources of sports venues still restrict the proportion of adults who meet the minimum LTPA recommendation. In addition to the low per capita supply of sports venues, other characteristics of sports venues have a significant impact on the national levels of LTPA [16]. These characteristics include number and area, distribution in different networks, opening status, weekly visits, types of sports venues, sports venues in urban parks and urban-rural divide. Nevertheless, there have been few studies conducted on the provision of sports venues in China and on their impacts on LTPA and national fitness policies. Therefore, the purpose of this study was to investigate the provision of sports venues in China, analyze the pertinent characteristics of sports venues associated with LTPA and discuss the development of sports venues and national fitness policies in the context of promoting LTPA and public health. The findings of this study will be useful in clarifying the gap between the provision of sports venues and national LTPA targets, which could support decisions on upgrading existing sports venues and the size, type and function of new sports venues, while also preparing proposed investments.

2. Methods

2.1. Data and Subjects

The Sixth National Sports Venues Census (NSVC), China's most recent one, was conducted in 2013. It included 84 types of sports venues (e.g., basketball courts, fitness stations and small-area sports courts) in various networks, industries and ownerships in China. The General Administration of Sports and the Ministry of Education led the national census. This census was divided into four hierarchical levels (national, provincial, prefecture and county), with data aggregated from the smallest to the largest spatial scale. Additional census details were obtained from the Data Compilation of the Sixth NSVC [17].

Sports venues refer to sports facilities dedicated to sports training, competitions and fitness activities, including functional affiliated buildings that are required for this purpose. Sites that are temporarily used for sports activities, such as conference rooms, auditoriums, warehouses, etc., are not included, and nor is construction. Major sports facilities used in this study included basketball courts (an outdoor stadium with fixed baskets for training and fitness, a minimum competition area 28 m long by 15 m wide and a surrounding, 2-m buffer extending from the competition area), fitness stations (facilities occupying small areas in communities, villages, parks, green areas, etc., consisting of a collection of outdoor fitness equipment, which are economical, practical and can be used free of charge), table tennis courts (an outdoor venue with fixed tables for playing table tennis, with a site area not under 40 sq. m), small-area sports courts (an outdoor stadium with a ring runway between 200 and 400 m), table tennis rooms (indoor sports venues for table tennis sports training and fitness use, at least 192 sq. m. Venues with temporary equipment and facilities in conference rooms, restaurants, auditoriums and corridors were excluded), track and field grounds (an outdoor stadium with a 400-m circular runway, no fixed stands or less than 500 seats), stadiums (a sports building with more than six standard 400-m runways, a soccer field in the center of the venue, a fixed grandstand and no less than 500 seats) and urban fitness trails (built in urban communities, residential areas, cultural and sports plazas, street gardens, waterfronts, parks and roadside green belts, free circular or non-circular trails open to the public for fitness).

The opening status of sports venues included membership (not open to the whole society, e.g., the sports venues of schools, enterprises and institutions are only open to the teachers and students of the school or the employees of the unit), part-time access (open to the public less than 8 h per day) and full-time access (open to the whole society every day for more than 8 h).

The criteria for regular PA in Chinese adults include three or more PA events per week of at least 30 min of moderate-intensity PA. The target heart rate during moderate intensity activities is about 64–76% of the maximum heart rate (The accepted method for obtaining the maximum heart rate is an estimation based on an individual's age subtracted from 220) [18].

2.2. Analysis

Two categories of data were selected to describe the provision of sports venues. One category contained the property information of sports venues, i.e., number, area, type, opening status and weekly visits. The other was related to the ownership and location information of sports venues. We chose differences in sports venues between 2000 and 2013 to analyze past trends in the venue number and area. The number and area of sports venues were investigated in three networks (sports, education and others), including the corresponding opening status and weekly visits for publicly accessible venues. We sorted and analyzed the top five types of sports venues in number and area with full-time access status, venues in urban parks and the urban-rural divide in both outdoor and indoor venues. These data were compiled to calculate accessibility ratios and per capita use and to chart trends over time. The data analysis was conducted in Microsoft Office Excel 2016. All data were entered by two research assistants to ensure the initial data accuracy. Descriptive statistics were calculated.

3. Results

3.1. Trends in Number and Area of Sports Venues

The number of sports venues increased constantly between 2000 and 2013, with an inflection point around the year 2008 in which the annual growth in both number and area suddenly increased (Figure 1a,b). At the end of 2013, there were 1.69 million sports venues, and the venue area was 1.99 billion m^2, including 0.17 million indoor venues and 1.52 million outdoor venues covering an area of 0.06 billion and 1.93 billion m^2, respectively. There were 12.45 venues for every 10,000 residents, and the per capita area was 1.46 m^2.

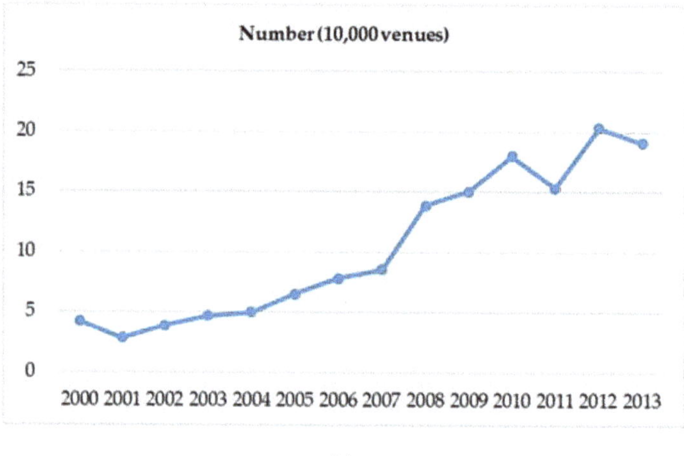

(a)

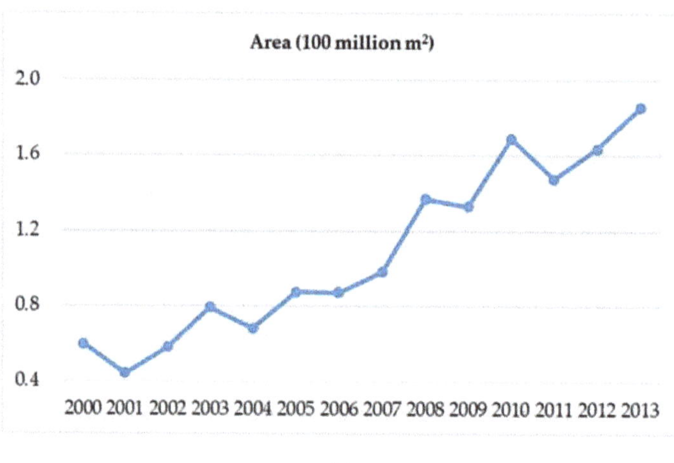

(b)

Figure 1. Annual growth in (**a**) number and (**b**) area of sports venues between 2000 and 2013.

3.2. Sports Venues in Different Networks, Their Opening Status and Weekly Visits

Sports venues were distributed disproportionately across the sports network, education network and others (Table 1). More than 90% of venues were included in the education network and others.

Under the education network, middle and elementary schools comprised 34.52% of the total venue number and 46.53% of the area.

Table 1. Distribution and opening status of sports venues in different networks.

	Number Ratio (%)	Area Ratio (%)	Full-Time Access	Part-Time Access	Membership
Sports network	1.43	4.77	15,656	4868	3798
Education network	38.98	53.01	56,135	153,072	451,314
Others	59.59	42.22	774,050	76,333	107,184
Total	100	100	845,841	234,273	562,296

In total, the percentages of full-time access, part-time access and membership venues were 51.5, 14.3 and 34.2%, respectively (Table 1). The sports network and others had 64.4 and 80.8% of full-time access venues. However, the education network had 68.3% of membership venues. Middle and elementary schools with a large number of sports venues had only 7.1 and 23.4% of full-time and part-time access venues. Among 1.08 million publicly accessible venues, 85.1, 11.7 and 3.2% of venues had less than 500, between 500 and 2500, and more than 2500 weekly visits, respectively.

3.3. Types of Sports Venues with Full-Time Access Status

The top five types of sports venues in number and area are shown in Tables 2 and 3. Basketball courts were the most popular venues, both high in number and area (36.31% and 18.34%). Fitness stations had the second highest number (22.41%), and small-area sports courts covered the largest area (22.68%), and the percentages of full-time access were over 90%.

Table 2. Top five types of sports venues in number.

	Number (10,000 Venues)	Ratio	Full-Time Access
Basketball court	59.64	36.31%	47.60%
Fitness station	36.81	22.41%	91.37%
Table tennis court	14.57	8.87%	47.04%
Small-area sports court	8.91	5.42%	7.79%
Table tennis room	4.87	2.97%	34.78%
Total	124.80	75.98%	

Table 3. Top five types of sports venues in area.

	Area (100 Million m^2)	Ratio	Full-Time Access
Small-area sports court	4.42	22.68%	7.79%
Basketball court	3.58	18.34%	47.60%
Ground track field	1.69	8.67%	12.42%
Stadium	1.05	5.39%	7.02%
Urban fitness trail	0.59	3.03%	96.68%
Total	11.33	58.13%	

3.4. Sports Venues in Urban Parks

In 2013, there were 21,013 sports venues in 12,401 urban parks, with an area of 367,962 ha, i.e., 1.69 venues per park and 6.37 venues per 100 ha park area. Urban parks had 119,750 facilities with fitness stations (3.32% of the total), and 2366 urban fitness trails (19.24%) with a total length of 6450 km (32.91%).

3.5. Sports Venues in Urban and Rural Areas

In total, 58.62% and 68.60% of the number and area were located in urban areas, and 41.38% and 31.40% were in rural areas (Table 4). In terms of indoor venues, however, urban areas had 82.50% and

91.53% in number and area, and rural areas only had 17.50% and 8.47%. On average, urban and rural areas had 13.17 and 10.80 venues per 10,000 persons, and 1.83 m^2 and 0.97 m^2 per capita.

Table 4. Number and area of sports venues in urban and rural areas.

	Urban Area		Rural Area	
	Number	Area	Number	Area
Indoor venues	12.87	0.54	2.73	0.05
Outdoor venues	83.40	12.83	65.24	6.07
Total	96.27	13.37	67.97	6.12
Indoor venues	12.87	0.54	2.73	0.05
Outdoor venues	83.40	12.83	65.24	6.07
Total	96.27	13.37	67.97	6.12

Notes: Number in 10,000 venues; Area in 100 million m^2.

4. Discussion

This study represents the first analysis of the pertinent characteristics of sports venues associated with LTPA in China. The number and area of sports venues are the fundamental indicators for the overview of the characteristics. Both indicators increased dramatically between 2000 and 2013. The turning point occurred around the year 2008 when the Summer Olympic Games were held in Beijing, China. However, numbers were still small compared with the United States and Japan [19]. Every Japanese resident had 19 m^2 of venue space on average in 2010 [20], but the Chinese per capita area was only 1.46 m^2 in 2013. China's Sports Development Five-year Plan is the national guideline for the growth goal of sports venues. The goals of the 12th Five-year Plan (2011–2015) were more than 1.20 million venues and 1.5 m^2 per capita by the end of 2015. There were 1.69 million sports venues at the end of 2013, and the area per capita reached 1.5 m^2 at the end of 2014 [21]. The new goal of the 13th Five-year Plan (2016–2020) was 1.8 m^2 per capita. By 2030, the Healthy China Initiative aims to obtain 2.3 m^2 per capita of sports venues. Indeed, a lack of sports venues can exacerbate use conflicts and potentially lead to social issues in China. For example, a group of elderly square dancers fought with young basketball players for the use of a basketball court in Luoyang city, in central China's Henan province. Local police were finally called to stop the fight [22]. Media reported that similar incidents also happened in Fujian Province and Shanghai. A study led by the Central China Normal University found that Chinese teenagers' increasing use of internet gaming can possibly be related to the low per capita area of sports venues in the country [23]. It is urgent to find paths for the sustainable development of sports venues in the context of China's increasingly tense people-land relationship.

The opening status of sports venues can limit residents' utilization of PA spaces. Our findings show that only half of sports venues were fully open to the public, meaning that the realized number and area per capita can be even lower. A lack of sports venues forces people who want to engage in LTPA to occupy other urban spaces that are not planned and designed for LTPA. For example, fast-walking groups have become a popular phenomenon that includes dozens, sometimes hundreds, of urban residents who use roadways to walk for exercise. Unfortunately, in 2017, a taxi crashed into a fast-walking group, causing one death and two injuries [24]. However, it is worth noting that 85% of publicly accessible sports venues received less than 80 daily visits, and only 18.4% of adults had access to the public sports venues, while nearly 40% chose to use vacant places that were not PA-oriented to engage in LTPA, such as in public open spaces or along streets [13].

Sharing campus sports venues can increase public PA opportunities, especially for nearby neighborhoods and communities [25,26]. More than one third of the number of and almost half of the area of sports venues were located at middle and elementary schools. Nearly 80% of children and youths aged between 6 and 19 engaged in regular PA, while only 14.7% of adults aged 20 and above were active [13]. This is likely because about 70% of sports venues in middle and elementary schools were not open to the public. Adults are generally no longer affiliated to middle or elementary schools, and thus they become a public that has a limited access to PA facilities within middle and elementary schools. In China, a higher level of education can lead to a higher percentage of participation in regular PA [13]. In addition to an awareness of exercise, this trend is likely related to school PA facilities (including high schools and universities) because a higher level of education means a longer access to campus PA facilities where it is much easier to develop an exercise habit.

The concept of shared use has been treated as an efficient and effective approach since social ecological models were adopted to develop PA opportunities [27]. The Global Action Plan on Physical Activity 2018–2030, released by the World Health Organization to reduce physical inactivity worldwide, proposed measures to encourage and strengthen the policy of shared use of school facilities with the strategic objective of creating active environments. The US "Healthy People 2020" strategy also established a similar objective to increase access to PA spaces and facilities in public and private schools. Japan's Basic Act on Sport clearly states that Japanese national and public primary and secondary schools' sports facilities need to be open to community residents [28]. As early as 2009, National Fitness Regulations released by the State Council of China proposed that public schools should actively create conditions to open exercise facilities to the public. However, many issues have accompanied these openings, such as how to maintain student safety, responsibility for accidental injuries of users, paying for increased management costs, etc. Lei divided the risk types of opening school facilities into charge management, personal safety, property security, environmental safety and liability [29]. In 2017, the Ministry of Education and General Administration of Sport in China jointly issued Implementation Opinions on Promoting the Opening of School Sports Facilities to the Society, stating that schools are required to open exercise facilities to students in their free time and on public holidays. By 2020, schools that meet opening conditions should also improve their level of openness and efficiency. The Implementation Opinions provided guidance on opening hours, potential users, the charging standard and the security mechanism. Although some cities were indeed promoting the opening of school exercise facilities, e.g., Shanghai and Hangzhou, the media often reported that residents complained that they had difficulty entering schools to use exercise facilities. As of 2018, there were 18 national policies on the opening of school exercise facilities [30], but the effect of the policies has yet to be tested.

The types of sports venues should match the needs of residents' LTPA. Of all the sports venues, basketball courts accounted for the largest number and the second largest area. The 2014 CFSR shows that the out-of-school PA programs for children and adolescents aged 6 to 19 years included active party games (22.2%), long-distance running (18.0%) and basketball (11.2%), while for adults aged 20 and above basketball was rarely played [13]. The number of fitness stations was over 20% of the total, and over 90% were open full-time to the public. The users of fitness stations are largely middle-aged and elderly [31]. Because the first fitness stations to have been built are now approximately 25 years old (the first was built in 1996 at the Tianhe Sports Center of Guangzhou), many problems related to maintenance and the changing needs of a growing population need to be solved. These include the location, site layout, management and maintenance of fitness stations [32,33]. Small-area sports courts had the largest area, but only 7.79% of them had a full-time access, largely because most of these were found in middle and elementary schools. Li investigated 32 elementary schools within the downtown area of Dalian and found that 84.4% had small-area sports courts [34]. If the policy of opening school exercise facilities to society is to be well-implemented, the opening ratio of small-area sports courts must greatly increase in order to meet the growing needs for activities of adults aged 20 and older whose most popular activities are fitness walking (54.6%) and jogging (12.4%) [13]. Urban fitness trails can be another ideal venue for these activities. There are 12,299 urban fitness trails (nearly 19.6 million meters) nationwide, and more than 95% of the trails have a full-time access. However, Lin and Qiu studied the urban fitness trails in Fujian Province and found that about 40% of the trails received less than 500 visitors per week. This was, in part, because the trail venues were often located in less-populated areas or were built long enough ago to not meet modern exercise needs [35]. The Implementation Plan for the Million-kilometer Fitness Trail Project released in 2018 strives to construct about 300 km of fitness trails in each county-level administrative unit by 2020 [36]. The city of Suzhou in Jiangsu Province has been working on the country's first urban fitness trail system plan [37].

Urban parks in developed countries that mix physical structures and green spaces are considered to be ideal environments explicitly built to promote LTPA and public health [38,39]. On average, US residents visit local parks and recreational facilities 29 times annually, and 52% of park users who were surveyed said that PA is a key factor in their decision to access parks and recreational facilities [40]. However, the importance of urban parks in promoting LTPA has not been fully valued or has even been ignored in the development of Chinese sports venues [41]. Only about 1.3% of sports venues are located in urban parks that are usually open to the public for free. In contrast, one observational park-based PA study shows that more than 50% of Chinese park users are able to engage in moderate-to-vigorous PA [42]. Additionally, 26.4% of people appeal for the building of sports venues close to urban parks [13], indicating that the public has a strong willingness to conduct physical activities in urban parks. In China, park-based running has become part of the urban lifestyle (increasing development in the process) [43], and fitness trails in urban parks play a vital role in this trend. For example, Beijing's Olympic Forest Park has become China's most popular running park, mainly because of its 13-km long, 3-m wide and 10-cm thick plastic fitness trails. The number and length of urban fitness trails in urban parks were nearly 20% and 30% of the total at the end of 2013.

China's society has long been in a state of "urban-rural dual structure", and the urban-rural gap of sports venues exactly embodies some aspects of this duality [19]. The number and area of urban sports venues were 1.42 and 2.18 times those of rural areas, and the number and area of urban indoor venues were as high as 4.71 and 10.8 times those of rural areas. In terms of the per capita number and area, the difference was at 1.22 and 1.89 times for total venues, and at 4.06 and 9.31 times for indoor venues. There is still a clear gap between urban and rural areas, particularly for indoor venues. The health risks associated with exposure to air pollution likely outweigh the benefits of outdoor physical activities in China [44,45]. Because most indoor pollutants can be efficiently reduced or removed by air purifiers [46], indoor venues become ideal places for healthy exercise without the negative effects of air pollution. The paucity of indoor venues in rural areas deepens this urban-rural divide. Residents' rates of regular PA in urban and rural areas were 22.2% and 14.3%, respectively [47], a huge difference (7.9%) closely related to the aforementioned gap. Therefore, the National Fitness Program of China (2016–2020) stresses and promotes the extension of basic public exercise services to rural areas. China has made significant progress since the Fifth NSVC in 2003, when only 8.18% of sports venues were distributed in rural areas.

Our study was limited in several ways. First, the data used in this study were from 2013, the year of the latest decadal national census. However, sports venues have changed dramatically since 2013. For example, by the end of 2017, the overall number of sports venues exceeded 1.96 million, and the per capita area reached 1.66 m^2, according to a newly published report that only described the overall trend [21]. Second, detailed data on the use of sports venues was unavailable, such as users' gender, age and socioeconomic status. In addition, the incompatibility of statistical indicators in the Sixth NSVC and the 2014 CFSR made it challenging to conduct extended analyses. Third, non-standard sports venue data were not available. These venues are mostly found in rural areas, and the growth in the number of rural residents who regularly participated in PA was higher than that of urban residents [13]. Since non-standard sports venues usually use a small area and their investment cost is only one third that of standard venues, they may be more suitable for large cities with limited land resources or crowded historical quarters. Finally, urban parks play an important role in LTPA, yet information on park-based venues was lacking.

5. Conclusions

Public sports venues are physical spaces that accommodate fitness for all and are a basic resource for national public health. Accordingly, promoting the construction of public sports venues has always been one of the key contents of numerous sports developments, as well as of national fitness and public health policies. Promotion measures for sports venues should focus on both new and existing resources. On the one hand, China's society aims to effectively expand new fitness resources to ensure that the per capita area continues to increase, particularly increasing popular fitness venues such as multipurpose small-area sports courts and urban fitness trails. More public open spaces, such as city parks, and urban vacant places, such as old factories, warehouses, old commercial facilities, etc., should be retrofitted as sports venues. However, the NIMBY (Not in My Backyard) issue occasionally occurs when selecting the location of sports venues [48]. The administration needs to further revitalize existing resources and pay close attention to the use, management and upgrading of sports venues and facilities. In particular, it is important to ensure that public and school sports venues that meet open conditions can be fully open to local residents. Finally, in the policy making process, Chinese governments need to pay attention to the potential impact of related, external factors such as the gap between the urban and the rural and the potential advantage of indoor venues against summer heat and air pollution.

Author Contributions: Conceptualization, K.W.; Data curation, K.W.; Formal analysis, K.W.; Funding acquisition, K.W.; Methodology, X.W.; Writing—original draft, K.W.; Writing—review & editing, X.W. All authors have read and agreed to the published version of the manuscript.

Funding: This study was funded by Ministry of Education in China Project of Humanities and Social Science (No. 17YJCZH166) and Northwest A&F University (No. 2452015280).

Acknowledgments: Thanks go to Sijia Chen for formatting the manuscript.

References

1. Guthold, R.; Stevens, G.A.; Riley, L.M.; Bull, F.C. Worldwide trends in insufficient physical activity from 2001 to 2016: A pooled analysis of 358 population-based surveys with 1.9 million participants. *Lancet Glob. Health* **2018**, *6*, e1077–e1086. [CrossRef]

2. Bray, F.; Ferlay, J.; Soerjomataram, I.; Siegel, R.L.; Torre, L.A.; Jemal, A. Global cancer statistics 2018: GLOBOCAN estimates of incidence and mortality worldwide for 36 cancers in 185 countries. *CA Cancer J. Clin.* **2018**, *68*, 394–424. [CrossRef]

3. *IDF Diabetes Atlas*, 8th ed.; International Diabetes Federation: Brussels, Belgium; Available online: http://www.diabetesatlas.org (accessed on 20 September 2019).

4. State Council's Opinions on Implementing Healthy China Action. Available online: http://www.gov.cn/zhengce/content/2019-07/15/content_5409492.htm (accessed on 20 September 2019).

5. Zhang, J.; Chaaban, J. The economic cost of physical inactivity in China. *Prev. Med.* **2013**, *56*, 75–78. [CrossRef] [PubMed]

6. Ng, S.W.; Howard, A.G.; Wang, H.J.; Su, C.; Zhang, B. The physical activity transition among adults in China: 1991–2011. *Obes. Rev.* **2014**, *15*, 27–36. [CrossRef] [PubMed]

7. Jiang, B.; Liang, S.; Peng, Z.R.; Cong, H.; Levy, M.; Cheng, Q.; Wang, T.; Remais, J.V. Transport and public health in China: The road to a healthy future. *Lancet* **2017**, *390*, 1781–1791. [CrossRef]

8. Wang, L.; Ning, P.; Yin, P.; Cheng, P.; Schwebel, D.C.; Liu, J.; Wu, Y.; Liu, Y.; Qi, J.; Zeng, X.; et al. Road traffic mortality in China: Analysis of national surveillance data from 2006 to 2016. *Lancet Public Health* **2019**, *4*, e245–e255. [CrossRef]

9. Bedimo-Rung, A.L.; Mowen, A.J.; Cohen, D.A. The significance of parks to physical activity and public health: A conceptual model. *Am. J. Prev. Med.* **2005**, *28*, 159–168. [CrossRef] [PubMed]

10. Rachel Morgan. Importance of Leisure & Recreation for Health. Available online: https://healthfully.com/importance-of-leisure-recreation-for-health-7692320.html (accessed on 7 July 2020).

11. Raza, W.; Krachler, B.; Forsberg, B.; Sommar, J.N. Health benefits of leisure time and commuting physical activity: A meta-analysis of effects on morbidity. *J. Transp. Health* **2020**, *18*, 100873. [CrossRef]

12. Ng, S.W.; Popkin, B.M. Time use and physical activity: A shift away from movement across the globe. *Obes. Rev.* **2012**, *13*, 659–680. [CrossRef]

13. General Administration of Sport of China. Available online: http://www.sport.gov.cn/n16/n1077/n1422/7300210.html (accessed on 14 September 2019).

14. Gao, W.; Cao, K. Increasing public access to sports venues to promote physical activity among older adults in China. *J. Aging Phys. Activ.* **2018**, *26*, 169–170. [CrossRef]
15. Guo, M.; Liu, C.; Liu, M.; Wang, J. The process and enlightenment of the development of sports facilities in China. *J. Beijing Sport Univ.* **2009**, *32*, 12–16.
16. Wu, S.; Luo, Y.; Qiu, X.; Bao, M. Building a healthy China by enhancing physical activity: Priorities, challenges, and strategies. *J. Sport Health Sci.* **2017**, *6*, 125–126. [CrossRef] [PubMed]
17. Department of Sports Economy, General Administration of Sport of China. The Data Compilation of the Sixth National Sports Venues Census. Available online: http://www.sport.gov.cn/pucha/index.html (accessed on 12 August 2018).
18. Heart Rate Guidelines for Adults and Children. Available online: https://www.ncsf.org/enew/articles/articles-heartrateguidelines.aspx (accessed on 11 July 2020).
19. Li, G.; Sun, Q. The empirical study of the development of the sports ground since the 21st century: Based on the Fifth and Sixth National Sports Ground Survey data statistical analysis. *J. Xi'an Phys. Educ. Univ.* **2016**, *33*, 164–171.
20. Chen, Z. Research on Public Sports Services in Japan. In *Annual Report on Development of Public Sports Services in China (2013)*; Dai, J., Ed.; Social Sciences Academic Press: Beijing, China, 2013.
21. Han, H.; Zheng, J. Logic, connotation and value of "People-centered Sports Development Outlook" based on study of General Secretary Xi Jinping's speeches on sports. *J. Wuhan I. Phys. Educ.* **2018**, *52*, 5–11.
22. China Focus: Square-dance Basketball Battle Causes Furor. Available online: http://www.xinhuanet.com/english/2017--06/07/c_136347068.htm (accessed on 12 November 2019).
23. Chinese Teenagers' Addiction to Internet Gaming Can Possibly Be Related to the Low Area Per Capita of Sports Venues in the Country. Available online: http://edu.people.com.cn/n/2014/0915/c1053-25664967.html (accessed on 2 October 2019).
24. Fast-Walker Killed after Taxi Strikes Group in E China. Available online: http://en.people.cn/n3/2017/0712/c90882--9240647.html (accessed on 2 December 2019).
25. Booth, M.; Okely, A. Promoting physical activity among children and adolescents: The strengths and limitations of school-based approaches. *Health Promot. J. Austr.* **2005**, *16*, 52–54. [CrossRef] [PubMed]
26. Kanters, M.A.; Bocarro, J.N.; Moore, R.; Floyd, M.F.; Carlton, T.A. Afterschool shared use of public school facilities for physical activity in North Carolina. *Prev. Med.* **2014**, *69*, 44–48. [CrossRef] [PubMed]
27. Sallis, J.F.; Cervero, R.B.; Ascher, W.; Henderson, K.A.; Kraft, M.K.; Kerr, J. An ecological approach to creating active living communities. *Annu. Rev. Public Health* **2006**, *27*, 297–322. [CrossRef] [PubMed]
28. Xu, P. Optimization Design Research of Primary and Secondary Schools' Gymnasium and Surrounding Facilities under Open Policy. Master's Thesis, Dalian University of Technology, Dalian, China, 2017.
29. Lei, Y. Study on the Risks and Risk Response Measures of the Public Opening Stadiums in Shanghai Primary and Middle Schools. Master's Thesis, Shanghai University of Sport, Shanghai, China, 2017.
30. Jin, Y.; Wang, W.; Wu, X. The policies of opening to the public of school sports facilities in China based on the text analysis of policies. *J. Jilin Sport Univ.* **2018**, *34*, 83–89.
31. Wang, C. The investigation of the Changchun City national fitness path. *J. Jilin I. Architect. Civil Eng.* **2013**, *30*, 111–114. Available online: http://en.cnki.com.cn/Article_en/CJFDTotal-JLJZ201302032.htm (accessed on 16 November 2019).
32. Yang, C. An Investigation on the Setting and Using of Comprehensive Fitness Equipment in Changchun. Master's Thesis, Jilin University, Changchun, China, 2010.
33. Cai, L. The Status Research of the National Fitness Path Configuration in the Urban Districts of Harbin. Master's Thesis, Harbin Normal University, Harbin, China, 2015.
34. Li, G. Investigation and Countermeasure Research on the Current Situation of the Big Class Break Sports Activities in the Primary School of Dalian City. Master's Thesis, Liaoning Normal University, Dalian, China, 2016.
35. Lin, L.; Qiu, G. Present and future of urban exercise trails in Fujian Province. *J. Xiamen Univ. Technol.* **2016**, *24*, 105–110.
36. Multi-department Notice on Issuing the "Implementation Plan for the Million-kilometer Walking Trail Project". Available online: http://www.gov.cn/xinwen/2018-03/16/content_5274663.htm (accessed on 2 December 2019).
37. Chen, S. Planning and Construction of Fitness Trails in Suzhou. In Proceedings of the 2017 China Urban Transportation Planning Conference, Shanghai, China, 7–9 July 2017.

38. McKenzie, T.L.; Cohen, D.A.; Sehgal, A.; Williamson, S.; Golinelli, D. System for Observing Play and Recreation in Communities (SOPARC): Reliability and feasibility measures. *J. Phys. Act. Health* **2006**, *3*, 208–222. [CrossRef] [PubMed]

39. Wang, K.; Liu, J. The spatiotemporal trend of city parks in Mainland China between 1981 and 2014: Implications for the promotion of leisure time physical activity and planning. *Int. J. Environ. Res. Public Health* **2017**, *14*, 1150. [CrossRef] [PubMed]

40. National Recreation and Park Association. NRPA Americans' Engagement with Parks Survey. Available online: https://www.nrpa.org/globalassets/research/engagement-survey-report.pdf (accessed on 12 December 2019).

41. Zhang, M.; Wang, K.; Liu, J. Analysis and enlightenment of the classification system of American urban parks for physical activity needs. *Planners* **2018**, *34*, 148–154.

42. Tu, H.; Liao, X.; Schuller, K.; Cook, A.; Fan, S.; Lan, G.; Lu, Y.; Yuan, Z.; Moore, J.B.; Maddock, J.E. Insights from an observational assessment of park-based physical activity in Nanchang, China. *Prev. Med. Rep.* **2015**, *2*, 930–934. [CrossRef] [PubMed]

43. China Urban Research Report, 2017 Q3. Available online: http://huiyan.baidu.com/reports/2017Q3.html (accessed on 12 December 2019).

44. Xu, J.; Gao, C.; Lee, J.K.; Zhao, J. PM2.5: A barrier to fitness and health promotion in China. *J. Sport Health Sci.* **2017**, *6*, 292. [CrossRef]

45. Qin, F.; Yang, Y.; Wang, S.T.; Dong, Y.N.; Xu, M.X.; Wang, Z.W.; Zhao, J.X. Exercise and air pollutants exposure: A systematic review and meta-analysis. *Life Sci.* **2019**, *218*, 153–164. [CrossRef]

46. Huang, Y.; Su, T.; Wang, L.; Wang, N.; Xue, Y.; Dai, W.; Lee, S.C.; Cao, J.; Ho, S.S. Evaluation and characterization of volatile air toxics indoors in a heavy polluted city of northwestern China in wintertime. *Sci. Total Environ.* **2019**, *662*, 470–480. [CrossRef]

47. National Health Commission of the People's Republic of China. *Status Report on Nutrition and Chronic Diseases of Chinese Residents 2015*; People's Medical Publishing House: Beijing, China, 2015.

48. Why is it difficult and expensive for large and medium city residents to exercise? Venue opening is not enough. Available online: http://www.chinanews.com/sh/2018/05-08/8508308.shtml (accessed on 3 December 2019).